UNDERSTANDING ELECTRICITY AND ELECTRONICS

FOURTH EDITION

Peter Buban
and
Marshall L. Schmitt

McGraw-Hill Book Company

New York   St. Louis   San Francisco   Dallas   Auckland
Bogotá   Düsseldorf   Johannesburg   London   Madrid
Mexico   Montreal   New Delhi   Panama   Paris
São Paulo   Singapore   Sydney   Tokyo   Toronto

ABOUT THE AUTHORS

Peter Buban teaches electricity and electronics at the Quincy (Ill.) Area Vocational Technical Center. He received his M.A. degree from Northeast Missouri State University. His many years of practical military and industrial experience in the electrical and electronics fields include such jobs as electronics technician, chief radioman, and electrician. Mr. Buban is the coauthor of *Technical Electricity and Electronics.*

Marshall L. Schmitt is a Senior Program Specialist, Division of Educational Replication, U.S. Department of Education. He has responsibilities for industrial arts and technology programs in this division. He received his doctorate from Pennsylvania State University. He has taught courses in electricity and electronics for U.S. Navy personnel and at the high school level in New York State. He also taught in the Industrial Arts Department at North Carolina State University, Raleigh, N.C. Dr. Schmitt is the coauthor of *Technical Electricity and Electronics.*

Editor: Hal Lindquist
Contributing Editor: Ronald Grigsby Kirchem
Coordinating Editor: Linda Richmond
Design Supervisor: Valerie Scarpa
Production Supervisor: Salvador Gonzales

Cover Photograph: Werner H. Müller/Peter Arnold, Inc.

This book was set in 10 pt. Korinna by York Graphic Services, Inc.

Library of Congress Cataloging in Publication Data

Buban, Peter.
  Understanding electricity and electronics.

  (McGraw-Hill publications in industrial education)
  Includes index.
  Summary: A textbook of the theory and practical applications of electricity and electronics.
  1. Electric engineering—Juvenile literature.
2. Electronics—Juvenile literature. [1. Electric engineering. 2. Electronics] I. Schmitt, Marshall Langdon, date. II. Title. III. Series.
TK146.B845 1982   621.3   81-2450
ISBN 0-07-008678-8   AACR2

3 4 5 6 7 8 9 10 DODO 90 89 88 87 86 85 84 83 82

Editor's Foreword

The McGraw-Hill Publications in Industrial Education constitute a comprehensive series of textbooks covering important areas of industrial education. *Understanding Electricity and Electronics*, a volume in this series, has proved to be one of the most widely used textbooks in its field. To remain effective in such a rapidly changing and expanding field, a textbook must be updated, improved, and expanded. This fourth edition incorporates all the vital aspects of theory and practice necessary to keep pace with the latest technology and pedagogy.

This book reflects the authors' many years of teaching experience at various educational levels and their broad knowledge of the practical side of the field. Their ingenuity and resourcefulness are constantly being tested and refined in the classroom and laboratory. The result is a textbook that is a superior source of information for the junior and senior high school student, as well as for the technical student taking electricity and electronics for the first time.

A two-color format has been used throughout this book to highlight outstanding concepts. A highly readable typeface and an open, pleasing overall design sustain the student's interest. Discussion topics, learning-by-doing exercises, individual-study activities, and construction projects are all designed to challenge the theoretical and practical minds of boys and girls having a wide range of backgrounds, interests, and abilities.

The authors, editor, and publisher hope that the fourth edition of *Understanding Electricity and Electronics* will stimulate the interest of students in this most important area. Perhaps the inspiration fostered by study of this fascinating field of energy and communication will result in satisfying the need for future electrical and electronics specialists. Indeed, it might stimulate the imaginations of potential electrical engineers and technicians, those who will contribute to the society of tomorrow.

Chris H. Groneman

McGraw-Hill Publications in Industrial Education

Chris H. Groneman, Consulting Editor

Books in Series

GENERAL INDUSTRIAL EDUCATION Groneman and Feirer
GENERAL METALS Feirer
GENERAL POWER MECHANICS Crouse, Worthington, Margules, and Anglin
GENERAL WOODWORKING Groneman
GETTING STARTED IN DRAWING AND PLANNING Groneman and Feirer
GETTING STARTED IN ELECTRICITY AND ELECTRONICS Groneman and Feirer
GETTING STARTED IN METALWORKING Groneman and Feirer
GETTING STARTED IN WOODWORKING Groneman and Feirer
TECHNICAL ELECTRICITY AND ELECTRONICS Buban and Schmitt
TECHNICAL WOODWORKING Groneman and Glazener
UNDERSTANDING ELECTRICITY AND ELECTRONICS Buban and Schmitt

Contents

v

Learning by Doing Activities

Preface

Understanding Electricity and Electronics is intended for a basic course in electricity and electronics. While the organization of the book has been especially designed to adapt easily to varying course lengths, sufficient material has been included for a full-year course.

Two publications are correlated with the text to provide a complete basic electricity and electronics learning program. They are a *Student Laboratory Manual* and a *Teacher's Resource Guide.*

The *Student Laboratory Manual* contains procedures for performing 50 experiments. Special attention has been given to the use of readily available and relatively low-cost components and materials. A unique feature of the laboratory program that this manual presents are instructions for the construction of the circuit board, the power supply, and the coils that are to be used in performing the experiments.

The *Teacher's Resource Guide* provides suggestions for using the pupil text and for conducting the laboratory program. It also contains reproducible Student Learning Guides and answer keys to the Learning Guides and to the Laboratory Manual. The Student Learning Guides consist of objective statements that are directly related to 46 of the 48 textbook units. These guides are a convenient way to check progress and to identify those topics that require further attention.

Early in the fourth edition of this updated text the student is introduced to the fundamental theories on which later discussions of practical applications are based. This logical development of the subject is particularly effective in relating the technological advances in electricity and electronics to basic concepts in physics, mechanics, and chemistry, to which the student is introduced. This approach is strengthened by a judicious use of basic mathematics, primarily in problem-solving situations that involve fundamental electrical relationships.

Classroom experience with the previous editions and many useful suggestions from users of the text have helped shape the general organization and content of this edition. The career and avocational opportunities in electricity and electronics open to both men and women are explored in detail. The discussions of materials, devices, and processes reflect the latest developments in their respective areas. In keeping with the objective of relating the subject to the real world, basic consumer information on electrical products and their use is included when pertinent. The unit on other sources of energy looks toward the future, touching on energy problems likely to affect the lives of everyone. This edition also includes new information in the electricity and electronics field, such as high-efficiency motors, wire wrapping, and light-wave communications.

The text consists of ten main sections that present a total of forty-eight units of instruction and suggestions for the construction of several projects. End-of-unit materials include learning-by-doing exercises, self-tests, review and discussion topics, and individual-study activities. Each is designed to achieve a particular objective in motivating, evaluating, and reinforcing the student's comprehension of what has been studied. *Learning-by-doing* exercises provide practical "hands-on" experiences in the use of circuit diagrams, tools, and test equipment. They may be used as individual or group assignments or as demonstrations by the instructor. Each of the exercises has been designed to use inexpensive materials usually available in the school shop or laboratory. The *self-test* sections provide a way for students to assess their own progress through each unit in a challenging and independent fashion. The *review and discussion* topics may be used effectively by the teacher to stimulate classroom discussions of specific topics, to highlight particular points, and to develop topics more fully. In this manner the instructor can very easily expand the treatment of certain subject matter as the needs and interests of the class demand. The *individual-study* activities are particularly suited as project assignments for those students showing special interests or aptitudes. Small groups of two, three, or four students can also be assigned these activities. The primary purpose is to encourage students to make use of school, community, and industry resources to prepare a report and then to use their oral or written skills to communicate the information to their class. The pursuit of these objectives can serve as an effective method of extending the scope of the student's learning experiences beyond the classroom.

Section Ten is devoted to suggestions for projects that will provide practical experiences in the design and construction of useful electrical and electronic devices. These projects are free of detailed step-by-step instructions since the authors feel such details are best left to instructors, who are familiar with the facilities available to them and the abilities of their students.

As in the previous edition, metric units are included with the customary American units. With the United States destined to join the rest of the industrial world in the adoption of the metric system of measurements, it is important that students obtain a working knowledge of these units early in their industrial education.

The authors wish to express their sincere appreciation to Ruth Buban and Doris L. Schmitt, who have provided encouragement and assistance in the preparation of this text. The authors also wish to thank John W. Gouldin, electronics instructor, West Springfield High School, Springfield, Va., for his special assistance in obtaining photographs of his students in his electronics laboratory.

They also wish to acknowledge the companies, organizations, and individuals who generously contributed material for use in this book. Specific reference to these sources is given, where appropriate, throughout the text. Comments and suggestions for improvement of this volume are most welcome.

Peter Buban & Marshall L. Schmitt

Introduction

Why study electricity and electronics? Try to imagine living today without using electricity and electronics. You would have no electric lights, no telephone, no television, and no doorbell to announce your friends. You could not buy a radio or a flashlight. There would be no electricians or electrical engineers. Electricity and electronics have made life not only easier but also more interesting.

Because electricity has had such an impact on our lives, we need to know something about its uses, its dangers, and its potential for the future. This is true today more than ever before. As you grow older, you will probably be called on as a citizen to help make many decisions. One of these decisions will, no doubt, deal with the fuels used to *generate*, or produce, electric energy. Unit 44 discusses the generating plants that change various forms of energy to electric energy. The advantages and disadvantages of the different generating methods are also presented. Studying this information will help you make informed judgments today and in the future about the fuels that produce electric energy.

In a more personal way, we live with an ever-increasing variety of electrical and electronic products. If we are to choose and use them wisely, we should know something about them. For example, we should know how they are designed, manufactured, and made more efficient. Also, we should know what their limits are and to what uses they can best be put. The study of electricity and electronics can open new avenues of interest for you. This study may suggest career or hobby opportunities you may like. However, you will need further study to develop your interests more fully.

Only by expanding your knowledge can you hope to deal with living in this fast-moving world of ours. Studying this text will help give you this base.

Unit 1 Atoms, Electrons, and Electric Charges

Before we discuss the basics of electricity and electronics, we need to understand what matter is. *Matter* is anything that occupies space and has *mass*, or weight. Wire, rubber, and glass are examples of matter. Scientists study the properties of matter in order to know how it works. Scientists are also concerned with the study of energy. *Energy* is the capacity for doing work. Energy has different forms, such as heat energy and electric energy. Electric energy results from the motion of tiny bits of matter called *electrons*. Energy can make changes in materials, such as putting them together, taking them apart, or simply moving them from one place to another.

The basic unit of matter is the *atom*. In studying electricity and electronics, it is important for you to understand the atom, because the electron is one of its parts. In this unit, the structure of the atom and its electrical properties are discussed.

STRUCTURE OF THE ATOM

An atom is made up of tiny particles. Two of these particles, the *electron* and the *proton,* are important to our studies.

Electrons move around the center, or *nucleus*, of an atom in paths. These paths are usually called *shells* (Fig. 1–1). An atom can have several shells around its nucleus. Each of these shells can have only up to a certain number of electrons. This number is called the *quota* of a shell. When every shell of the atom contains its quota of electrons, the atom is said to be in a stable condition. The nucleus of the atom is made up of particles called *protons* and *neutrons.* These are held together tightly by a binding energy.

Fig. 1–1. This is the inside of an atom of oxygen gas. It contains eight electrons that move around the nucleus in two shells.

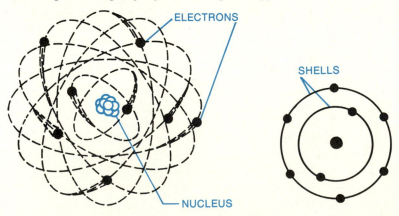

ELECTRONS

SHELLS

NUCLEUS

All electrons are alike, and all protons are alike. Thus, atoms differ from one another only in the number of electrons and protons they contain (Fig. 1–2). The number of protons in the nucleus is the *atomic number* of that atom. Neutrons weigh about the same as protons. The term *atomic weight* refers to the total number of particles (both protons and neutrons) in the nucleus of an atom.

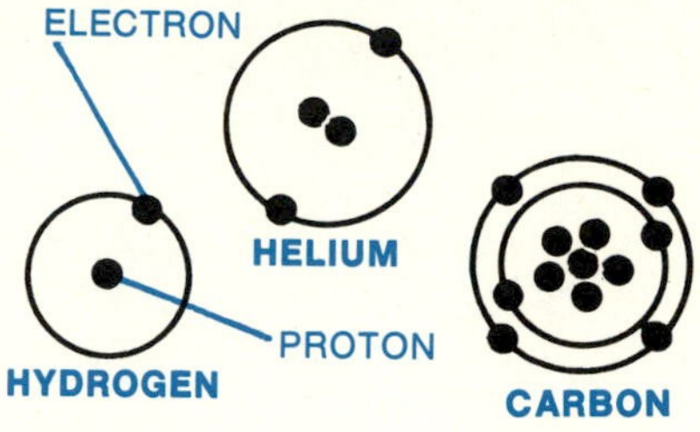

Fig. 1–2. Atoms of hydrogen, helium, and carbon

ELEMENTS, COMPOUNDS, AND MOLECULES

When all the atoms in a substance are alike, the substance is called an *element.* Copper, iron, and carbon are among the more than 100 different elements known to exist. Different elements can combine to form a substance called a *compound.* Water, sugar, and plastic materials are examples of compounds.

The smallest particle of a compound that has all the properties of that compound is called a *molecule.* A molecule contains atoms of each of the elements that form the compound (Fig. 1–3).

CHARGES

Electrons and protons have tiny amounts of energy known as *electric charges.* Electrons have negative (−) charges. Protons have positive (+) charges. Neutrons have no electric charge. Thus, they are neutral. The amount of the negative charge of each electron is equal to the amount of the positive charge of each proton. These opposite charges attract each other. This attraction helps hold the atom together.

Under normal conditions, the negative and the positive charges in an atom are equal in value. This is because the atom has an equal number of electrons and protons. An atom in this condition is said to be electrically neutral (Fig. 1–4).

Fig. 1–3. Molecules of water and sulfuric acid. A molecule is made up of a number of atoms.

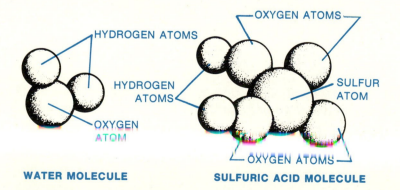

WATER MOLECULE   SULFURIC ACID MOLECULE

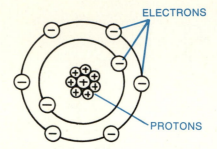

Fig. 1-4. An electrically neutral atom

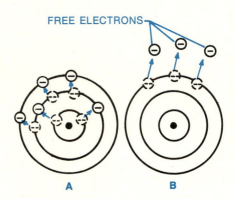

Fig. 1-5. Some electrons within an atom may move to higher energy levels within the atom or leave the atom as the result of the absorption of energy.

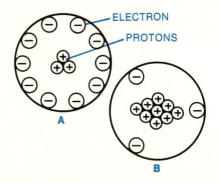

Fig. 1-6. Ions: (A) negatively charged ion; (B) positively charged ion

VALENCE ELECTRONS

Electrons in the outermost shell of an atom are called *valence electrons*. In the study of electricity and electronics, we are concerned mostly with the behavior of valence electrons. They can, under certain conditions, leave their "parent" atoms. The number of valence electrons in atoms also determines important electrical and chemical characteristics of the substance.

ENERGY LEVELS AND FREE ELECTRONS

The electrons in any shell of an atom are said to be located at certain energy levels. These are related to the distance of the electrons from the nucleus of the atom. When outside energy such as heat, light, or electricity is applied to certain materials, the electrons within the atoms of these materials gain energy. This may cause the electrons to move to a higher energy level. Thus, they move farther from the nuclei of their atoms (Fig. 1-5 at A).

When an electron has moved to the highest possible energy level (or the outermost shell of its atom), it is least attracted by the positive charges of the protons within the nuclei of the atom. If enough energy is then applied to the atom, some of the outermost-shell, or valence, electrons will leave the atom. Such electrons are called *free electrons* (Fig. 1-5 at B).

IONS

An *ion* is a charged atom. If a neutral atom gains electrons, there are then more electrons than protons in the atom. Thus, the atom becomes a negatively charged ion (Fig. 1-6 at A). If a neutral atom loses electrons, protons outnumber the remaining electrons. Thus, the atom becomes a positively charged ion (Fig. 1-6 at B). Ions with unlike charges attract one another. Ions with like charges repel one another. The process by which atoms either gain or lose electrons is called *ionization.*

ELECTRIC CHARGES IN ACTION

A simple way of generating an electric charge is by friction. For example, if you rub a rubber balloon briskly with a wool cloth, electrons will move from the cloth to the balloon, causing the balloon to become negatively charged. If you then put the balloon against a wall, the balloon's negative charge will repel electrons from the surface of the wall (Fig. 1-7).

This will, in turn, cause the surface of the wall to become positively charged. The attraction between the opposite charges of the balloon and a small surface area of the wall is strong enough to hold the balloon in place.

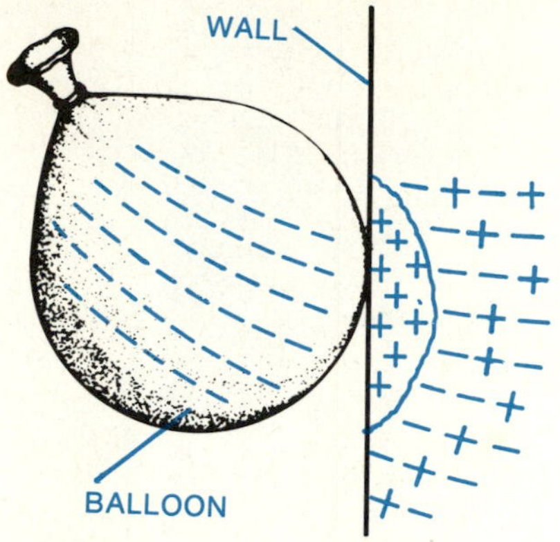

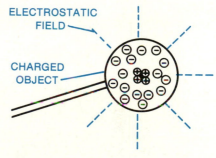

Fig. 1–7. Opposite electrical charges attract each other and thus cause a rubber balloon to "stick" to a wall.

STATIC ELECTRICITY

The attraction between the charged balloon and wall represents work done by *electrostatic energy,* which is often called *static electricity.* An *electrostatic field* is the energy that surrounds every charged object (Fig. 1–8). In this kind of electricity, there is no movement of electrons between the balloon and the wall. Thus, the electricity is said to be *static,* or at rest.

Static electricity is sometimes thought of as a nuisance. However, devices such as capacitors and air cleaners and industrial processes such as the manufacture of abrasive paper make use of it.

USEFUL APPLICATIONS OF STATIC ELECTRICITY

Figure 1–9 shows how static charges are used in making abrasive paper. *Sandpaper* is a common term for abrasive paper. Such paper is used to smooth the surfaces of wood or metal. A conveyor belt carries the abrasive particles over a negatively charged plate. The abrasive particles are negatively charged by contact. At the same time, the paper, which is tacky from an adhesive coating, is moving under a posi-

Fig. 1–8. An invisible electrostatic field is present around every charged object.

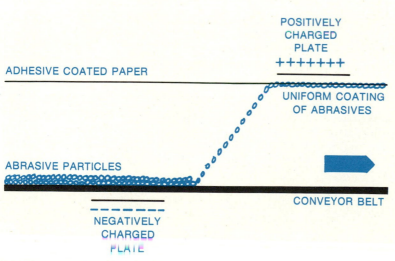

Fig. 1–9. Electrostatics used in making abrasive paper

5

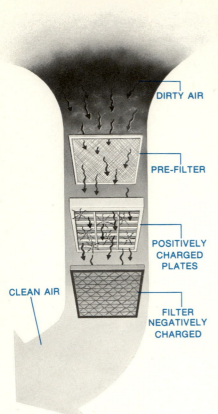

Fig. 1–10. Electrostatic air cleaner

tively charged plate. This makes the paper positive. Since opposite charges attract, the abrasive particles are attracted to the paper. These particles form a very uniform and dense abrasive surface on the paper.

Static electricity can also be used in an air cleaner (Fig. 1–10). This device can be used in home heating systems to clean the air as it circulates through the furnace. A similar device can be used by industry to reduce air pollution. Electrostatic air filters are much more efficient than simple cloth or paper filters. Those remove only large particles from the air. An electronic filter can remove tiny particles. As shown in Fig. 1–10, the dirty air passes through a paper prefilter that removes large pieces of dirt and lint. The air then moves through a series of plates that give a positive charge to the dirt and dust particles. Finally the air moves through a negatively charged filter. The positive particles are attracted to the negative filter. They become trapped there. After the filters become very dirty, the system must be shut down and cleaned.

Another practical use of static electricity is shown in Fig. 1–11. Spray painting is a common way of producing beautifully finished surfaces. However, it is wasteful. Much of the paint never sticks to the surface. It stays in the air as

Fig. 1–11. Electrostatic spray painting

spray dust. Spray dust is not only wasteful but also dangerous to health. Most of the spray dust can be eliminated by positively charging the object to be painted and negatively charging the paint. Moreover, since the charge decreases where the paint is applied, a uniform thickness of paint is produced.

LEARNING BY DOING

1. Making an Electroscope. An *electroscope* is a device used to show the presence of an electric charge. It can also be used to demonstrate the laws of electrical attraction and repulsion.

MATERIALS NEEDED

2 6-in. (152-mm)* pieces of bare copper wire, no. 6 AWG (4.0 mm)
2 strips of tin or aluminum foil, ¼ × 1¼ in. (6 × 32 mm)
1 wood disk, ¾ in. thick × 2½ in. in diameter (19 × 64 mm)
toy rubber balloon, medium size
glass rod
wool cloth
silk cloth
1-pint (0.47-liter) glass jar

Procedure

1. Cut the edges of the wood disk at an angle to form the jar stopper (Fig. 1–12 at A). Drill holes through the disk.
2. Shape the strips of foil (Fig. 1–12 at B).
3. Complete the electroscope assembly as shown in Fig. 1–12 at C.
4. Blow up the balloon. Stroke it briskly in one direction with the wool cloth. Then press the balloon against the electroscope terminals, or wires. What happens? Explain this action.
5. Rub the glass rod briskly in one direction with the silk

*Wherever appropriate, metric equivalents of U.S. customary units are given in parentheses immediately following the customary units. In most cases, metric values are not exact but are rounded off for convenience. In some cases, the closest standard metric sizes are given for standard U.S. sizes (for example, wire and pipe sizes).

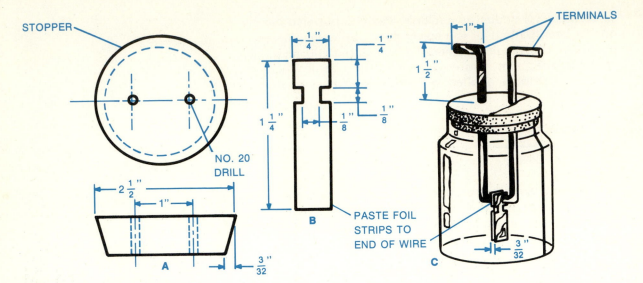

Fig. 1–12. Electroscope assembly

cloth. Then press it against the terminals of the electroscope. Explain what happens when this is done.

6. Recharge the balloon by stroking it again with the wool cloth. Hold one of the terminals with your fingers. Bring the balloon near the other terminal. What happens? Can you explain this action?

7. Recharge the balloon once again. Have a classmate rub the glass rod with the silk cloth. As your classmate touches one of the terminals with the rod, bring the balloon near the other terminal. Why are the strips of foil now attracted to each other?

SELF-TEST

Test your knowledge by writing, on a separate sheet of paper, the word or words that most correctly complete the following statements:

1. _____ is anything that occupies space and has mass.

2. The capacity for doing work is called _____.

3. Electrons move around the nucleus of an atom in paths that are usually called _____.

4. The nucleus of an atom is made up of particles called _____ and _____.

5. Atoms differ from one another only in the number of _____ and _____ they contain.

6. The number of protons in the nucleus of an atom is known as the atomic _____ of that atom.

7. When all the atoms within a substance are alike, the substance is called an _____.

8. Common examples of elements are _____, _____, and _____.

9. Different elements can combine to form a substance called a _____.

10. A _____ is the smallest particle of a

compound that has all the properties of that compound.

11. Electrons are _____ charges. Protons are _____ charges.
12. An electrically neutral atom is one that has the same number of _____ and _____.
13. The electrons in the outermost shell of an atom are called _____ electrons.
14. The energy _____ of an electron is related to its distance from the nucleus of an atom.
15. An ion is an atom that is either _____ charged or _____ charged.
16. If a neutral atom gains electrons, it becomes a _____ ion.
17. If a neutral atom loses electrons, it becomes a _____ ion.
18. Ions with unlike electric charges _____ each other. Ions with like electric charges _____ each other.
19. The process by which atoms either gain or lose electrons is called _____.
20. A simple way of generating an electric charge is by _____.
21. The attraction between two opposite electric charges is an example of _____ electricity.
22. A charged object is surrounded by an _____ field.

FOR REVIEW AND DISCUSSION

1. Define the term *energy*.
2. Identify two different forms of energy. Can you list more?
3. Describe the makeup of an atom.
4. How do atoms differ from one another?
5. What is an element? A compound? A molecule?
6. Define the atomic number of an atom.
7. Explain what is meant by an electrically neutral atom.
8. What are valence electrons?
9. Explain the relationship between electron energy levels and free electrons.
10. How does an atom become a negative ion? A positive ion?
11. State the laws of electrical attraction and repulsion.
12. Describe a simple way of generating an electric charge.
13. Explain static electricity. Give an example of its presence.
14. What is meant by an electrostatic field?

INDIVIDUAL-STUDY ACTIVITY

Prepare a written or an oral report on electron theory as it relates to the study of electricity.

Unit 2   Electric Circuits

An *electric circuit* is a combination of parts connected to form a complete path through which electrons can move. The purpose of a circuit is to make use of the energy of moving electrons. Therefore, a circuit is also a system of parts, or components, by which electric energy can be changed into other forms of energy, such as heat, light, or magnetism.

PARTS OF A CIRCUIT

A basic complete circuit has four parts: (1) the energy source, (2) the conductors, (3) the load, and (4) the control device (Fig. 2–1).

Energy Source. The *energy source* in a circuit produces the force that causes electrons to move. It is like a pump that forces water through a pipe. In electricity, this force is called *voltage,* or *electromotive force.* The basic unit of force is the *volt.* The flow of electrons is called *current.* The most common energy sources used in electric circuits are chemical cells and electromechanical generators. These devices do the work needed to move electrons through the parts of the circuit.

Conductors. The *conductors* in a circuit provide an easy path through which electrons can move through the circuit. Copper is the most commonly used conductor material. It is formed into wire, bars, or channels (Fig. 2–2). Copper wire may be bare or covered with some kind of insulating material.

Fig. 2–1. A basic electric circuit

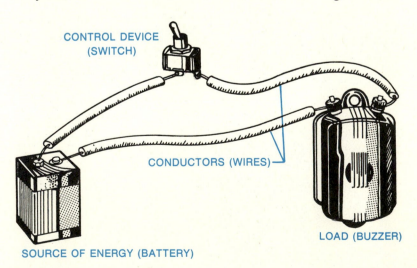

CONTROL DEVICE
(SWITCH)

CONDUCTORS (WIRES)

SOURCE OF ENERGY (BATTERY)

LOAD (BUZZER)

COPPER ROD

"DRAWN" WIRE

Fig. 2–2. Wire is manufactured by pulling copper rods through dies. (*Rome Cable Corporation*)

The insulation prevents electrons from moving outside the wire (Fig. 2–3).

In some circuits, metal objects other than copper conductors form the conducting paths. In an automobile, for example, the entire car frame serves as a conductor. It completes a number of circuits that connect the voltage source (the car battery) to various electrical and electronic devices (the loads) (Fig. 2–4).

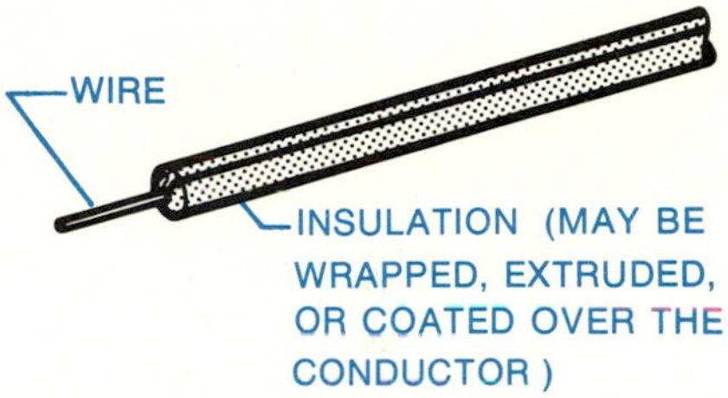

Load. The *load* is that part of a circuit that changes the energy of moving electrons into some other useful form of energy. A light bulb is a very common circuit load. As electrons move through the filament of the lamp, the energy of the electrons in motion is changed into heat energy and light energy (Fig. 2–5).

Control Device. A simple circuit-control device is the common mechanical wall switch. It lets us open or close a circuit

WIRE

INSULATION (MAY BE WRAPPED, EXTRUDED, OR COATED OVER THE CONDUCTOR)

Fig. 2–3. Insulated wire. Some common insulating materials are neoprene, rubber, nylon, polyethylene, Teflon, vinyl, cotton, asbestos, paper, and enamel.

Fig. 2–4. An automobile frame (*Ford Motor Company*)

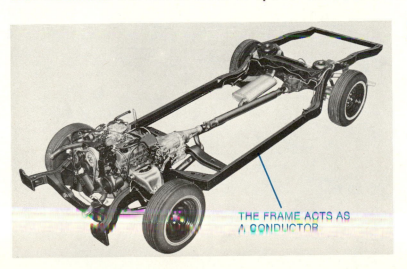

THE FRAME ACTS AS A CONDUCTOR

SYMBOL

Fig. 2–5. A typical electrical load. The light bulb converts electric energy into heat and light energy. (*General Electric Company*)

(Fig. 2–6). When the switch is "on," it acts as a conductor to keep electrons moving through the circuit. The circuit is said to be *closed*. When the switch is "off," the circuit path is interrupted. Electrons can no longer move through the circuit. The circuit is said to be *open*.

In addition to the familiar on-off switches, other devices can provide a switching action and control the flow of electrons in a circuit. Examples are electromagnetic relays, diodes, transistors, and electron tubes, all discussed later in this book.

KINDS OF CIRCUITS

Loads can be connected into a circuit in *series*, in *parallel*, or in *series-parallel* combinations. A series circuit provides only one path through which electrons can move from one terminal of the energy source to the other (Fig. 2–7 at A). In a parallel circuit, there may be two or more different paths through which electrons can flow (Fig. 2–7 at B). A series-parallel circuit is shown in Fig. 2–7 at C. Note that series and parallel circuits are combined to form one circuit with several paths.

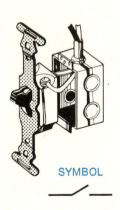

SYMBOL

Fig. 2–6. A typical on-off switch

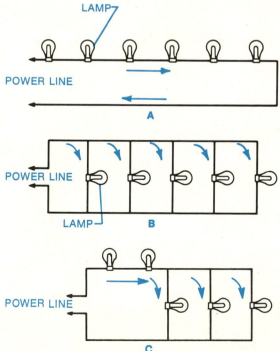

Fig. 2–7. Circuit connections: (A) series circuit; (B) parallel circuit; (C) series-parallel circuit

Test your knowledge by writing, on a separate sheet of paper, the word or words that most correctly complete the following statements:

1. An electric circuit makes it possible to use the energy of moving _____.
2. In a circuit, electric energy is changed into other forms of energy, such as _____, _____, or _____.
3. The four basic parts of a complete circuit are the _____, the _____, the _____, and the _____.
4. Common sources of energy used in electric circuits are _____ and _____.
5. The most common conductor material is _____.
6. That part of a circuit that changes electric energy into another form of energy is the _____.
7. The series circuit provides only _____ path through which electrons can move.
8. In a _____ circuit, there may be two or more electron paths.

FOR REVIEW AND DISCUSSION

1. What is an electric circuit?
2. Name and define the four parts of a basic complete circuit.
3. Describe basic series and parallel circuits.
4. What is the main feature of a series-parallel circuit?

INDIVIDUAL-STUDY ACTIVITY

Prepare a written or an oral report describing familiar electric circuits you use in everyday activities. Identify the parts of the circuits and tell what each does.

Unit 3   Circuit Diagrams and Symbols

Once you have developed a circuit on paper to do a task, you should plan how to build it. In electrical and electronics work, the basic part of such a plan is usually a diagram. A *diagram* shows how the different parts are connected to form the complete circuit. Developing the diagram is an important step in designing circuits. The diagram will guide those who will maintain and repair the circuit. The diagram also serves as a record for those who may want to copy or study it or for those who wish to change the circuit.

This unit discusses several common kinds of diagrams. It also offers much useful information about reading and drawing diagrams.

PICTORIAL DIAGRAMS

If you build a project from a do-it-yourself kit, one of the items included in the construction manual will be a *pictorial diagram*. This is a drawn picture of the parts of the circuit. It

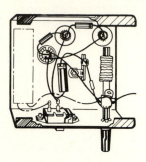

Fig. 3–1. Pictorial diagram of a one-transistor radio

shows how the parts are connected and where they are to be located within the assembly (Fig. 3–1). The pictorial diagram also shows the wiring devices, such as lugs and sockets, that are to be used. The pictorial diagram is usually drawn *to scale.* That means that the *relationships* of the sizes and locations of the parts are accurate whether they are shown full-size, smaller, or larger.

Pictorial diagrams are easy to follow. For this reason, they are used with do-it-yourself kits and in manufacturing to show what finished products will look like.

One disadvantage of a pictorial diagram is that it does not give clear electrical information about the circuit. It simply shows what the circuit will look like after it is put together. Such diagrams do not show the electron path or the way the parts relate to one another electrically. Pictorial diagrams usually take a lot of time to prepare and often take up much space.

SCHEMATIC DIAGRAMS

The *schematic diagram* is the standard way of communicating information in electricity and electronics. On this diagram, the components are shown by *graphic symbols.* These are letters, drawings, or figures that stand for something. In electricity and electronics, a graphic symbol shows the operation of a part in a circuit. Because the symbols are small, the diagrams can be drawn small. The symbols and related lines are used to show how the parts of a circuit are connected and how they relate to one another.

Examine the schematic diagram of the two-transistor radio circuit shown in Fig. 3–2. If you read this diagram from left to right, you see the components in the order in which they are used to convert radio waves into sound energy. The antenna (at the left) collects the radio waves (energy), and the headset (at the right) converts this energy into sound. By using the diagram, you can trace the operation of the circuit from beginning to end. Because of this, schematic diagrams are used widely by engineers and technicians. They are used in all kinds of circuit design, construction, and maintenance activities.

Identification of Components. As shown in Fig. 3–2, the components on a schematic diagram are also identified by *letter symbols.* These include R for resistors, C for capacitors, and Q for transistors. You will learn later what these are and what they do. For now, you should simply be able to recognize their symbols. The letters are further combined with numbers, such as R1, R2, and R3, to distinguish different components of the same kind. The numerical values of components are often shown directly on the schematic diagram, together with the component identifications. When these are not given in this way, they are stated in the parts list in the notes that come with the diagram.

Other Symbols. A number of other common symbols used with schematic diagrams are shown in Fig. 3–3. One of the organizations responsible for standardizing these and other symbols is the American National Standards Institute (ANSI). Complete listings of the symbols used in electricity

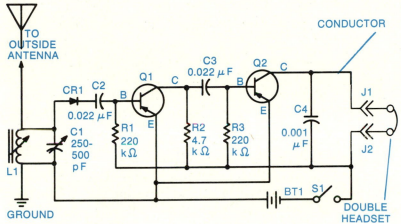

Fig. 3–2. Schematic diagram of a two-transistor radio

Identification of components

BT1	Battery
C1	Variable capacitor
	(The arrow crossing the symbol indicates the value of the component can be varied.)
C2, C3, and C4	Fixed capacitors
CR1	Crystal diode
J1 and J2	Insulated tip jacks
L1	Antenna coil (variable ferrite-core inductor)
Q1 and Q2	p-n-p transistors
R1, R2, and R3	Carbon-composition resistors
S1	Single-pole–single-throw switch

NOTE: The letters B, C, and E near the transistor symbols indicate the Base, Collector, and Emitter leads of the transistors.

Conductor symbols:

Conductors crossing but not connected.

Conductors electrically connected (the dot is often omitted when the connection is obvious).

Fig. 3–3. Here are some common symbols used on schematic diagrams. Other symbols are shown throughout the book.

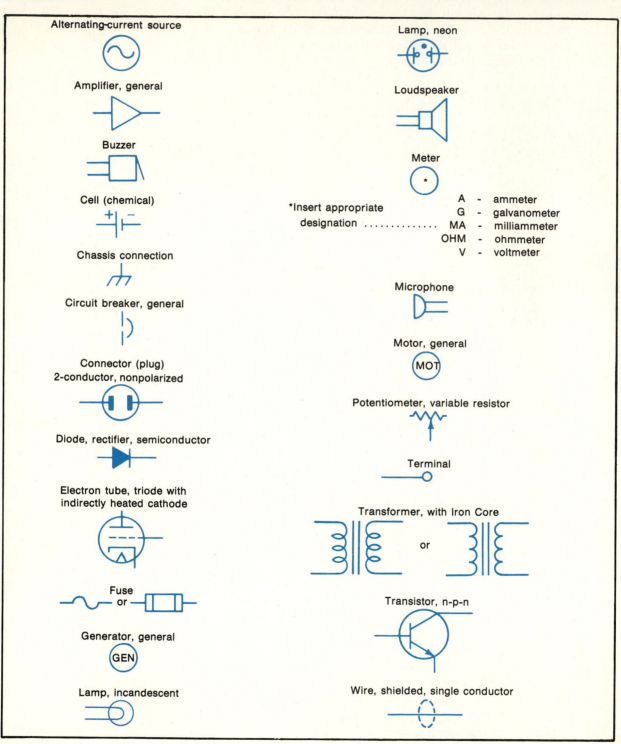

Alternating-current source

Amplifier, general

Buzzer

Cell (chemical)

Chassis connection

Circuit breaker, general

Connector (plug)
2-conductor, nonpolarized

Diode, rectifier, semiconductor

Electron tube, triode with
indirectly heated cathode

Fuse
or

Generator, general
GEN

Lamp, incandescent

Lamp, neon

Loudspeaker

Meter
*

*Insert appropriate
designation ....

A   -   ammeter
G   -   galvanometer
MA   -   milliammeter
OHM   -   ohmmeter
V   -   voltmeter

Microphone

Motor, general
MOT

Potentiometer, variable resistor

Terminal

Transformer, with Iron Core
or

Transistor, n-p-n

Wire, shielded, single conductor

and electronics are available from this organization. Standards used in other technical fields are also available from the ANSI.

READING THE SCHEMATIC DIAGRAM

The ability to read schematic diagrams is important. Most circuits you may want to study, build, or repair are probably illustrated by such a diagram. Reading a diagram is not so hard as it first may appear. With some practice, you will find that reading a schematic is convenient and saves time.

To begin to learn how to read schematic diagrams, look at the pictorial and the schematic diagrams of the same circuit shown in Fig. 3–4. Read from left to right. Find and identify the same components in both. Trace the wire connections. Some connections in the pictorial diagram may be covered by other components. Try to find these wires in the schematic diagram.

Unlike a pictorial diagram, a schematic diagram does not show where the parts or connecting wires are located on the *chassis*. The chassis is the frame or base on which the parts are mounted. Also note that the schematic diagram does not show any wiring devices such as lugs and sockets. Whether and where to use other wiring devices depends on the sizes

Fig. 3–4. (A) Pictorial (bottom view) and (B) schematic diagrams of a rectifier circuit

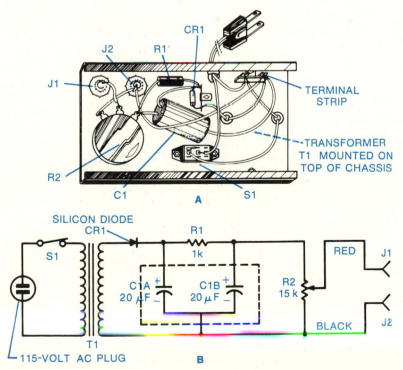

17

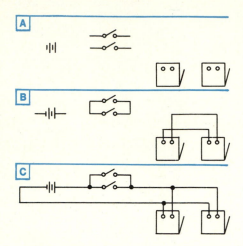

Fig. 3–5. Drawing the schematic diagram

and shapes of the components and the space available. Wiring also depends on the electrical characteristics of the parts of the circuit. They must be wired according to the schematic diagram.

WIRING FROM A SCHEMATIC DIAGRAM

To wire a circuit from a schematic diagram, you must know how to read the diagram. You also need to know about kinds and sizes of wires, about how to make connections, and about wiring tools and devices. These are discussed in Section 5, "Electrical Wiring—Materials, Tools, and Processes."

DRAWING A SCHEMATIC DIAGRAM

Sometimes you will wire a circuit without using a schematic diagram. At other times, you may have to work on a wired circuit for which no diagram is available. In such cases, you may often need or want to make a schematic diagram directly from the wired circuit. The following suggestions will help you make a neat, readable diagram:

1. Use standard symbols for all the components. If there is no standard, adapt a symbol for your use. Note its meaning on the diagram and in a separate symbol list.
2. Locate the symbols so that the lines that will show the conductors connecting them will not be too close together. Use as few line crossings and bends as possible (Fig. 3–5 at A.)
3. Do not draw the symbols so close to one another that they crowd the diagram and make it hard to read.
4. Use unruled paper for schematic diagrams to keep from confusing drawn lines with the lines printed on the paper.
5. After drawing the symbols, connect them with straight vertical and horizontal lines (Fig. 3–5 at B and C).
6. Use the dot symbol when it is necessary to show that conductors are electrically connected.

WIRING, OR CONNECTION, DIAGRAMS

Wiring diagrams are used to show circuit-system connections in a simple way. They are very commonly used with home appliances and automobile electrical systems.

The typical wiring diagram shows the components of a circuit in a pictorial manner. The components are usually identified by name (Fig. 3–6). Such a diagram also often shows the positions of the components in relation to each other. A color code may identify certain wires.

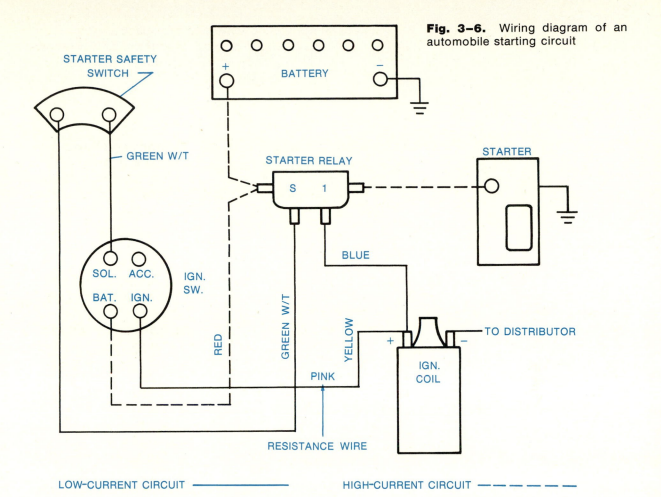

Fig. 3-6. Wiring diagram of an automobile starting circuit

LOW-CURRENT CIRCUIT ———————   HIGH-CURRENT CIRCUIT — — — — —

ARCHITECTURAL FLOOR-PLAN DIAGRAMS

Architects, electrical designers, and contractors use floor-plan diagrams to show where the parts of a building's electrical system are to be located. These include receptacle outlets, switches, lighting fixtures, and conduit and other wiring devices. These are shown with symbols (Fig. 3–7). A basic floor-plan diagram sometimes omits the conduit system and associated wiring. These and their sizes are shown, however, on the contractor's working drawings. It is from these that the electrician does the wiring job.

LEARNING BY DOING

2. **Drawing Schematic Diagrams.** Using the right symbols, draw schematic diagrams of the circuits shown in the picto-

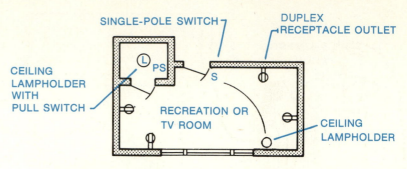

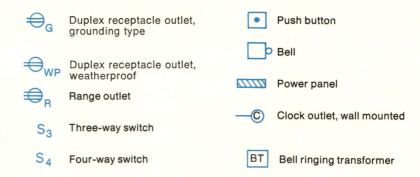

Other symbols used with architectural floor-plan diagrams

$\ominus_G$	Duplex receptacle outlet, grounding type	⊡	Push button
$\ominus_{WP}$	Duplex receptacle outlet, weatherproof		Bell
$\ominus_R$	Range outlet		Power panel
S_3	Three-way switch	─©─	Clock outlet, wall mounted
S_4	Four-way switch	BT	Bell ringing transformer

Fig. 3–7. A simple architectural floor-plan diagram and some standard symbols used on these diagrams

rial diagrams in Fig. 3–8. You should draw these on unruled $8\frac{1}{2} \times 11$ inch (216 × 279 mm) paper. You should first draw rough, freehand diagrams. Then make finished diagrams with a straightedge.

SELF-TEST

Test your knowledge by writing, on a separate sheet of paper, the word or words that most correctly complete the following statements:

1. A _____ shows how the different parts are connected to form the complete circuit.
2. The pictorial diagram is usually drawn to _____.
3. On a schematic diagram, the components are shown by _____.
4. You read most schematic diagrams from _____ to _____.
5. Using a schematic diagram makes it possible to trace the _____ of a circuit from beginning to end.
6. Examples of letter symbols used to identify components on a schematic diagram are _____ for resistors, _____ for capacitors, and _____ for transistors.
7. The dot symbol is used to show that wires are electrically _____ at that point.
8. A wiring diagram often shows the _____ of the components in relation to each other.
9. Architectural floor-plan diagrams show where the parts of a building's _____ system are located.

FOR REVIEW AND DISCUSSION

1. Explain the purposes of a circuit diagram.
2. Describe a pictorial diagram.

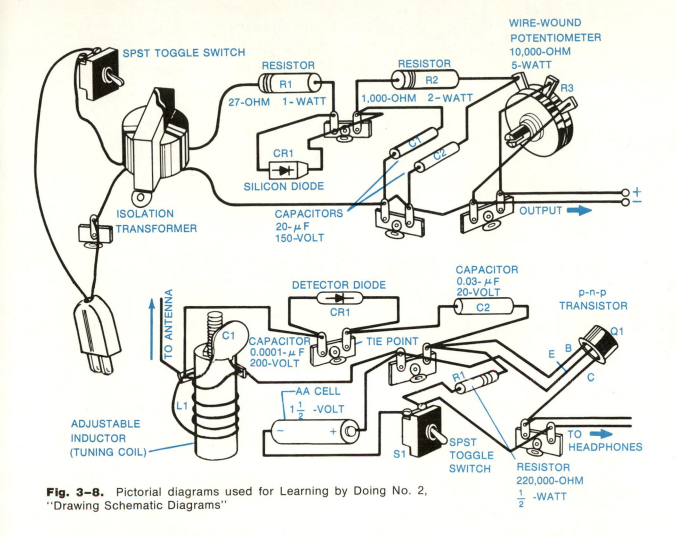

Fig. 3-8. Pictorial diagrams used for Learning by Doing No. 2, "Drawing Schematic Diagrams"

3. Name some disadvantages of the pictorial diagram.
4. Describe a schematic diagram. Tell what information is given by such a diagram.
5. How are the different components identified on a schematic diagram?
6. Draw and identify the symbols for 10 electrical and/or electronic components with which you are familiar.
7. List several practices to be followed when drawing a schematic diagram.
8. Describe a wiring diagram.
9. What is the purpose of an architectural floor-plan diagram?

INDIVIDUAL-STUDY ACTIVITIES

1. Prepare a written or an oral report on the development of standard electrical symbols in this country.
2. Obtain large-size pictorial and schematic diagrams. Explain them to the students in your class. Tell about the advantages of each.
3. Obtain an architectural floor-plan diagram from a local architect, contractor, or electrician. Explain it to your class.
4. Prepare an architectural floor-plan diagram of one room in your home.

Unit 4   Safety Rules, Practices, and Devices

Safety is everybody's business. Developing safe work habits depends on the right attitude—a feeling that you *want* to work safely. With this attitude, you will try to understand the safety rules and practices for your activities. And you will find yourself following these rules and practices. You will want to do these things, not only for your own safety but also for the safety of others.

Failing to follow electrical safety rules and practices can injure you or others and destroy property. Shock and burns may result when the body conducts electricity. Property may be damaged by electrical fires started by overheated wires or sparks. These must be prevented if we are to use electric energy safely and efficiently.

ELECTRIC SHOCK

An *electric shock* is a physical sensation caused by the reaction of the nerves to electric current. In minor cases, there is only a harmless "jerking" of the affected muscles. In more severe cases, the breathing and heart muscles become paralyzed. If the muscles are permanently damaged, the result is often death by *electrocution.* That means "to be killed by electricity." If the muscles are not permanently damaged, they can often be restored to normal operation by *artificial respiration.* Mouth-to-mouth artificial respiration is an effective way to restore normal breathing. In this method, you periodically blow air into the victim's mouth. At the same time, you pinch the victim's nostrils shut. Call a medical doctor at once. If a doctor is not available, call the nearest rescue squad.

The amount of current that can cause serious damage to the muscles of a person's body depends on his or her physical condition. Records show that people have been electrocuted by very low values of current. Serious muscular damage is likely to result if an excessive current passes through the chest area. This happens when the conducting path goes from hand to hand or from one hand to the feet (Fig. 4-1).

Under normal conditions, the epidermis, or outer layer of skin, presents a high resistance to current. If the epidermis becomes wet from sweat or another liquid, its resistance is greatly lowered. Under this condition, a voltage lower than 120 volts can cause a dangerous amount of current to pass through the body.

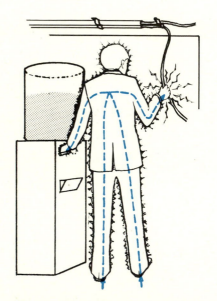

Fig. 4-1. Current paths through the chest region of the body

BURNS

In addition to shock, an excessive current passing through the body can cause severe burns. Such burns result from heat produced by friction between electrons and body tissues. Electrical burns often occur inside the body along the path followed by current. Burns of this kind can be very painful and difficult to heal.

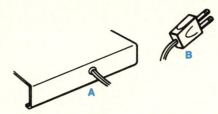

Fig. 4—2. Points at which electrical cord defects most often occur: (A) where cords enter appliances; (B) where cords enter attachment plugs

GENERAL SAFETY RULES AND PRACTICES

Understanding the causes and effects of electric shocks and burns is important. Safety rules and practices are designed to prevent such accidents. No set of rules and practices can cover all cases. However, the list that follows will help you use and work with electricity safely.

1. Always treat electrical wires in a circuit as being potentially dangerous.
2. Always unplug the power cord of a piece of equipment before removing the case or cabinet in which the wiring is contained.
3. Do not touch a water or gas pipe, a sink, a bathtub, or any wet surface while handling the metal parts of electrical equipment that is plugged in.
4. Never place any plugged-in equipment where it can fall into a sink or a bathtub containing water.
5. Be sure to read and follow all instructions supplied with electrical or electronic equipment before attempting to use it.
6. Make it a habit to periodically inspect all the electrical appliances and tools you use. In this way, you will find any potentially dangerous conditions, such as loose wires, undesirable bare wires, and frayed cords (Fig. 4-2).
7. Always unplug or disconnect equipment that does not seem to be working normally.
8. Do not work on any equipment unless you know the right procedures to follow. It is always a good practice to call in an expert when you are unsure about the safe and proper way to handle any equipment.
9. If you must work with the exposed parts of a circuit that is plugged in, handle the parts with one hand only. Keep the other hand behind your back or in a pocket. This will keep the current from passing through your chest. Do not wear rings or watches when touching live electric circuits.
10. It is always a safe practice to use only equipment that has the Underwriters' Laboratories label (Fig. 4-3). This

Fig. 4—3. The Underwriters' Laboratories label

shows that the item has been made according to strict safety standards.

11. Find and learn to work the main power switch in your home and in other places where you work with electricity. This will let you quickly turn off all the circuits in an emergency.

12. During an electric shock, the muscles will often cause the fingers to hold the point of contact tightly. Use all available means to separate from a point of electrical contact as quickly as possible. The possibility of serious muscular damage and burns increases the longer current passes through your body.

13. If an electric shock causes breathing to stop, a doctor should be called immediately. Artificial respiration must be started right after the electrical contact has been broken. Extreme care must be taken in removing a person from the point of contact. Otherwise, the path of current can pass through the rescuer, causing her or him to suffer severe shock also.

14. Although an electric shock may not be dangerous, it will almost always startle or frighten you. This may, in turn, cause you to fall or to bump against hard surfaces, resulting in serious injury.

GROUNDING FOR PERSONAL SAFETY

The *ground,* or earth, usually is a good conducting point for electrons when a path exists between it and a *live,* or charged, wire. Because of this, putting any part of your body between the ground and such a wire presents the danger of severe shock. *Grounding* wires and devices are designed to reduce this hazard.

If a bare wire of an electrical appliance comes into contact with the bare metal of its cabinet, the cabinet becomes an extension of the wire. If the cabinet is *ungrounded,* or not connected to the ground, the line voltage will appear between the cabinet and ground. In most homes, the line voltage is about 115 volts. Thus, there is a danger of severe shock if a person makes contact between the cabinet and any surface that is in close contact with the ground.

Grounding Wires, Plugs, and Outlets. Metal cabinets, cases, and frames of appliances, tools, and machines are usually grounded by a separate grounding wire. This grounding wire is part of their power cords. The grounding wire is connected to the cabinet, case, or frame of the device and to the round prong of the cord's plug. It is not connected to any

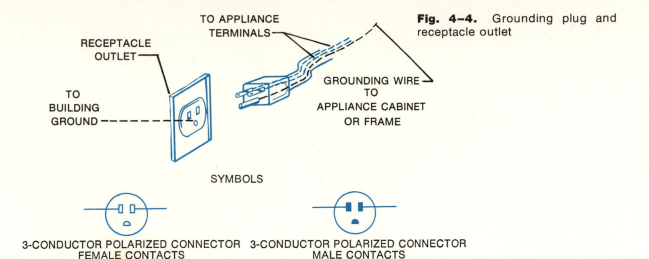

RECEPTACLE OUTLET

TO APPLIANCE TERMINALS

Fig. 4—4. Grounding plug and receptacle outlet

TO BUILDING GROUND

GROUNDING WIRE TO APPLIANCE CABINET OR FRAME

SYMBOLS

3-CONDUCTOR POLARIZED CONNECTOR FEMALE CONTACTS

3-CONDUCTOR POLARIZED CONNECTOR MALE CONTACTS

part of the device's electric circuit. When the plug is put into a grounded outlet, the grounding wire of the cord is automatically connected to the building ground. This connection is made through a grounding wire in the cord leading to the outlet (Fig. 4–4). Since the cabinet, case, or frame is then at the same voltage as the ground, a shock hazard cannot exist between it and any grounded surface.

The three-pronged grounded plug is called a *polarized plug.* Its prongs will fit into the outlet only when they are lined up properly. For example, the round prong connected to the equipment ground wire can fit only into the round hole in the outlet. When this happens, the other two prongs automatically line up with their slots in the outlet. One is connected to the "hot" wire. The other is connected to the "neutral" wire. For a description of the hot and neutral wires, see Unit 31.

Individual Grounding. Metal cabinets of appliances and frames of machines do not always use grounding plugs and outlets. These devices are grounded by connecting them directly to water pipes or to ground rods (Fig. 4–5). Because plastic plumbing is often used, the plumbing system must be inspected to make sure the water pipe is an adequate ground. Gas pipes should never be used for grounding.

Green Wire. Some appliances have cords that have a two-pronged plug but still contain three wires. One of these, a grounding wire, is green. It extends a short distance from the side of the plug. A *spade lug* is connected to this end of the green wire. This lug is used to connect the green wire to a grounding point. The other end of the green wire is attached

Fig. 4–5. Grounding metal cabinets of appliances: (A) grounding to a water pipe; (B) use of a ground rod

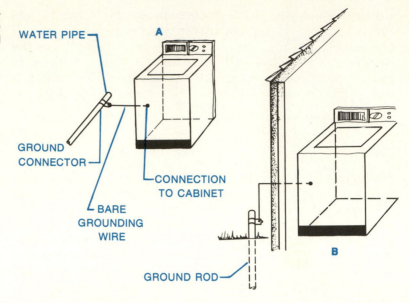

WATER PIPE

A

GROUND CONNECTOR

BARE GROUNDING WIRE

CONNECTION TO CABINET

GROUND ROD

B

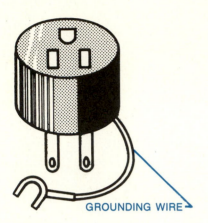

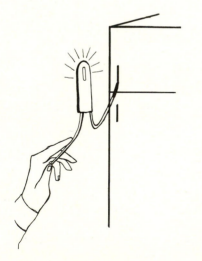

GROUNDING WIRE

Fig. 4–6. A typical two-to-three-wire adapter

to the metal case of the electrical appliance or tool. When connected properly, the green wire provides adequate protection against electric shock.

Two-to-three-wire Adapter. A two-to-three-wire *adapter* is a handy device. It makes it possible to use a grounding plug in a two-socket outlet (Fig. 4–6). Such an adapter does not automatically ground the device. It must be connected, using the grounding wire, to the ground point.

NEON-LAMP HANDITESTER

An inexpensive device called a *neon-lamp handitester* can be used to see whether an exposed wire is touching the metal cabinet or case of an electrical appliance. This condition presents, of course, the danger of severe shock. To use a handitester, simply press one tip firmly against a bare metal surface of the cabinet or case under test and hold the other tip in your hand (Fig. 4–7). The neon lamp of the tester will glow if a dangerous voltage exists between the cabinet and ground. If it does, the appliance or the tool should be unplugged at once and fixed as soon as possible.

WIRED-TO-CHASSIS CIRCUITS

One wire of the line cord of some radios and television sets is connected directly to the chassis. This then serves as a circuit conductor. This presents a definite shock hazard. A

Fig. 4–7. Using a neon-lamp handitester to check for the presence of voltage between an appliance cabinet and ground

dangerous voltage may exist between the chassis and ground. Check the chassis with a neon-lamp handitester. If the test lamp glows, unplug the line cord. Then reverse the plug by turning it halfway around. Put the plug back in. In most cases, this will eliminate the dangerous condition. If not, unplug the device at once. Have it checked by a service technician.

DOUBLE, OR REINFORCED, INSULATION

Many appliances and portable motor-driven tools now offer protection against shock by means of *double*, or *reinforced, insulation*. In these devices, the wiring is kept from grounding by special kinds of insulation between the wiring and the equipment cabinet or case. To further protect against shock, certain gears and screws that hold the assembly together are often made of nylon or some other insulating material. Very often, the entire case is made of strong plastic material. This kind of construction makes it unnecessary to ground the cabinet or case.

OPEN AND SHORT CIRCUITS

When a wire becomes broken, burned out, or disconnected at some point, its circuit becomes open. When this occurs, the current will stop flowing (Fig. 4–8 at A). An open circuit can be dangerous. The voltage of the circuit will appear across the point at which the circuit is broken. Wires or equipment with a broken connection must be handled carefully if the equipment is still plugged in.

If two bare live wires come in contact with one another in such a way as to bypass the load, an excessive amount of current will flow in the circuit. This condition is known as a *short circuit* (Fig. 4–8 at B). Short circuits are very dangerous. They are dangerous because the excessive current is usually high enough to damage the insulation of the wire and perhaps start a fire. In most circuits, fuses and circuit breakers are used to automatically open the circuit if there is excessive current flow. These devices are discussed below.

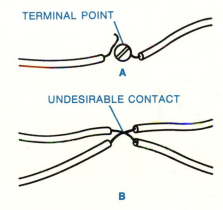

Fig. 4–8. Examples of (A) open circuit; (B) short circuit

FUSES

A *fuse* is a safety device that works as a switch to turn a circuit off when the current goes over a specified value. Too much current may be due to a short circuit within the load. It may result from a condition known as *overloading*. Overloading happens when a circuit delivers more current to the

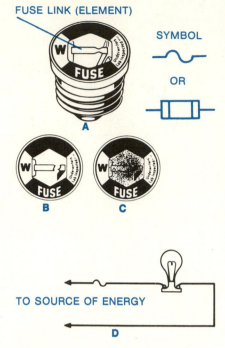

FUSE LINK (ELEMENT)

SYMBOL

OR

A

B   C

TO SOURCE OF ENERGY

D

Fig. 4–9. Medium-base plug fuse: (A) fuse in good condition; (B) blown or broken fuse link; (C) blown fuse link indicated by darkened "window"; (D) method of connecting a fuse to a circuit (*Bussmann Manufacturing Division, McGraw-Edison Company*)

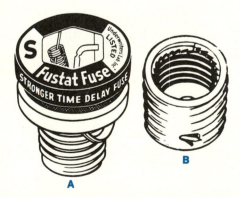

A   B

Fig. 4–10. Tamperproof fuse: (A) plug fuse; (B) fuse adapter (*Bussmann Manufacturing Division, McGraw-Edison Company*)

load than the conductors can carry safely. Then the circuit becomes overheated. Overloading generally happens because too many appliances are plugged into the outlets served by the circuit.

Plug Fuses. Plug fuses are commonly used in household circuits. Such a fuse has a strip of metal called a *fuse link,* or *element* (Fig. 4–9). The link is usually made of zinc. It is designed to melt or burn out when more than the rated value of current passes through it. The electrical size of a fuse is equal to this value of current. It is given in a unit called an *ampere.* The number of amperes is usually printed on the body of a fuse, for example, 15 A or Amp. This means that the fuse will open if the current goes over 15 amperes. Plug fuses are put into a circuit by screwing them into special sockets. Some plug fuses have a metal screw base like the base of an ordinary light bulb. This kind of plug fuse is known as an *Edison,* or *medium, base.* The use of medium-base plug fuses is generally being discontinued. Most electrical building codes now prohibit their use. However, they are still quite common in older buildings.

Tamperproof Fuse. The tamperproof fuse is designed to prevent a fuse from being replaced with one of a different electrical size. For this reason, it is also called a noninterchangeable plug fuse. A socket is used that will accept only certain sizes of fuses. For example, a 25-A or a 30-A fuse will not fit into a socket for 15-A and 20-A fuses. In order to use these fuses in a medium-base socket, an adapter must be installed (Fig. 4–10). The adapter is first screwed into the fuse socket. Then the fuse is screwed into the adapter. Once in place, the adapter cannot be easily removed.

Dual-element Plug Fuse. A dual-element, or time-delay, plug fuse is used in many homes. It is designed to withstand a current larger than its rated value for a short time. If the excessive current continues after this time, the fuse blows out just like an ordinary plug fuse. It must then be replaced.

The construction of a dual-element plug fuse is shown in Fig. 4–11. When too much current passes through the thermal cutout of the fuse, the solder holding the fuse link to the thermal cutout becomes hot. If this current continues, the solder is softened. This allows the spring to pull the fuse link out of the thermal cutout and break its connection.

Dual-element plug fuses are most often used in motor circuits. A motor draws much more current on starting than it does after reaching full speed. A dual-element plug fuse will withstand this short period of high current. It then continues to protect the motor circuit after the current has decreased to its running value. An ordinary plug fuse would probably burn out during the starting period. The dual-element plug fuse is slow to act under normal conditions. However, a sudden, heavy flow of current will cause its fuse link to quickly melt and open the circuit just like the ordinary plug fuse.

Cartridge Fuses. Like plug fuses, cartridge fuses are available in either ordinary or dual-element versions (Fig. 4–12 at A). Some of these fuses are the renewable-link kind (Fig. 4–12 at B). A burned-out link in such a fuse can be replaced. The cartridge case can be used over and over again.

Cartridge fuses are mounted in clip holders (Fig. 4–12 at C). Small-size glass cartridge fuses are sometimes enclosed in spring-loaded fuse holders (Fig. 4–12 at D).

FUSE SAFETY RULES

Fuses will protect circuits only if they are used in the right way. The following rules will help you use fuses correctly and handle them safely:

1. Never replace a fuse of the proper size with a larger one.
2. Do not try to repair any blown fuse unless it is the renewable-link kind.

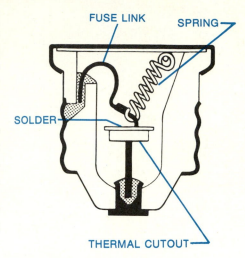

Fig. 4–11. Cutaway view showing construction of a dual-element plug fuse (*Bussman Manufacturing Division, McGraw-Edison Company*)

Fig. 4–12. Cartridge fuses and holders: (A) typical fuse cases; (B) cutaway view of renewable-link fuse; (C) clip-style fuse holders; (D) spring-loaded fuse holder (*Bussmann Manufacturing Division, McGraw-Edison Company*)

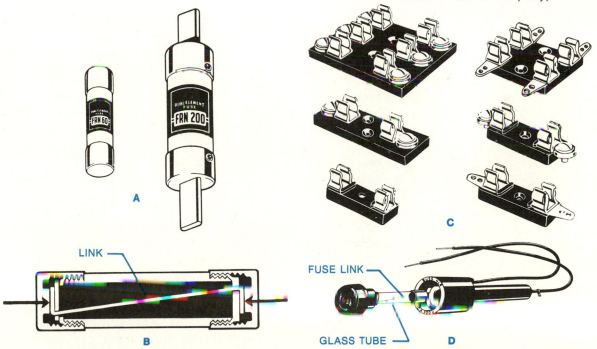

3. Call an electrician or some other qualified person if you are in doubt about the right-size fuse to use in a circuit.
4. Never put a coin or any other object in a fuse socket.
5. Whenever possible, disconnect the main switch at the load center or switch box before replacing a fuse that is blown.
6. When replacing a fuse in a load center or switch box while the main switch is on, stand on a dry surface (preferably wood or a rubber mat). Use one hand to make the replacement. Keep the other hand behind your back or in a pocket.
7. To prevent overheating and sparking, make sure that a fuse is firmly inserted into its socket or fuse holder.
8. Remember that even an ordinary fuse of the right size will often blow out if used in a circuit with a motor in it. Use a dual-element fuse instead.
9. When a fuse of the right size blows out repeatedly, this is an indication that the circuit is overloaded or that there is a short circuit somewhere in the circuit. Such conditions must be corrected before the circuit is put back into use.
10. Never try to bypass a fuse by connecting a wire between the terminals of its socket or holder. Always keep a supply of spare fuses of the right sizes on hand in case of emergency.

CIRCUIT BREAKERS

A *circuit breaker* is a mechanical device that performs the same protective function as a fuse. Unlike a fuse, however, it can also serve as an ordinary on-off switch. Some circuit breakers have a switch mechanism that works by *thermal,* or heat, control. Others use a magnetic switch mechanism. Larger circuit breakers have switchlike mechanisms that are both thermally and magnetically controlled (Fig. 4–13). Thermal control involves the use of a bimetallic element, discussed in Unit 33. Magnetic control involves the use of a solenoid, discussed in Unit 14.

Like fuses, circuit breakers are rated according to the amount of current in amperes that can pass through them before they are *tripped,* or opened. However, unlike a fuse, a circuit breaker can be reset after being tripped. Some circuit breakers have adjustable trip sizes. Others have only a single trip value.

When a circuit breaker of the kind shown in Fig. 4–13 trips, its lever moves to or near the "off" position. To reset it, simply move the lever back to the "on" position.

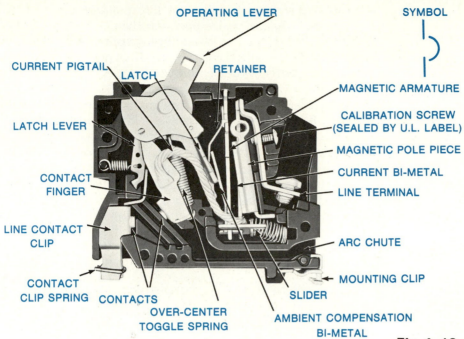

OPERATING LEVER

SYMBOL

CURRENT PIGTAIL

LATCH

RETAINER

MAGNETIC ARMATURE

CALIBRATION SCREW
(SEALED BY U.L. LABEL)

LATCH LEVER

MAGNETIC POLE PIECE

CURRENT BI-METAL

CONTACT
FINGER

LINE TERMINAL

LINE CONTACT
CLIP

ARC CHUTE

MOUNTING CLIP

CONTACT
CLIP SPRING   CONTACTS

SLIDER

OVER-CENTER
TOGGLE SPRING

AMBIENT COMPENSATION
BI-METAL

Fig. 4–13. Construction of a typical circuit breaker (*Cutler-Hammer*)

GROUND-FAULT CIRCUIT INTERRUPTER

The *ground-fault circuit interrupter* (GFCI) is a device that gives added personal protection against electric shock. Devices such as fuses, equipment grounds, and double insulation are used for personal and equipment protection, but they have shortcomings. An electric current of about 60 milliamperes (mA) can cause the heart action of adults to become irregular or to stop altogether. A 15-ampere fuse or circuit breaker disconnects the circuit only when the current exceeds 15 amperes. This amount is much greater than 60 milliamperes.

Figure 4–14 shows a portable ground-fault circuit interrupter. They are also available built into circuit breakers and duplex outlets. It is small enough to fit into a pocket. This GFCI can be plugged into a common duplex receptacle. However, it must be grounded properly to the receptacle. Before connecting an electrical cord of an appliance to a GFCI, you should test the GFCI to see if it is working properly. Do this by following the specific instructions for the GFCI being used. Ground-fault circuit interrupters work on the basic electrical principle that the current flowing in the hot wire to the device should equal the current flowing in the neutral wire from the same device. Any difference in the currents in these wires indicates that the current is taking another path to complete the circuit. The other path could be an unwanted

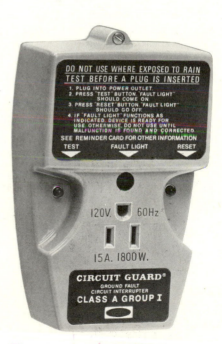

Fig. 4–14. A portable ground-fault circuit interrupter (GFCI) (*Harvey Hubble, Inc.*)

ground in the electrical equipment. Ground-fault circuit interrupters sense the difference between the currents in the hot and neutral wires. If there is a difference of more than 5 milliamperes, a safety switch trips in the GFCI and interrupts the circuit. This protects the person who is operating the electrical equipment from a serious electric shock. The GFCI does not completely eliminate the feeling of an electric shock. However, it does open the circuit quickly enough to prevent injury to a person of normal health. Thus, a GFCI provides protection against dangerous currents that do not overload 15- or 20-ampere fuses or circuit breakers. Nevertheless, do not use a GFCI in place of a separate grounding wire. Moreover, a GFCI does not protect a person who accidentally touches bare hot and neutral wires at the same time.

SAFETY GLASSES

The use of safety glasses is strongly recommended in electrical and electronics shops. Safety glasses should be worn when soldering, cutting wires, handling chemicals or television picture tubes, and using machines such as the drill press, portable electric drill, or grinder. All these activities could injure the eyes. It is always good practice to wear safety glasses while doing them.

OSHA

As a future worker, you should become familiar with the *Occupational Safety and Health Act*, commonly called OSHA. The U.S. Congress passed this act in 1970. Its basic purpose is to ensure that every worker in the nation has safe and healthful working conditions.

The act does several things. It encourages employers and employees to reduce hazards in the work place. It establishes responsibilities and rights for employers and employees. It develops required job safety and health standards and enforces them.

Both the worker and the employer have responsibilities under the act. The worker must read the OSHA posters at the job sites. The worker must obey all OSHA standards that apply. The worker must follow all the employer's safety and health rules and regulations. The employer must tell all employees about OSHA. The employer must provide a hazard-free work place. The employer must make sure employees have and use safe tools and equipment. It is not possible to describe OSHA in detail in this textbook. However, much information about OSHA is available from the U.S. Department of Labor. That is the government agency that administers the act.

3. Making a Neon-lamp Handitester. This device can be used to conveniently check for the presence of voltage at receptacle outlets and terminals and between ungrounded appliance and tool housings and ground. A voltage of at least about 60 volts is needed to light the lamp. A direct-current (dc) voltage will cause only the electrode of the neon lamp connected to the negative terminal to glow. Thus, the handitester can be used to test dc-voltage polarity.

MATERIALS NEEDED

1 clear, hard plastic (Lucite or vinyl) tube, $\frac{1}{2}$-in. (12.7-mm) OD, $\frac{1}{4}$-in. (6.35-mm) ID, 4 in. (101.6 mm) long
1 neon lamp, type NE-2
1 carbon-composition or film resistor; 220,000 ohms; $\frac{1}{2}$ watt
12-in. (305-mm) test-prod wire, black, no. 18 AWG (1.00 mm)
12-in. (305-mm) test-prod wire, red, no. 18 AWG (1.00 mm)
1 test solderless prod, black, 4-in (102-mm) handle
1 test solderless prod, red, 3-in. (76-mm) handle
plastic insulating tubing (spaghetti)
epoxy cement
clear liquid casting plastic (optional)

Procedure

1. Make the neon-lamp, resistor, and test-prod-wire connections shown in Fig. 4–15. Make sure that all solder

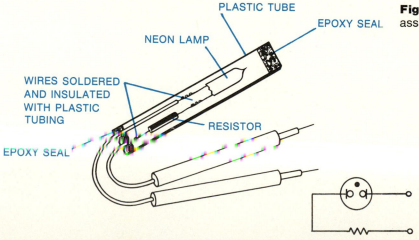

PLASTIC TUBE
EPOXY SEAL
NEON LAMP
WIRES SOLDERED AND INSULATED WITH PLASTIC TUBING
RESISTOR
EPOXY SEAL

Fig. 4–15. Neon-lamp handitester assembly and circuit diagram

joints and bare wires of the assembly are completely insulated.
2. Carefully put the assembly into the plastic tube.
3. Seal the ends of the tube with epoxy cement or with some other suitable sealing compound. Put the cement or sealing compound on the lamp end of the tube a little at a time to keep it from covering a part of the lamp.

NOTE: A more attractive handitester can be made by sealing the test-prod-wire end of the tube with clear liquid casting plastic after the parts have been put into it. You then place the tube upright and fill it with the casting plastic from the lamp end.

SELF-TEST

Test your knowledge by writing, on a separate sheet of paper, the word or words that most correctly complete the following statements:

1. An electric shock is a physical sensation caused by the reaction of the nerves to _____.
2. The danger of severe shock is greatly increased if the epidermis becomes _____.
3. In addition to shock, an excessive current through any part of the body can cause severe _____.
4. Electrical wires, wherever they may be found in a circuit, should always be treated as being potentially _____.
5. It is always safe practice to read and thoroughly understand all available operating _____ supplied with electrical and electronic products.
6. You should make it a habit to periodically _____ all electrical and electronic equipment you use.
7. A plug-in device should be immediately _____ if it does not seem to be working normally.
8. An _____ Laboratories label on a product shows that it has been made according to strict _____ standards.
9. Putting any part of your body between the _____ and a live wire presents the danger of severe shock.
10. A grounding wire is used to prevent a shock hazard from existing between a metal cabinet and _____.
11. Grounding plugs and outlets automatically connect power cords to the building _____.
12. The color of the grounding wire in the power cords of tools and appliances is _____.
13. The use of _____, or _____, insulated products significantly lessens the danger of severe electric shock.
14. A fuse is a safety device that works as a _____ to turn a circuit _____ when there is too much current.
15. When a circuit is overloaded, it becomes _____.
16. The electrical size of an ordinary plug fuse is given in _____.
17. Dual-element fuses are usually needed in _____ circuits.
18. Never replace a fuse of the proper size with a _____ one.
19. A _____ is a mechanical device that, like a fuse, protects a circuit from excessive current.
20. Ground-fault circuit interrupters sense the difference between the currents in the _____ and _____ wires.
21. OSHA stands for _____.

FOR REVIEW AND DISCUSSION

1. Describe an electric shock.
2. Why is it especially important not to allow the body to become a conductor of current from hand to hand or from hand to foot?
3. Explain why moisture on the skin greatly increases the danger of severe electric shock.
4. What causes electrical burns?
5. Discuss several general safety rules and practices.
6. Why is it important to ground metallic cabinets of appliances and metallic housings of portable electric tools?
7. Discuss grounding plugs and outlets.
8. For what purpose is a two-to-three-wire adapter used?
9. Explain the use of a neon-lamp handitester as a safety aid.
10. Why does a wired-to-chassis circuit present a dangerous condition when operated at 115 volts?
11. What is meant by double, or reinforced, insulation construction?
12. For what purpose is a fuse used? Explain the operation of this device.
13. Name and describe four common kinds of fuses.
14. In what kinds of circuits are dual-element plug fuses commonly used? Why?
15. Discuss several fuse safety rules.
16. What is a circuit breaker?
17. How is a typical circuit breaker reset after it has tripped?
18. What is the major advantage of using a ground-fault circuit interrupter?
19. Name several activities that require wearing safety glasses.
20. Identify three responsibilities OSHA requires of a worker and of an employer.

INDIVIDUAL-STUDY ACTIVITIES

1. Prepare a written or an oral report about electric shock.
2. Prepare a list of the electrical safety practices that could be improved or that should be begun in your home.
3. Prepare a written or an oral report about the purposes and the policies of the Underwriters' Laboratories.
4. Give a demonstration showing various kinds of fuses and their general uses.
5. Obtain publications on artificial respiration from your local chapter of the American Red Cross or other area health agencies. Discuss the steps involved in mouth-to-mouth respiration.
6. Obtain examples of polarized plugs. Discuss the different kinds that are available.
7. Discuss how the Occupational Safety and Health Act (OSHA) affects your school.

Unit 5 Occupations in Electricity and Electronics

The rapidly growing fields of electricity and electronics provide jobs for a large portion of the nation's workers (Table 5-1). According to those who study employment trends, many more workers will be needed in these fields in the future. This is due to the development and the use of new systems, products, and processes. Some of these new systems are now in operation and some are in the planning stage.

Perhaps you, too, will be one of the ever-increasing number of young persons entering these areas of employment. Your participation in an electricity and electronics course is a good start in this direction. Such a course can provide you with many interesting and useful educational experiences. It can serve to stimulate your interest in preparing for a job in the field. An introduction to these occupations at this time may suggest some career ideas. These will help make the rest of the course more meaningful. This information may also help you plan a school schedule. You should study those subjects that are recommended for an introduction to and/or preparation for entry into a particular occupation.

Table 5-1.   Approximate Number of Persons Employed in Various Electrical and Electronics Occupations

Occupation	Number of Persons Employed in U.S.
electrical and electronic engineers	329,000
physical scientists	215,000
engineering and science technicians	985,000
radio operators	53,000
telephone operators	311,000
electricians	590,000
household appliance and accessory installers and mechanics	145,000
office machine repairers	63,000
radio and television repairers	131,000
electric power and cable installers and repairers	111,000
telephone installers and repairers	297,000
telephone line installers and repairers	77,000

EDUCATION AND TRAINING

A high school education can and does provide the necessary background for entry into many electrical and electronics jobs. For those who go into jobs immediately after graduation, continued education and training generally is needed. This training is often acquired in on-the-job apprenticeship programs. Attendance at special training sessions and evening classes is very helpful, too. Individualized programs such as approved correspondence courses can meet special training needs. In general, the more education you have, the better are your employment opportunities.

The basic education and training needed for entry into professional and skilled occupations is usually obtained through formal training programs (Fig. 5–1). These include programs at both public and private 2- and 4-year colleges, technical institutes, and trade schools. In addition to these programs, participation in technical activities in the military services can often qualify a person for entry into many different jobs. Military experience also may prepare you for jobs that require education and training beyond the high school level.

Whatever training program you may choose, you should begin at once to make the necessary plans for entering such a program. Much information about this can be obtained from your school instructor and guidance counselor. College and school representatives, labor unions, and those who are working at jobs can provide additional information.

ELECTRICAL OR ELECTRONIC ENGINEERING

Engineers are creators of ideas and problem solvers. They understand the principles of electrical theory, mathematics,

Fig. 5–1. Future engineers in a college program studying the properties of light waves (*U.S. Department of Labor*)

Fig. 5–2. Electrical engineer designing control systems for manned flight operations (*National Aeronautics and Space Administration*)

physics, and chemistry. With this knowledge, they design and develop new systems and products (Fig. 5–2). They also improve the performance of systems and products already in use.

Most engineers are graduates of colleges or universities. Courses leading to a bachelor of science degree in electrical engineering (EE) are usually completed in 4 years. However, some engineers continue their education in specialized areas for which advanced study is needed.

In addition to the regular electrical engineering degree, another 4-year bachelor of science engineering degree is offered now by many colleges. This is the electronics engineering technology (EET) degree. Graduates of this program are highly trained in the fields of science and mathematics. Their training also includes much work in the more practical areas of electricity and electronics. This training involves the improvement of systems and products already in use. In addition, the installation and maintenance of complicated electrical systems and equipment are studied.

ELECTRIC POWER AND LIGHTING DESIGNERS

Electrical designers are responsible for the design of electrical systems. Electrical systems are found in all kinds of indoor and outdoor facilities, such as buildings, highways, bridges, and stadiums. The systems electrical designers design provide power for air-conditioning equipment, business machines, pumps, elevators, and lighting. In fact, they design

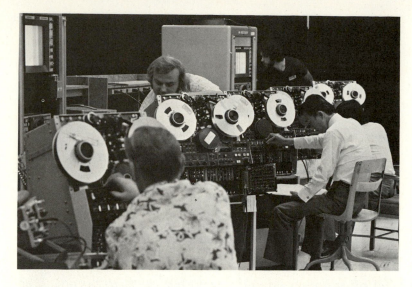

Fig. 5–3. Electronic technicians give final systems tests to professional videotape recorders. (*Ampex Corporation*)

almost any equipment connected to a power supply. Electrical designers are usually employed by engineering companies in the construction field or by architectural firms. Many designers work for the contractors who install the systems designed. A designer normally works under the direction of an engineer. The engineer takes legal responsibility for the designer's work by officially signing the plans and specifications produced. A designer's formal education may include a full 4-year engineering course of study. Some designers have only a 2-year technical education or on-the-job training. These persons begin as electrical *drafters.* A drafter's work includes drawing detailed plans for the entire electrical system. Electrical designers must be extremely knowledgeable about the various building codes. These codes govern the design and installation of electrical systems in the area in which the facilities will be built.

ENGINEERING TECHNICIANS

Engineering technicians are skilled workers whose jobs require a broad knowledge of scientific and mathematical principles. They apply this knowledge to activities involving the design, construction, operation, installation, and maintenance of electrical equipment. It is also important for engineering technicians to be skilled in writing and in public speaking. This is because their duties include selling, preparing reports, and speaking before various groups. This is particularly true of technicians who work directly with engineers in developing and producing new products (Fig. 5–3). Many engineering technicians advance to their jobs after a long pe-

Fig. 5-4. Radio and television service technician testing the performance of a television receiver

riod of on-the-job training. Another route to becoming an engineering technician is to complete an approved course of study at a community college or technical institute. Many of these schools offer an associate in science degree in electronic engineering technology. This degree usually requires successful completion of a 2-year program of study.

RADIO AND TELEVISION SERVICE TECHNICIANS

Skilled radio and television service technicians are experts in the use of test instruments and hand tools. These technicians use troubleshooting techniques and adjustment procedures to repair, maintain, and install electronic appliances and devices (Fig. 5-4). Most of these technicians specialize in home radio and television sets. However, many of them also work on a wide variety of other electronic equipment.

Technical skills are not the only requirement for successful radio and television service technicians. These technicians work both in the shop and in customers' homes. Because of this, they must be courteous, neat, and businesslike in their conduct with other people. Many employers consider this an important quality when selecting new people for permanent jobs.

The employment of radio and television service technicians is expected to increase sharply during the coming years. There will also be a marked increase in the complexity of the systems with which the technicians will work. Any plan for becoming a service technician should include attending a technical or vocational school. Many schools offer programs that emphasize this kind of training. These programs normally last from 9 to 18 months. Regardless of training and

Fig. 5–5. Technical director monitoring programs at television broadcasting station (*WRC-TV, Washington, D.C.*)

experience, service technicians must stay up to date. Special courses and lectures about new products and techniques are offered at technical schools. These courses often are taught by factory personnel. Attending these courses provides an effective way for radio and television technicians to improve their skills.

BROADCAST TECHNICIANS

Broadcast technicians set up, operate, and maintain the equipment used to record and transmit radio and television programs. This includes transmission consoles, microphones, disc and tape recording and playback equipment, television cameras, videotape recorders, motion-picture projectors, and lighting and sound-effects devices (Fig. 5–5).

Broadcast technicians receive their basic training on the job or by taking courses in the technical aspects of broadcasting. In addition to the technical skills needed, any person who operates broadcast transmitters is required by federal law to have a first-class radiotelephone license issued by the Federal Communications Commission (FCC). Applicants for the license must pass a series of written examinations. These examinations test knowledge of transmitting and receiving equipment and electromagnetic wave theory. In addition, the examinations include questions about governmental and international regulations and practices concerning broadcasting. Preparation for the license examinations is available in special school programs and correspondence courses.

Fig. 5–6. Telephone central-office-equipment installer at work on a distributing frame (*American Telephone and Telegraph Company*)

TELEPHONE WORKERS

A large number of skilled technicians and craft workers are needed in the telephone industry. Telephone workers install, maintain, and repair the vast amount of wiring and equipment used in the telephone industry. Telephone companies frequently employ young, inexperienced men and women with the proper technical aptitude for this work. Then they are trained for some of these jobs.

Central-office-equipment Installers. These workers assemble, adjust, and wire the switching and dialing equipment in central offices of telephone companies (Fig. 5–6). Their jobs may involve installing a completely new central-office system. They may also add equipment to an existing system or replace outdated equipment. New employees receive their training both on the job and by classroom instruction. Because of the complexity of the work, this training continues as new techniques and equipment are developed and introduced.

Telephone Installers. The principal responsibility of these workers is to install, rearrange, and remove telephones and related equipment in homes and offices. New telephone installers usually begin their training with classroom instruction. This is followed by work with an experienced installer for on-the-job training. In many telephone companies, especially the smaller ones, an installer may work on private branch exchange (PBX) equipment. Private branch exchange systems are often found in private companies. These systems generally have special requirements. For example, in some private companies, incoming night calls must be switched automatically to different telephones. These telephones are often on the desks of persons who are working overtime or late shifts.

Repair Personnel. Repair personnel and installers, together, make up the largest group of craft workers employed by the telephone industry. Repair personnel locate and correct troubles that occur in customer telephones and related equipment (Fig. 5–7). Those who work with private branch exchange equipment test and maintain switchboards, batteries, relays, and power plants. Others maintain the equipment used for radio and television broadcasts, radiotelephones, and teletypewriters. They may also repair and maintain equipment such as electric-signal systems, public-address systems, and time clocks.

Line Workers and Cable Splicers. Line workers install wires and cables and do other jobs associated with telephone transmission lines. Although new installations are a very important part of their work, line workers also inspect, repair, and maintain existing telephone lines.

Cable splicers make line connections between cables after they have been installed by line workers underground or on poles. Splicers work on aerial platforms, underground, and in basements of large buildings. The principal duty of cable splicers is to connect the individual wires that run in large cables. Cable splicers also rearrange wire connections at cable splices, whenever telephone line circuits are changed, and maintain and repair cables.

APPLIANCE SERVICE WORKERS

Appliance service workers install, repair, and rebuild large and small electrical appliances. They work on appliances such as ranges, refrigerators, washing machines, clothes dryers, irons, and toasters (Fig. 5–8). In small shops, they work on all kinds of appliances. In larger shops, they usually specialize in one general kind.

Appliance service workers are generally employed by the service departments of stores, wholesale houses, and appliance manufacturers. Others work in electrical-repair shops or operate their own shops. As the number of appliances grows, so will the need for qualified service workers.

Most workers who service appliances begin as helpers and learn their skills through practical experience. These workers

Fig. 5–7. An apprentice repairer checking a line in a telephone circuit system (*U.S. Department of Labor*)

Fig. 5–8. Appliance service worker repairing an electric coffee pot (*U.S. Department of Labor*)

often attend training sessions sponsored by appliance manufacturers. They also meet with factory representatives to learn about new products and the general procedures for servicing them.

The repair of simple appliances can be learned by on-the-job training in several months. However, to become competent in servicing a variety of complicated appliances takes several years of study and training. This training period can be shortened if the beginning service worker has some knowledge of electricity. Knowing about electrical theory and the use of basic test instruments provides a good foundation for training. In addition, an understanding of the characteristics and operation of electrical and electronic devices is helpful.

BUSINESS-MACHINE SERVICE WORKERS

Business-machine service workers maintain and repair the rapidly increasing numbers and kinds of office equipment. This includes electric typewriters, calculating machines, cash registers, electronic computer systems, dictating and transcribing machines, duplicating and copying machines, and microfilm equipment. All these are combinations of mechanical and electrical or electronic systems. Business-machine service workers must have a strong background in electrical and electronic theory and applications. They must also have the ability to analyze a mechanical system. This is essential to understand the system's relationship to the circuits and devices that control it. Much of the work of business-machine service workers is done in offices during regular business hours. Therefore, it is important that they be neat and cooperative in addition to being technically able. Many business-machine service workers begin as inexperienced helpers. Then, after several years of on-the-job training, they become qualified to work on some machines. Because of the nature of this work, a person who plans to become a business-machine service worker needs a good background in both mechanical and electrical or electronic technology.

CONSTRUCTION ELECTRICIANS

The occupation of construction electrician is one of the most important in the skilled building trades. These workers lay out, assemble, install, connect, and test electric circuits, fixtures, devices, machinery, and control equipment. They work in houses, factories, hospitals, schools, and offices. Since most of their work is done indoors, year-round employment is more common for electricians than it is for other workers in the building trades. Most construction electricians are employed by electrical or building contractors. Others are self-employed.

Construction electricians usually begin their careers as apprentices. During this time, they work under the close supervision of experienced workers (Fig. 5–9). Apprenticeship training also involves classroom instruction in all phases of wiring and installation techniques. The beginner should have some previous training in basic mathematics, electrical theory, and the use of common hand and power tools. Technical and vocational schools often offer courses in electrical wiring, machinery, and control devices.

MAINTENANCE ELECTRICIANS

Maintenance electricians form the largest group of skilled electrical workers. They inspect, troubleshoot, repair, and maintain electrical wiring, machinery, motors, generators, and other electrical equipment. Most of this work is done in industrial plants. Machinery in factories must be constantly maintained and repaired to prevent costly interruptions in production. Maintenance electricians also work in hospitals, commercial and institutional buildings, and large apartment developments (Fig. 5–10).

AUTOMOTIVE SERVICE

The electrical system of the modern automobile must be periodically checked and tested. In small garages, this is usually done by an automotive mechanic who has experience with the electrical systems (Fig. 5–11). In larger garages, automotive electrical work is done by specialists. These people are trained to interpret the readings of a variety of electronic testing instruments and to make the needed repairs.

ELECTRICAL UTILITY LINE WORKERS

Line workers install and repair transmission and distribution cables and equipment of electrical utility companies. These companies supply power to residential, commercial, and industrial users. Line work is performed on both low-voltage and high-voltage systems. Work is also performed overhead on tall towers or underground with duct and conduit systems (Fig. 5–12). High-voltage line workers are highly skilled. They are especially trained and experienced in the problems connected with voltages that may be as high as 750,000 volts.

Fig. 5–9. A construction electrician wiring an electric outlet box (*U.S. Department of Labor*)

Fig. 5–10. A maintenance electrician repairing a motor from an air conditioning unit (*U.S. Department of Labor*)

Fig. 5-11. Making electrical connections for a "jump start" on an automobile (*Ford Motor Company*)

Most line workers receive their training on the job. However, theoretical and practical training in general electricity may be obtained in high schools, special technical institutes, and community colleges.

PREPARING FOR THE FUTURE

What you have just read are brief introductions to some of the important electrical and electronic occupations. These occupations will be open to young women and men in the years ahead. There is much more to be said about each of them. In addition, there are many more occupations from which to choose (Fig. 5-13). If you plan to enter an electrical or electronic occupation, taking an electricity and electronics course is an important step in that direction. Do not forget that your prospective employer will be interested in both your technical skills and aptitude and your personal qualities. Will you be interested in your job and in becoming better at it? Will you be willing to cooperate with coworkers? Are you trustworthy and dependable? Developing these qualities can be your best preparation for the future.

SOURCES OF INFORMATION

Much useful information about many different kinds of occupations can be found in the *Occupational Outlook Handbook*, published by the U.S. Department of Labor, Bureau of Labor Statistics. This handbook is available at your school or local library. Valuable occupational information can also be

Fig. 5-12. Utility line worker

Fig. 5-13. Scientist working in the space program shown with a model of the Orbiting Solar Observatory (*National Aeronautics and Space Administration*)

obtained from your school counselors. In addition, discuss your job interests with local technicians and with representatives of colleges, technical schools, and community colleges.

SELF-TEST

Test your knowledge by writing, on a separate sheet of paper, the word or words that most correctly complete the following statements:

1. There are approximately a million _____ and _____ technicians employed in the United States.
2. It usually takes _____ years to complete the formal education necessary to become an electrical engineer.
3. Persons who draw the plans for electrical systems are called _____.
4. _____ installers are likely to be found working on private branch exchange (PBX) systems.
5. Construction _____ are persons who install, connect, and test electrical circuits in buildings.
6. A U.S. Department of Labor publication entitled _____ provides much useful information on occupations.

FOR REVIEW AND DISCUSSION

1. Discuss why a high school education has become standard for American workers.
2. Name several high school courses that are helpful in preparing for technical occupations.
3. Name several kinds of post–high school education and training programs that are useful in preparing for technical occupations.
4. What personal characteristics do you consider to be important qualifications for entering a technical occupation?

INDIVIDUAL-STUDY ACTIVITIES

1. Prepare a written or an oral report telling about your possible choice of a future occupation and the steps you feel you will have to take to achieve this goal.
2. Visit a local electrical or electronics service shop or manufacturing plant and tell the class about your experiences.

Electric Circuits and Devices

Unit 6 Voltage, Current, and Power

Static electricity is electrical energy at rest. Although this form of electricity can be very useful, it cannot operate a load such as lamps, heaters, motors, and other devices. This takes dynamic, or active, electricity.

Dynamic electricity involves the transfer of energy from a source to a load. This is done by the movement of electrons through a circuit. The energy that forces the electrons through a circuit is the *electromotive force.* The electromotive force, abbreviated emf, is measured in units called *volts.* Because of this, it is often referred to as *voltage.* When you hear or read the term *voltage,* you will know that it is the force that moves electrons through a circuit.

As electrons flow through the load, their energy is changed into some other form of energy. In an electric heater, for example, the electrons passing through the heater wire, or element, give up their energy in the form of heat. This is how moving electrons can be used to transfer energy from a source to a load.

Because electrons are negatively charged, they are attracted by positive charges and repelled by negative charges. If two charged objects are connected by a conducting material such as a wire, a *current* of electrons will flow from the negative object to the positive object. To produce a continuous electric current in the wire, energy must be supplied continuously (Fig. 6–1). In a circuit, this energy is provided by a source such as a *dry cell,* a *battery,* or a *generator.*

Electrons are not used up as they move through a circuit. Therefore, the number of electrons that returns to the positive (+) terminal of a source of energy is the same as the number of electrons that leaves the negative (−) terminal of the source of energy (Fig. 6–2).

THE VOLT AND THE AMPERE

The basic unit of voltage is the volt. It is named after Alessandro Volta. He was an Italian professor who lived from 1745 to 1798. Most of the electrical and electronic equipment found in homes operates at about 120 volts. A common flashlight dry cell produces 1.5 volts. A modern automobile battery produces 12 volts. The voltage used to operate a television picture tube may be as high as 27,500 volts. The letter symbol for voltage is E.

The basic unit of current is the *ampere.* It is named after André Marie Ampère. He was a French physicist and mathematician who lived from 1775 to 1836. One ampere of current equals the movement of 6,280,000,000,000,000,000 electrons past any point in a circuit during one second of time. By using *scientific notation,* such a large number can be written as 6.28×10^{18}.

A 100-watt light bulb requires about 0.8 ampere of current to operate. A $\frac{1}{4}$-horsepower electric motor needs about 4.6 amperes of current to operate. The cranking motor of an automobile may use over 200 amperes when the starter switch is turned on. The letter symbol for current is I.

VOLTAGE AND CURRENT REQUIREMENTS

If an electrical device is to work properly, the source of energy must be able to do two things. First, it must supply the voltage. Second, it must deliver the current for which the device was designed. For example, you can connect eight flashlight dry cells together in such a way as to form a battery that produces 12 volts. However, if you tried to start an automobile engine with this battery, it would not work. This is because it does not have the ability to deliver the large amount of current needed to operate the automobile's electric cranking motor. To do that, a larger battery, also producing 12 volts but having a much larger current-delivering capacity, must be used. It is important, therefore, to know both the voltage and current requirements of electrical appliances and tools. These requirements are often given on the nameplates attached to these products. They may also be found in the technical literature that accompanies the products.

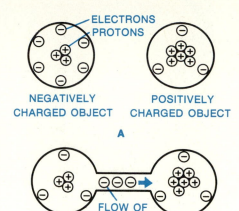

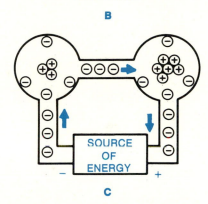

Fig. 6–1. (A) Two charged objects; (B) electrons are attracted to the positively charged object; (C) a source of energy provides a continuous supply of electrons.

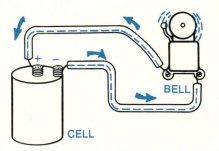

Fig. 6–2. Electrons are not consumed as they move through a circuit.

Table 6-1. Common Numerical Prefixes Used with Metric and Other Units

Prefix and Symbol	Numerical Equivalent	Powers of Ten
giga (G)	1 000 000 000	$= 10^9$ (thousand-millions*)
mega (M)	1 000 000	$= 10^6$ (millions)
kilo (k)	1 000	$= 10^3$ (thousands)
milli (m)	0.001	$= 10^{-3}$ (thousandths)
micro (μ)	0.000 001	$= 10^{-6}$ (millionths)
nano (n)	0.000 000 001	$= 10^{-9}$ (billionths or thousand-millionths)
pico (p)	0.000 000 000 001	$= 10^{-12}$ (trillionths or million-millionths)

*In the United States we call this number a billion; in England a million-million (1,000,000,000,000) is called a billion.

UNIT PREFIXES

Very small and very large values of voltage and current are usually expressed in a shorthand way. This is done by using a system of *decimal prefixes*. The most common of these prefixes used in electricity and electronics and the numerical values they represent are given in Table 6-1. As shown in this table, the prefix *milli-* (m) means one-thousandth (0.001). A current of 0.001 ampere can then be expressed as a current of 1 milliampere, or more simply, 1 mA. Likewise, a voltage of 1000 volts can be expressed as 1 kilovolt, or 1 kV.

DIRECT CURRENT

Direct current (dc) is produced in a circuit by a steady voltage source. That is, the positive and negative *terminals*, or poles, of the voltage source do not change their charges over time. These terminals are said to have *fixed polarity*. Therefore, the direction of the current does not change over time. Such a voltage is provided by electric cells, batteries, and dc generators. Direct current may be constant, or steady, in value (Fig. 6–3 at A). It may vary, or change, in value (Fig. 6–3 at B). The current also may be pulsating, or interrupted (Fig. 6–3 at C). The applied voltage and the nature of the load determine the kind of direct current supplied.

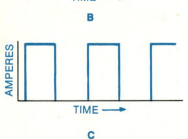

Fig. 6–3. Direct current: (A) steady direct current; (B) varying direct current; (C) pulsating direct current

ALTERNATING CURRENT

Alternating current (ac) is produced by a voltage source whose polarity changes, or alternates, with time. This causes the current in the circuit to move in one direction and then in the other (Fig. 6–4). The most common source of alternating voltages is alternating-current generators, or alternators.

In addition to changing direction, most kinds of alternating current change in value with time. For example, the variation of current with time may follow the form of a sine wave. This is called *sinusoidal alternating current.* Most electric utilities in this country supply sinusoidal alternating current and voltage to their customers.

Sine Wave. A graph called a *sine wave* shows the direction and the value of the current that passes through a given point in a circuit during a certain period of time (Fig. 6–5). One complete wave is a *cycle* of alternating current. The time it takes to complete one cycle is the *period* of the wave.

In Fig. 6–5, points on the vertical line AB represent current values. Base line CD is the time line. The heavy curved line, the sine wave, shows how the current changes in value during one cycle. That part of the sine wave above the base line represents the movement of current in one direction. The part of the sine wave below the base line represents the movement of current in the other direction.

During one alternation, or one-half, of the sine-wave cycle, the current moving in one direction increases from zero to a maximum value, and then returns to zero. At that time, the current begins to increase again, but in the opposite direction. It again increases to a maximum value and then decreases to zero. This completes one cycle.

Frequency. The number of times that a cycle of alternating current is repeated during one second of time is called the *frequency* of the current. The basic unit of frequency is the *hertz* (Hz). It is named after Heinrich Rudolph Hertz. He was a German physicist who lived from 1857 to 1894. The hertz has replaced cycles per second (cps) as the unit of frequency. However, cps still appears in some older books and other literature.

In this country, the frequency supplied by power companies is generally 60 Hz, although 50 Hz is used in some areas. Frequencies in the thousands and millions of hertz are used in radio and television broadcasting and microwave communications. They are usually expressed in terms of kilohertz (kHz), megahertz (MHz), and gigahertz (GHz) units (see Table 6–1).

USING DIRECT OR ALTERNATING CURRENT

Electric energy is most often supplied in the form of alternating current. Therefore, most electrical and electronic prod-

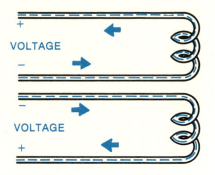

Fig. 6–4. The polarity of the voltage in an ac circuit alternates, or changes, at regular intervals.

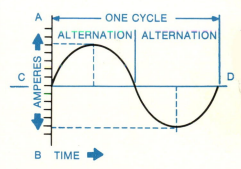

Fig. 6–5. A sine wave describes the variation in the value of current with time in the usual ac circuit.

ucts are designed to operate with it. However, many of these products will work with either alternating or direct current. This is true, for example, of common lamps and almost all equipment with heating elements, such as toasters and irons. The right voltage must, of course, be applied to these devices.

Direct current is needed to operate devices such as transistors and electron tubes. Radios that are not battery-operated must be provided with a source of direct current. A special circuit in the radio changes the power-line alternating current into the necessary direct current. This special circuit is called a *rectifier circuit*. Direct current must also be used for certain electrochemical processes such as charging batteries and electroplating.

Direct-current power supplies are designed to provide direct-voltage outputs that can usually be varied from zero to a certain maximum voltage. These power supplies are used wherever a controlled variable voltage is needed, such as in experimental work and testing of circuits.

Alternating current must be used to operate the induction motors used in appliances such as washing machines and refrigerators. Other motors, called *universal motors*, can be operated with either alternating or direct current. Transformers must be operated with alternating current or with varying or pulsating direct current. Inverters are used to change direct current to alternating current.

NONSINUSOIDAL ALTERNATING CURRENTS

Alternating voltages and currents may have waveforms that are not like sine waves. These are *nonsinusoidal waveforms* (Fig. 6–6). Television sets, electronic computers, and other devices may have circuits in which these voltage waveforms are present.

ELECTRIC POWER AND ENERGY

Power is the time rate of doing work. In an electric circuit, power may also be defined in two other ways. First, it is the rate at which electric energy is delivered to a circuit. Second, it is the rate at which an electric circuit does the work of converting the energy of moving electrons into some other form of energy. The basic unit of power is the *watt* (W). It is named after James Watt. He was a Scottish inventor who lived from 1736 to 1819. Much of the electrical equipment we use is rated in terms of the power or watts it consumes. Thus, a light bulb is rated as 60 W, 100 W, 200 W, and so on.

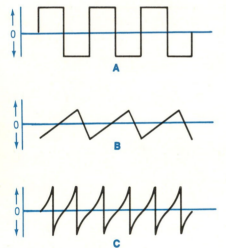

Fig. 6–6. Nonsinusoidal ac waveforms: (A) square waveform; (B) sawtooth waveform; (C) peaked waveform

Toasters, irons, clothes dryers, and other appliances also have a certain wattage rating. Since electric power is the rate at which energy is delivered to the circuit, it is easy to calculate. You simply multiply the power by some time unit. This will give you the total energy delivered to the circuit. The time unit usually used is the hour. The power unit is usually given in terms of kilowatts (1,000 watts). Therefore, the energy unit on which most electric companies base their bills is the kilowatthour (kWh).

SELF-TEST

Test your knowledge by writing, on a separate sheet of paper, the word or words that most correctly complete the following statements:

1. Electromotive force causes _____ to move through a circuit.
2. Electromotive force is measured in units called _____.
3. The flow of electrons in a circuit is called an electric _____.
4. Electrons flow through a circuit from the _____ terminal of the energy source to the _____ terminal of the energy source.
5. A source of energy must be able to supply the _____ and deliver the _____ for which the load was designed.
6. Direct current moves through a circuit in one _____ only.
7. The action of a sinusoidal current is best described by what is called a _____ wave.
8. In this country, the frequency of the alternating current in power lines is usually _____ Hz.
9. Examples of devices or products that can be operated with either direct or alternating current are _____ and _____.
10. Direct current must be used for the operation of _____ and _____ tubes.
11. A _____ circuit is one that changes alternating current into direct current.
12. Electric companies base their bills on a unit called the _____.

FOR REVIEW AND DISCUSSION

1. Define voltage and current.
2. Name the basic units of voltage and current.
3. What are the letter symbols for voltage and current?
4. Explain why it is impossible to start a 12-volt automobile cranking motor with a typical 12-volt dry-cell battery.
5. Write the numerical values of the prefixes *mega-, kilo-, milli-,* and *micro-.*
6. Describe steady direct current, varying direct current, and pulsating direct current.
7. Give two important characteristics of sinusoidal alternating current.
8. Draw a sine wave. Explain how it can be used to describe a cycle of sinusoidal alternating voltage or current.
9. Define frequency. Name the basic unit of frequency.
10. Name three kinds of loads that must be operated with direct current.
11. What is the basic unit of power?
12. What is the common unit used for electric energy?

INDIVIDUAL-STUDY ACTIVITIES

1. Prepare a written or an oral report telling about the numerical prefixes used with metric and other units. Discuss prefixes not included in Table 6-1.
2. Give a demonstration that will be helpful in explaining sinusoidal waveforms.

Unit 7   Conductors and Insulators

Since moving electrons have energy, they can be made to do work for us. What electrons do and how they can be controlled in a circuit mostly depends on the kinds of materials through which they flow. There are also certain materials through which electrons cannot flow easily. Knowing about these materials will help you understand better the many ways in which electric energy is used.

CONDUCTORS

A *conductor* is a material through which electrons flow easily. In such materials, valence electrons of the outermost shell can be quite easily removed from their parent atoms by the force of voltage. To put it another way, a conductor is a material having many free electrons.

Three good electrical conductors are silver, copper, and aluminum. In fact, metals generally are good conductors. Certain gases also can be used as conductors under special conditions. For example, neon gas, argon gas, mercury vapor, and sodium vapor are used in lamps.

How Electrons Flow. The basic way electrons flow through a wire is shown in Fig. 7–1. The action begins as the positive terminal of the cell attracts a valence electron away from atom 3. Atom 3 is now positively charged, having lost part of its negative charge. Thus, it attracts an electron from atom 2. Atom 2, in turn, attracts an electron from atom 1. As atom 1

Fig. 7–1. In a conductor material, voltage causes some electrons to be removed from their atoms. These ''free'' electrons then move through the conductor.

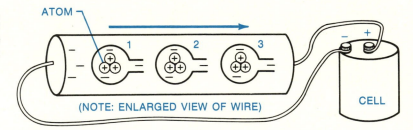

(NOTE: ENLARGED VIEW OF WIRE)

loses its electron, an electron leaves the negative terminal of the cell and is attracted to atom 1.

In a circuit, this action continues among very large numbers of atoms within the conductor. It produces a steady flow of free electrons through the conductor. The direction of this electron flow is always from the negative to the positive terminal of the source of energy.

Electrical Impulse. Individual electrons flow through a conductor at a relatively slow speed, usually less than 1 inch (25.4 mm) per second. However, an *electrical impulse,* or force, travels through a conductor at the speed of light, 186,000 miles (300 000 km) per second. This means that after a voltage is applied to a circuit, it produces a flow of free electrons through all points in that circuit at the speed of light.

It might be hard for you to imagine that an electrical impulse travels at the speed of light. Maybe this example will help. Suppose you could build an electric circuit 186,000 miles (300 000 km) long. This is more than seven times the distance around the Earth at the equator. One second after you closed the switch on such a circuit, the electrical impulse would take place throughout it.

RESISTANCE

As electrons flow through a conductor, they collide with other electrons and with other atomic particles. These collisions tend to reduce the number of electrons that flow through the conductor. This is like the difficulty that water has in passing through a hose or pipe that is partly filled with dirt or sand.

The opposition to current that results from electronic collisions within a conductor is called *resistance*. The basic unit of resistance is the *ohm*. It is named after Georg Simon Ohm. He was a German physicist who lived from 1787 to 1854. The letter symbol for resistance is R.

One ohm of resistance is that amount of resistance that will limit the current in a circuit to 1 ampere when 1 volt is

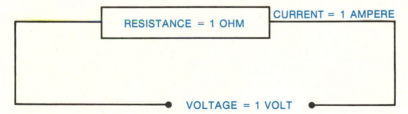

RESISTANCE = 1 OHM

CURRENT = 1 AMPERE

VOLTAGE = 1 VOLT

Fig. 7–2. One ohm is the amount of resistance that will allow one ampere of current to pass through a circuit to which a voltage of one volt is applied.

applied to the circuit (Fig. 7–2). Larger values of resistance are often expressed by using the prefixes *kilo-* and *mega-*. For example, a resistance of 2,000 ohms can be written as 2 kilohms, and a resistance of 2,000,000 ohms can be written as 2 megohms. The Greek capital letter omega, Ω, is used as an abbreviation for ohms after a numerical value, for example 25 Ω. A resistance of 2 kilohms can also be written as 2 kΩ. A resistance of 2 megohms can be written as 2 MΩ.

RESISTANCE OF METAL CONDUCTORS

The resistance of a metal conductor depends mainly on four things: (1) the kind of metal from which it is made, (2) its temperature, (3) its length, and (4) its cross-sectional area.

Different metals have different properties of electrical resistance. The resistances of several common metals compared with that of copper are given in Table 7-1.

In general, as a metal gets hotter, its resistance increases. A hot wire has more resistance than the same wire when it is cold. For example, the resistance of the tungsten wire filament in a 100-watt light bulb is about 10 ohms when the bulb is off. When the bulb is lit, the resistance of the white-hot filament increases to about 100 ohms. As in any conductor, heat causes atoms within the wire to move about much more

Table 7-1. Relative Resistance

Metals	Relative Resistance Compared with Annealed Copper*
aluminum (pure)	1.70
brass	3.57
copper (hard-drawn)	1.12
copper (annealed)	1.00
iron (pure)	5.65
silver	0.94
tin	7.70
nickel	6.25–8.33

*For example, a silver wire has only 0.94 times as much resistance as the same-size copper wire, whereas an aluminum wire has 1.70 times, or 170%, the resistance of copper.

rapidly than usual. This increases the number of collisions between free electrons and other atomic particles within the wire. As a result, the resistance to the flow of electrons through the wire also increases.

Just as resistance increases with an increase in temperature, a decrease in temperature will produce a decrease in resistance. If the temperature is reduced to absolute zero ($-273.16°$C, or $-459.69°$F), the resistance of the conductor will be zero. This is so because all molecular activity in the material will stop at absolute zero. There will thus be nothing to resist the flow of electrons. This condition is called *superconductivity*. Temperatures very close to absolute zero have been produced in the laboratory. This has made it possible to operate circuits having almost no resistance.

The resistance of a wire increases as its length increases. If

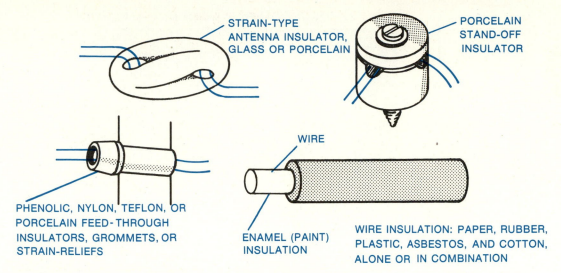

STRAIN-TYPE ANTENNA INSULATOR, GLASS OR PORCELAIN

PORCELAIN STAND-OFF INSULATOR

WIRE

PHENOLIC, NYLON, TEFLON, OR PORCELAIN FEED-THROUGH INSULATORS, GROMMETS, OR STRAIN-RELIEFS

ENAMEL (PAINT) INSULATION

WIRE INSULATION: PAPER, RUBBER, PLASTIC, ASBESTOS, AND COTTON, ALONE OR IN COMBINATION

Fig. 7–3. Common insulating materials and their uses

two wires of different lengths have the same cross-sectional area, the longer wire will have the greater resistance. There are two reasons for this. First, electrons must move a greater distance in the longer wire. Second, they collide with more particles in the longer wire.

If two wires are of the same length, the wire with the larger cross-sectional area will have less resistance. This is because the thicker wire has more free electrons. It also has more space through which these electrons can move.

INSULATORS

An electrical *insulator* is a material that does not easily conduct an electric current. Such materials contain valence electrons that are tightly bound to the nuclei of their atoms. As a result, it takes an unusually high voltage to produce significant numbers of free electrons in them. Such materials are also called *nonconductors* and *dielectrics*.

Common insulators are glass, porcelain, mica, rubber, plastics, paper, and wood. These materials are used to separate conductors electrically so that the currents they carry will flow in the right paths (Fig. 7–3).

There is no sharp line dividing conductors from insulators. All insulating materials will conduct electric current if a high enough voltage is applied across them. For example, air is usually thought of as being a fairly good insulator. However, during a thunderstorm, the huge voltages generated between clouds and the earth cause air to conduct current in the form of lightning.

The ability of a material to insulate is known as its *dielectric strength*. The dielectric strengths of several common insulators are given in Table 7–2.

Table 7-2. Dielectric Strengths of Common Insulating Materials

Material	Dielectric Strength Breakdown Voltage in volts per 0.001 in. (0.0254 mm)
Bakelite	300
Formica	450
glass (window)	200–250
mica	3,500–5,000
polystyrene	500–700
porcelain	50–100
Teflon	1,000–2,000
air	75
kraft paper (glazed)	150
wood	125–750

SELF-TEST

Test your knowledge by writing, on a separate sheet of paper, the word or words that most correctly complete the following statements:

1. A conductor is a material through which electrons can flow _____ .
2. In a conductor, there are many _____ electrons.
3. In addition to metals, certain _____ are also used as conductors.
4. Electrons move through a circuit conductor in a _____ to _____ direction.
5. The _____ to current that results from electronic collisions within a conductor is known as resistance.
6. The basic unit of resistance is the _____ .
7. The Greek capital letter _____ is used as an abbreviation for ohms after a numerical value. This letter is written as _____ .
8. In almost all metal conductors, the resistance _____ as the temperature increases.
9. With two wires of the same length, the wire with the larger cross-sectional area has _____ resistance.
10. All insulating materials will conduct current if a high enough _____ is applied to them.

FOR REVIEW AND DISCUSSION

1. Define an electrical conductor. Name at least three good conductor materials.
2. Describe the movement of electrons through a conductor.
3. What is meant by an electrical impulse?
4. Define electrical resistance.
5. Name and define the basic unit of resistance.
6. What are the four things that determine the resistance of a metal conductor?
7. Explain why, with most metal conductors, the resistance increases as the temperature increases.
8. Define an electrical insulator. Name five common insulating materials.
9. Under what condition can a material that is normally an insulator become a conductor?
10. What is meant by the dielectric strength of a material?

Unit 8   Resistance and Resistors

Resistance may or may not be useful in practical circuits. When too much current passes through a conductor, the resistance of the conductor may cause it to become hot. This, in turn, can create a fire hazard or cause the conductor to burn out. In this case, resistance is not desirable.

In other cases, resistance is deliberately put into a circuit. When this is done, the resistance is the circuit load. In an electric range, for example, the heating elements are made of a special wire called *resistance wire*. As current passes through this wire, the energy of moving electrons is changed into useful heat energy. Resistance is also put into many different circuits by the use of resistors.

Fig. 8–1.   Resistors are made in many sizes and shapes.

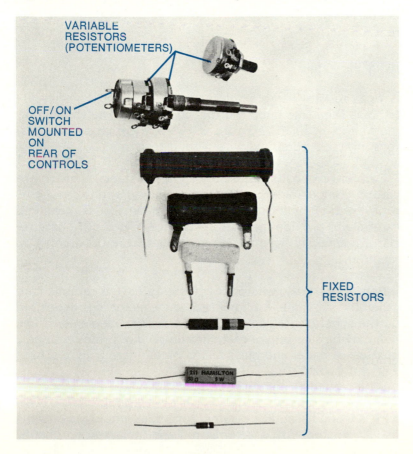

VARIABLE
RESISTORS
(POTENTIOMETERS)

OFF/ON
SWITCH
MOUNTED
ON
REAR OF
CONTROLS

FIXED
RESISTORS

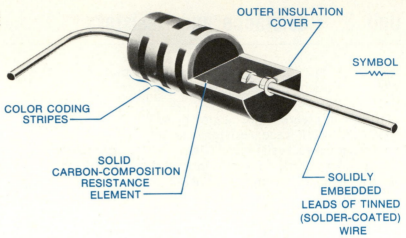

Fig. 8–2. The main elements of a fixed carbon-composition resistor (*Allen-Bradley Company*)

OUTER INSULATION COVER

SYMBOL

COLOR CODING STRIPES

SOLID CARBON-COMPOSITION RESISTANCE ELEMENT

SOLIDLY EMBEDDED LEADS OF TINNED (SOLDER-COATED) WIRE

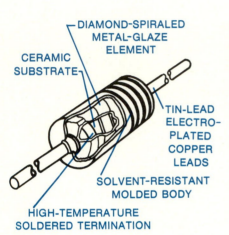

DIAMOND-SPIRALED METAL-GLAZE ELEMENT

CERAMIC SUBSTRATE

TIN-LEAD ELECTRO-PLATED COPPER LEADS

SOLVENT-RESISTANT MOLDED BODY

HIGH-TEMPERATURE SOLDERED TERMINATION

Fig. 8–3. Metal-glaze resistor

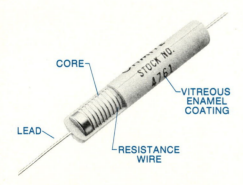

CORE

STOCK NO. 4761

VITREOUS ENAMEL COATING

LEAD

RESISTANCE WIRE

Fig. 8–4. Wire-wound resistor (*Ohmite Manufacturing Company*)

RESISTORS

A *resistor* is a device with a known value of resistance. Resistors are very common parts of many electric and electronic circuits (Fig. 8–1). They are used to control voltage and current.

FIXED RESISTORS

A *fixed resistor* has a single value of resistance. This value remains the same under normal conditions. The three common kinds of fixed resistors are carbon-composition resistors, film resistors, and wire-wound resistors.

Carbon-composition Resistors. The resistance element in a carbon-composition resistor is mainly graphite or some other form of solid carbon. The carbon material is measured carefully to provide the right resistance (Fig. 8–2). These resistors generally have resistance values from 0.1 ohm to 22 megohms.

Film Resistors. Resistors of this kind have a ceramic core called the *substrate*. A film of resistance material is deposited on the substrate. This serves as the resistance element. The film may be a carbon or metallic compound. It may also be a mixture of metal and glass called *metal glaze* (Fig. 8–3).

Wire-wound Resistors. The resistance element of a fixed wire-wound resistor is usually nickel-chromium wire. This wire is wound around a ceramic core. This whole assembly is usually coated with a ceramic material or a special enamel

(Fig. 8–4). Wire-wound resistors generally have resistance values from 1 ohm to 100 kilohms.

Tolerance. The actual resistance of a resistor may be greater or less than its rated value. This variation is called the *tolerance.* Carbon-composition, film, and wire-wound resistors each have a given tolerance. For example, common tolerances of carbon-composition resistors are ±5 percent, ±10 percent, and ±20 percent. This means that a resistor with a stated resistance of 100 ohms and a tolerance of ±5 percent can actually have a resistance of any value between 95 and 105 ohms. General-purpose wire-wound resistors usually have a tolerance of ±5 percent.

Resistors having tolerances as high as ±20 percent are used in many electric and electronic circuits. The advantage in using high-tolerance resistors is that they are less expensive to make than low-tolerance resistors. However, they can be used only in circuits where the variation is not important.

Precision Resistors. Some wire-wound and film resistors have actual values that are nearly equal to their rated values. These are called precision resistors. They are used in special circuits such as in test instruments.

Power Rating. The power rating of a resistor indicates how much heat a resistor can throw off before burning out. Since it is current that produces the heat, the power rating also gives some indication of the maximum current a resistor can safely carry. The power rating of a resistor is given in watts.

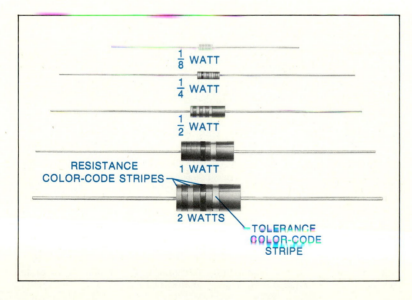

Fig. 8–5. Relationship between the physical size of carbon-composition resistors and their wattage ratings (*Allen-Bradley Company*)

Carbon-composition resistors have power ratings from $\frac{1}{16}$ watt to 2 watts. Wire-wound resistors have power ratings from 3 watts to hundreds of watts.

The physical size of a resistor has nothing to do with its resistance. A very small resistor can have a very low or a very high resistance. The physical size of a resistor does, however, suggest its wattage rating. For a given value of resistance, the physical size of a resistor increases as the wattage rating increases (Fig. 8–5). With experience, you can soon learn to tell the wattage ratings of resistors from their physical sizes.

RESISTOR COLOR CODE

The resistance and wattage values of wire-wound resistors are usually printed directly on them. The resistance values of fixed carbon-composition resistors and some kinds of film resistors are shown by a *color code.* The colors of the code and the numbers they stand for are given in Table 8–1. Three stripes are used for resistance values. A fourth stripe is often used to show the tolerance. A fifth stripe, when used, tells the failure rate. That is the amount the resistance will change over a period of time, such as 1,000 hours. The color code is read as follows:

1. The first, or end, stripe indicates the *first number* of the resistance value.
2. The second stripe indicates the *second number* of the resistance value.
3. The third stripe indicates the *number of zeros* that follow the first two numbers of the resistance value. If the third

Table 8–1. Resistor Color Code

Resistance Values First Three Stripes		Tolerance Values Fourth Stripe	
black	= 0	gold = ±5%	
brown	= 1	silver = ±10%	
red	= 2	none = ±20%	
orange	= 3		
yellow	= 4	**Failure Rate Fifth Stripe**	
green	= 5		
blue	= 6		
violet	= 7	brown = 1%	
gray	= 8	red = 0.1%	
white	= 9	orange = 0.01%	
gold	= divide by 10	yellow = 0.001%	
silver	= divide by 100		

(Fifth stripe, if present, indicates the failure rate, or the amount by which the resistance will change during a given period of time, usually 1,000 hours.)

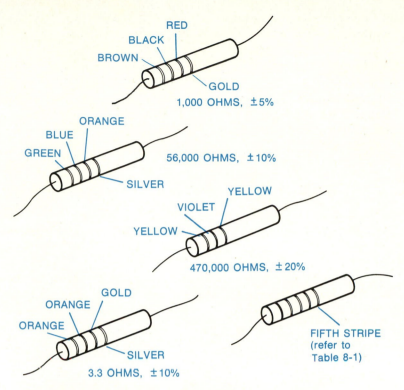

RED
BLACK
BROWN
GOLD
1,000 OHMS, ±5%

ORANGE
BLUE
GREEN
SILVER
56,000 OHMS, ±10%

YELLOW
VIOLET
YELLOW
470,000 OHMS, ±20%

ORANGE
GOLD
ORANGE
SILVER
3.3 OHMS, ±10%

FIFTH STRIPE
(refer to
Table 8-1)

Fig. 8–6. Reading the resistor color code

stripe is black, no zeros are added after the first two numbers, since black represents zero.

4. If the third stripe is gold, the number given by the first two stripes is divided by ten.

5. If the third stripe is silver, the number given by the first two stripes is divided by 100.

Several examples of resistance values given by the color code are shown in Fig. 8–6.

VARIABLE RESISTORS

Variable resistors are used when it is necessary to change the amount of resistance in a circuit. The most common variable resistors are called *potentiometers* and *rheostats*. Potentiom-

Fig. 8–7. Variable resistors: (A) potentiometer; (B) wire-wound rheostat (*Ohmite Manufacturing Company*)

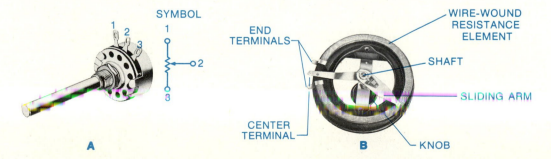

SYMBOL
1
2
3

END
TERMINALS

WIRE-WOUND
RESISTANCE
ELEMENT

SHAFT

SLIDING ARM

CENTER
TERMINAL

KNOB

A

B

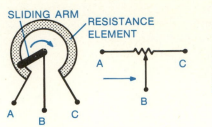

RESISTANCE BETWEEN B AND A INCREASES. RESISTANCE BETWEEN B AND C DECREASES.

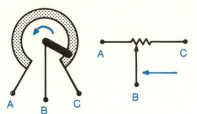

RESISTANCE BETWEEN B AND A DECREASES. RESISTANCE BETWEEN B AND C INCREASES.

Fig. 8–8. When the sliding arm of a variable resistor is moved, the resistance between the center terminals and end terminals changes.

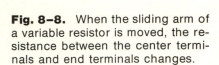

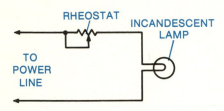

Fig. 8–9. Rheostat used to control current in a lamp circuit

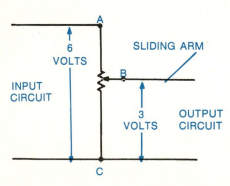

Fig. 8–10. Basic potentiometer action

eters generally have carbon-composition resistance elements. The resistance element in a rheostat is generally made of resistance wire. In both devices, a sliding arm makes contact with the resistance element (Fig. 8–7). In most variable resistors, the arm is attached to a shaft that can be turned in almost a full circle. As the shaft is turned, the point of contact of the sliding arm on the resistance element changes. This changes the resistance between the sliding-arm terminal and the terminals of the element (Fig. 8–8).

Rheostats are commonly used to control rather high currents, such as those in motor and lamp circuits. An example of how a rheostat is connected into a lamp circuit is shown in Fig. 8–9. Although similar to a potentiometer, a rheostat is usually larger because its resistance element carries greater currents and throws off greater amounts of heat.

A potentiometer can be used to vary the value of voltage applied to a circuit, as shown in Fig. 8–10. In this circuit, the input voltage is applied across terminals A-C of the resistance element. When the position of the sliding arm (terminal B) is changed, the voltage across terminals B-C will change. As the sliding arm moves closer to terminal A, the output voltage of the circuit increases. As the sliding arm moves closer to terminal C, the output voltage of the circuit decreases.

Potentiometers are commonly used as control devices in amplifiers, radios, television sets, and different kinds of meters. Typical uses include volume and tone controls; balance controls; linearity, brightness, and width controls; and zeroing adjustments.

The rating of a rheostat or of a potentiometer is the resistance of the entire resistance element. This resistance is measured from one end terminal to the other. The resistance and the wattage ratings of these devices are sometimes printed directly on them. They can also be found in manufacturers' specifications.

RESISTOR DEFECTS

Resistors are rugged devices. They seldom become defective unless too large a current passes through them. This may happen when there is a short somewhere in the circuit.

A carbon-composition or a film resistor that becomes overheated will often be burned completely apart. In other cases, overheating will cause such a resistor to be scorched, to crack, or to bulge out. The value of the resistance may then increase to many times normal.

When a wire-wound resistor becomes overheated, the resistance wire will often burn out at one point. This makes the resistor open. A carbon-composition, film, or wire-wound resistor also can become open if one of its wires is disconnected from the resistance element inside its body.

FUSIBLE RESISTORS

The most common kinds of *fusible resistors* are wire-wound devices that look like wire-wound resistors (Fig. 8–11). They

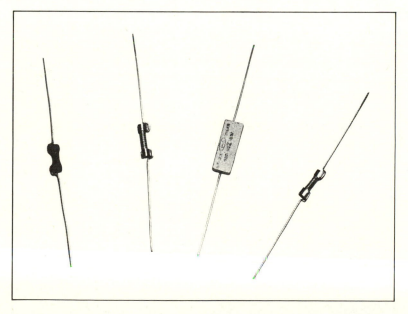

Fig. 8–11. Fusible resistors

are used often as fuses in amplifiers and in television sets to protect particular circuits. These fusible resistors usually have resistance values of less than 14 ohms. The resistance element, like the fuse link in a cartridge fuse, is designed to burn out when too much current passes through it.

LEARNING BY DOING

4. Using the Resistor Color Code. A color code is used with carbon-composition and film resistors. The following steps will help you practice reading this code.

MATERIALS NEEDED

10 different color-coded carbon-composition and/or film resistors.

Procedure

1. Make a table like Table 8–2 on a sheet of paper. Write the color code of each resistor in the right space in this table.
2. Write the resistance and the tolerance percentage of each resistor, as indicated by its color code, in columns 3 and 4 of the table.
3. Figure the tolerance range in ohms for each resistor and note this on the table. As an example, a 100-ohm resistor having a tolerance of 10 percent may have an actual resistance 10 percent greater than 100 ohms or 10 percent less than 100 ohms:

    10% of 100 = 10 ohms

    Therefore, the allowable range is (100 − 10) = 90 ohms to (100 + 10) = 110 ohms, and the range would be written as 90–110 ohms. Put the tolerance range in ohms for each resistor in column 5 of the table.

5. Installing a Loudspeaker "Fader" Control. The following circuit shows a very practical use of a potentiometer. It can be used to adjust the volume of two loudspeakers, such as the front and rear loudspeakers in an automobile.

MATERIALS NEEDED

potentiometer, 350 ohms, 2 watts
loudspeaker, if necessary. If this loudspeaker is to be added to an existing system, it should have the same electrical characteristics as the loudspeaker already in use.
hook-up wire, no. 20 AWG (0.8 mm), as needed

Table 8-2. Table for Learning by Doing No. 4, "Using the Resistor Color Code"

(1) Resistor	(2) Color Code	(3) Resistance	(4) Tolerance, %	(5) Tolerance Range, ohms
1				
2				
3				
4				
5				
6				
7				
8				
9				
10				

SAFETY
Many devices using loudspeakers, such as radios, televisions, and phonographs, *should not be expanded into multiple-speaker systems.* Often these devices are directly connected to the power lines. This means that the metal chassis *and the speaker wires* can be at a dangerous potential, or voltage, with respect to ground. Manufacturers of such equipment almost always disapprove of modifications. Check with a qualified technician before making any connections to these circuits.

Some equipment can be damaged by connecting the wrong speaker load. Again, check with a qualified technician or service center before attaching extra speakers to such equipment.

Procedure

1. Wire the circuit shown in Fig. 8–12. If the chassis of an automobile is used as one conductor of the loudspeaker circuit, the common wire is not necessary. In this case, one terminal of each loudspeaker is connected to a nearby part of the chassis as shown by the dotted lines.
2. Test the circuit by turning the shaft of the potentiometer first in one direction and then the other. As this is done, one of the loudspeakers should approach full volume while the volume of the other loudspeaker decreases. At the center position, the loudspeakers' volume should be equal.

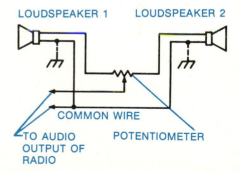
Fig. 8–12. Circuit used for Learning by Doing No. 5, "Installing a Loudspeaker 'Fader' Control"

Test your knowledge by writing, on a separate sheet of paper, the word or words that most correctly complete the following statements:

1. Resistors are used to put _____ into a circuit or part of a circuit.
2. Resistors are used to control _____ and _____.
3. A fixed resistor has a _____ single value of resistance.
4. The resistance element in a carbon-composition resistor is mainly _____ or _____.
5. In a film resistor, a film of _____ is deposited on a ceramic core.
6. The resistance element of a wire-wound resistor is made of _____ wire.
7. The amount by which the actual resistance of a resistor may vary from its indicated value is known as the _____ of the resistor.
8. A resistor that has a resistance that is very nearly equal to its indicated resistance is called a _____ resistor.
9. The _____ rating of a resistor indicates how much current the resistor can conduct before _____.
10. The physical size of a resistor has nothing to do with its _____.
11. A larger resistor of a given value has a greater _____ rating than a smaller resistor of the same value.
12. The first three stripes of the resistor color code indicate the _____ value of a resistor. The fourth stripe shows the _____ of the resistor.
13. The two most common kinds of variable resistors are called _____ and _____.
14. A potentiometer is most often used to change the value of the _____ applied to a circuit or to part of a circuit.
15. The sliding arm of a potentiometer or a rheostat is connected to the _____ of the device.
16. The resistance of a potentiometer or of a rheostat is measured between its _____ terminals.
17. The most common resistor defect is an open or burned-out condition caused by too large a _____ passing through.
18. The overheating of a carbon-composition or film resistor often causes the resistance of the resistor to _____.
19. A resistor that serves as a fuse is called a _____ resistor.

FOR REVIEW AND DISCUSSION

1. What is a resistor?
2. Name and describe three kinds of fixed resistors.
3. Define the tolerance of a resistor.
4. What is a precision resistor?
5. Explain the meaning of the wattage rating of a resistor.
6. Does the physical size of a resistor give some information about its resistance? Its wattage rating?
7. Describe the resistor color code and explain how it is used.
8. Describe the construction of a variable resistor. Name the two most common kinds of variable resistors.
9. For what purpose is a potentiometer most commonly used?
10. What condition most often causes a resistor to become defective?
11. For what purpose is a fusible resistor used?

INDIVIDUAL-STUDY ACTIVITIES

1. Prepare a written or an oral report on the uses of resistors in common household appliances and devices such as radio and television sets, dimming switches for lights, and so on. If possible, show examples of these resistors in the device discussed.
2. Explain to your class how the resistor color code is used.

Unit 9 Ohm's Law and Power Formulas

In any circuit where the only opposition to the flow of electrons is resistance, there are definite relationships among the values of *voltage, current,* and *resistance.* These relationships were discovered by Georg Simon Ohm in 1827. They are known as *Ohm's law.*

According to Ohm's law:

1. The voltage needed to force a given amount of current through a circuit is equal to the product of the current and the resistance of the circuit.
2. The amount of current in a circuit is equal to the voltage applied to the circuit divided by the resistance of the circuit.
3. The resistance of a circuit is equal to the voltage applied to the circuit divided by the amount of current in the circuit.

THE OHM'S LAW FORMULAS

By using the letter symbols for voltage (E), current (I), and resistance (R), the relationships given by Ohm's law can be expressed in the following formulas:

$$E = I \times R$$

By solving for I, this same formula can be written as

$$I = \frac{E}{R}$$

By solving for R, it can be written as

$$R = \frac{E}{I}$$

where E = voltage in volts
I = current in amperes
R = resistance in ohms

The arrangement of terms in a formula allows us to state certain general rules about the relationships of current, voltage, and resistance. For example, in the Ohm's law formula

$$I = \frac{E}{R}$$

the E in the numerator tells us that I (current) will change in step with E (voltage), providing the denominator R (resistance) does not change. Thus, if the voltage in a circuit dou-

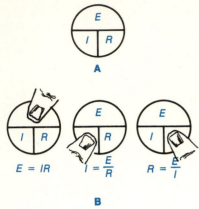
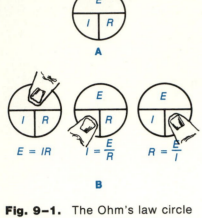

$E = IR$ $I = \dfrac{E}{R}$ $R = \dfrac{E}{I}$

B

Fig. 9–1. The Ohm's law circle

bles, the resulting current will be double its original value. If the voltage in a circuit is reduced to one-half its original value, the current will adjust itself to one-half of its original value. Thus, Ohm's law shows us that current is *directly proportional* to voltage.

In this formula, the R in the denominator tells us that with E (voltage) remaining the same, I (current) is *inversely proportional* to R (resistance). As resistance increases, current decreases in the same way. Similarly, a decrease in resistance results in an increase in current. For example, if the resistance in a circuit is increased to twice its original value, the current will decrease to one-half its original value. If the resistance is decreased to one-third its original value, the current will increase to three times its original value.

The Ohm's law formulas can be learned easily by using a circle, divided as shown in Fig. 9–1 at A. To use this circle, choose and cover one of the quantities (E, I, or R). The relationship of the other two quantities in the circle will show how the chosen quantity can be figured (Fig. 9–1 at B).

USING OHM'S LAW

Ohm's law is important in understanding the behavior of circuits. It is also important because it makes it possible to find the value of any one of the three basic circuit quantities (voltage, current, or resistance) if the values of the other two are known. Thus, circuits and their parts can be designed mathematically. This cuts down on measurements and experiments that would waste time and might damage equipment. The use of Ohm's law in solving practical circuit problems is shown in the following examples.

Problem 1: An electric light bulb uses 0.5 ampere of current when operating in a 120-volt circuit. What is the resistance of the bulb?

Solution: The first step in solving a circuit problem is to sketch a schematic diagram of the circuit itself. The second step is to label each of the parts and show the known values (Fig. 9–2).

In this problem, the values for I and E are known. To solve for R, we use the formula

$$R = \frac{E}{I}$$

or

$$R = \frac{120}{0.5}$$

$$= 240 \ \Omega$$

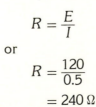

Fig. 9–2. Diagram for Problem 1

$I = 0.5\,A$

$E = 120\,V$ DC

LIGHT BULB $R = ?$

Problem 2: A bicycle horn has a current rating of 0.1 ampere printed on it. The resistance of the horn coil is known to be 15 ohms. Compute the voltage that must be applied to the horn circuit if it is to operate correctly (Fig. 9–3).

Solution: Since voltage is the unknown quantity, use the formula

$$E = I \times R$$
$$= 0.1 \times 15$$
$$= 1.5 \text{ volts}$$

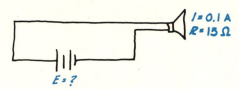

Fig. 9–3. Diagram for Problem 2

Problem 3: To use the right size fuse in an automobile circuit, it is necessary to find the current needed by a certain device. The device is to be connected to the 12-volt battery and has a resistance of 4.35 ohms (Fig. 9–4). Find the current.

Solution: Since current is the unknown quantity, use the formula

$$I = \frac{E}{R}$$

$$= \frac{12}{4.35}$$

$$= 2.76 \text{ amperes}$$

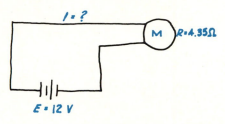

Fig. 9–4. Diagram for Problem 3

POWER FORMULAS

The power formulas show the relationships between electric power and voltage, current, and resistance in a dc circuit. The basic power formula is

$$P = E \times I$$

where P = power in watts
E = voltage in volts
I = current in amperes

From this formula, it is possible to get two other commonly used power formulas. For example, from Ohm's law we know that $E = I \times R$. By putting $I \times R$ in place of E in the basic power formula,

$$P = I \times R \times I$$
$$= I^2 R$$

Also, from Ohm's law we know that $I = E/R$. By putting E/R in place of I in the basic power formula,

$$P = E \times \frac{E}{R}$$

$$= \frac{E^2}{R}$$

USING THE POWER FORMULAS

The power formulas can be used to find the wattage ratings of circuit parts. They can be used to find the value of current in a circuit. And they can be used to find the cost of operating electrical and electronic products. The use of these formulas in solving practical circuit problems is shown in the following examples.

Problem 1: The current through a 100-ohm resistor to be used in a circuit is 0.15 ampere. What should the wattage rating of the resistor be?

Solution: Since the two known quantities in this problem are current and resistance, use the formula $P = I^2R$:

$$P = I^2R$$
$$= 0.15^2 \times 100$$
$$= 2.25 \text{ watts}$$

To keep a resistor from becoming overheated, its wattage rating should be about twice the wattage rating figured from a power formula. Thus, the resistor used in this circuit should have a wattage rating of about 5 watts.

Problem 2: Find the current used by a 60-watt incandescent lamp operating at 120 volts. Also find the current used by a 200-watt lamp and a 300-watt lamp operating at 120 volts.

Solution: In this problem, the power and voltage are known, and we wish to find the current. The simplest formula to use, therefore, is $P = E \times I$, from which we solve for I:

$$I = \frac{P}{E}$$

For the 60-watt, 120-volt lamp:

$$I = \frac{60}{120}$$
$$= 0.5 \text{ ampere}$$

For the 200-watt, 120-volt lamp:

$$I = \frac{200}{120}$$
$$= 1.67 \text{ amperes}$$

For the 300-watt, 120-volt lamp:

$$I = \frac{300}{120}$$
$$= 2.5 \text{ amperes}$$

In this problem, the voltage applied to each of the lamps is the same (120 volts). As the wattage of the lamps increases, the current in the circuit also increases. This means that there is a direct relationship between the power of a load and the current it uses.

Problem 3: The current in a house wiring system increases to twice the original value, from 2 amperes to 4 amperes. What effect does this have on the temperature of the wires in the circuit? Assume a resistance of 25 ohms.

Solution: According to the formula $P = I^2R$, if the resistance does not change, power is directly proportional to the square of the current. Thus, if the current is doubled, the power increases four times.

$$P = I^2R = 2^2 \times 25$$
$$= 4 \times 25 = 100 \text{ watts}$$

If the current is doubled from 2 amperes to 4 amperes,

$$P = 4^2 \times 25$$
$$= 16 \times 25 = 400 \text{ watts}$$

In this problem, doubling the current increases the power four times—from 100 watts to 400 watts. Power can be thought of as the rate at which the energy of moving electrons is changed into heat in the wires. If the current is doubled, the wires will become heated to four times the original temperature. This relationship between power and current is important to know. It means that even a small increase in current will cause a large increase in wire temperature. Overheated wires are a major cause of fires in houses and in other buildings.

KILOWATTHOUR FORMULA

The kilowatthour (kWh) is the unit of electric energy on which electric companies base their bills. Kilowatts multiplied by hours is equal to kilowatthours. For example, if a toaster rated at 1,000 watts is operated for 30 minutes (0.5 hour) the energy used is 1 kW × 0.5 hour = 0.5 kWh.

The amount of energy in kilowatthours used by an appliance or some other product can be figured by the following formula:

$$\text{kWh} = \frac{\text{wattage rating} \times \text{no. of hours used}}{1,000}$$

$$\text{kWh} = \frac{P \times h}{1,000}$$

where P = power in watts
h = time in hours

The electric energy in kilowatthours supplied by electric power companies is measured by a watthour meter. Such a meter is usually mounted on the side of a building. It is "read" at certain times. This reading is used to figure the electric bill for a period of time, usually 1 month.

The average price of electric energy in the United States is about 7¢ per kilowatthour. The average home in this country uses about 700 kWh of electric energy per month.

USING THE KILOWATTHOUR FORMULA

The kilowatthour formula can be used to find the cost of operating a product for any period of time if the cost per kilowatthour is known. This is shown in the problems below.

Problem 1: The average price of electric energy in a certain town is 6¢ per kWh. Find the cost of operating a 250-watt television set for 1.5 hours.

Solution:

$$\begin{aligned}
\text{kWh} &= \frac{P \times h}{1,000} \\
&= \frac{250 \times 1.5}{1,000} \\
&= 0.375
\end{aligned}$$

Since the energy costs 6¢ per kWh, the cost of operating the television set for 1.5 hours = 0.375 × 6, or 2.25¢.

Problem 2: At a price of 6.5¢ per kWh, what is the cost of operating a 1,200-watt clothes iron for 2 hours?

Solution:

$$\begin{aligned}
\text{kWh} &= \frac{P \times h}{1,000} \\
&= \frac{1,200 \times 2}{1,000} \\
&= 2.4
\end{aligned}$$

Total cost = 2.4 × 6.5, or 15.60¢

Problem 3: At a price of 7¢ per kWh, what is the cost of operating a 100-watt lamp for 6 hours?

Solution:

$$kWh = \frac{P \times h}{1,000}$$

$$= \frac{100 \times 6}{1,000}$$

$$= 0.6$$

Total cost $= 0.6 \times 7$, or 4.2¢

SELF-TEST

Test your knowledge by writing, on a separate sheet of paper, the word or words that most correctly complete the following statements:

1. Ohm's law states the relationships among _____, _____, and _____.
2. The voltage applied to a circuit is equal to the product of the _____ and the _____ of the circuit.
3. The current in a circuit is equal to the applied _____ divided by the _____.
4. The resistance of a circuit is equal to the applied _____ divided by the _____.
5. If the voltage applied to a circuit is doubled and the resistance remains constant, the current in the circuit will increase to _____ the original value.
6. If the voltage applied to a circuit remains constant and the resistance is doubled, the current will decrease to _____ the original value.
7. The three most commonly used power formulas are _____, _____, and _____.
8. If the current through a conductor is doubled and the resistance remains unchanged, the power consumed by the conductor will increase to _____ times the original amount.

9. The unit of electric energy on which electric companies base their bills is the _____.
10. The kilowatthours of energy used by an appliance can be calculated by multiplying the _____ of the appliance by the _____ it is used and dividing by 1,000.

FOR REVIEW AND DISCUSSION

1. State the relationships among voltage, current, and resistance as given by Ohm's law.
2. Using correct letter symbols, write the three forms of the Ohm's law formula.
3. A soldering iron uses 2 amperes of current when it is plugged into a 120-volt outlet. How much power does it use?
4. An automobile spotlight has a resistance of 2 ohms while using 6 amperes of current. What is the voltage of the car battery?
5. The heating element of an electric range has a resistance of 10 ohms. How much current does the element use when connected to a voltage of 115 volts?
6. The value of current through a 285-ohm resistor in a radio circuit is 0.3 ampere. What is the voltage across the resistor?
7. The voltage across an 8,000-ohm resistor

in a television circuit is 256 volts. How much current passes through the resistor?

8. What voltage is applied to a circuit with 2 amperes of current and a resistance of 57.5 ohms?

9. Using correct letter symbols, write the three basic power formulas.

10. Figure the current used by a 100-watt incandescent lamp that is operated at a voltage of 120 volts.

11. Figure the current used by a 1,000-watt toaster operated at 117 volts.

12. At a price of 7¢ per kilowatthour, what is the cost of operating a 300-watt television receiver for 6 hours?

13. At a price of 6¢ per kilowatthour, what is the cost of operating a 60-watt lamp for 24 hours?

14. Explain the general procedure for figuring an electric bill.

15. What happens to the amount of power used in a circuit when the voltage is decreased to one-half its original value?

INDIVIDUAL-STUDY ACTIVITIES

1. Prepare a written or an oral report telling about Georg Simon Ohm and his discovery of the relationships among voltage, current, and resistance.

2. Give a demonstration that will show your class at least one practical application of Ohm's law and one power formula.

3. Give a demonstration showing the relationship between the physical sizes of resistors and their wattage ratings.

4. Make a list of appliances and other products in your home that show the wattage rating on a nameplate or elsewhere. After finding out what the average price of electric energy per kilowatthour is in your community, figure the cost of operating each of the products for 1 hour.

Unit 10   Series and Parallel Circuits

When there is more than one load in a circuit, the loads can be connected in series or in parallel. It is important for you to know the characteristics of series and parallel circuits. Knowing them will help you understand how these circuits operate and how they are used.

Energy sources such as dry cells can also be connected in series or in parallel. These connections are also discussed in this unit.

SERIES CIRCUIT

In a *series circuit*, the parts are connected to the conductors one after the other (Fig. 10–1). Thus, there is only one path through which current can pass as it moves from one terminal of the energy source to the other. If such a circuit is broken or open at any point, the whole circuit is turned off. This feature is used to control and protect electrical systems. Such devices as switches, fuses, and circuit breakers are connected in series.

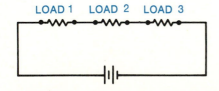

Fig. 10–1. A series circuit with three loads

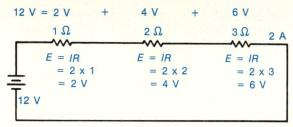

Other circuit loads connected in series include the filaments of electron tubes and ornamental (for example, Christmas-tree) lamps. In a radio or television system, many of the circuits are connected in series. Resistors or a combination of resistors and other devices make up these series circuits.

Current. Since a series circuit has only one electron path, as many electrons flow from any point of the circuit as flow to that point. This means that there is the same amount of current in all parts of a series circuit at the same time.

Voltage. The total voltage applied to a series circuit is automatically spread across the loads or devices in the circuit. The voltage across any load will be that amount needed to force the circuit current through the resistance of the load. This is also called a *voltage drop.* According to Ohm's law, the voltage drop across any load in the circuit is equal to the product of the current and the resistance:

$$E \quad = \quad I \quad \times \quad R$$
(Voltage drop across load)   (current through load)   (resistance of load)

In a series circuit, the sum of the voltage drops across the individual loads is equal to the total voltage applied to the circuit. This is known as Kirchhoff's voltage law. This can be proved by using Ohm's law, as shown in Fig. 10–2. If any one of the resistances is changed in such a circuit, all the voltage drops will change. However, the individual drops will still add up to the source voltage.

Resistance. The total resistance of a series circuit is equal to the sum of all the resistances in the circuit (Fig. 10–3). In circuits with short conductors, the resistance of the conductors usually does not add much to the total resistance. With long conductors, such as in power and telephone distribution systems, the resistance of the conductors is an important part of the total resistance.

Voltage Divider. A series voltage-divider circuit, or network, is often used when different values of voltage are

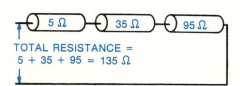

TOTAL RESISTANCE =
5 + 35 + 95 = 135 Ω

Fig. 10–3. Three resistors are hooked up in series. The total resistance of the circuit is the sum of the resistances.

77

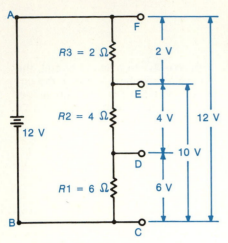

Fig. 10–4. A voltage-divider network

needed from a single energy source. A simple voltage divider is shown in Fig. 10–4. In this circuit, a voltage of 12 volts is applied to three resistors in series. The total resistance of these resistors limits the current through them to 1 ampere. The individual voltages are found as follows:

Total current

$$I = \frac{E}{R} = \frac{12}{2 + 4 + 6} = \frac{12}{12}$$
$$= 1 \text{ ampere}$$

Voltage drop across CD:
$$E = I \times R$$
$$= 1 \times 6 = 6 \text{ volts}$$

Voltage drop across DE:
$$E = I \times R$$
$$= 1 \times 4 = 4 \text{ volts}$$

Voltage drop across EF:
$$E = I \times R$$
$$= 1 \times 2 = \underline{2 \text{ volts}}$$

Total voltage drop CF:
$$= \quad 12 \text{ volts}$$

Voltage drop across CE:
$$E = I \times R$$
$$= 1 \times (6 + 4) = 1 \times 10$$
$$= 10 \text{ volts}$$

PARALLEL CIRCUIT

In a *parallel circuit,* the loads are connected "across the line." That is, they are connected between the two conductors that lead to the source of energy (Fig. 10–5). The loads and their connecting wires often are called the *branches* of the circuit. Parallel connections also are called *multiple connections* and *shunt connections.*

In a parallel circuit, the loads operate independently. Therefore, if one of the branches is disconnected or turned off, the remaining branches will continue to operate.

Voltage and Current. Note that all the branches of a parallel circuit have the same voltage (Fig. 10–6). The total current, however, is distributed among the branches. The amount of current in any branch can be found by using the Ohm's law formula $I = E/R$. The lower the resistance, the greater the current. Thus, while all the loads in a parallel circuit have the same voltage, they may or may not have the same amount of current. In a house wiring system, for example, lights and

Fig. 10–5. A parallel circuit with three branches

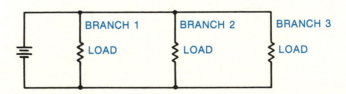

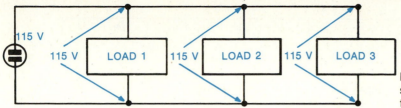

Fig. 10–6. In a parallel circuit, the same voltage is present across all the branches.

small appliances that operate at the same voltage require different amounts of current. Therefore, they are connected in parallel (Fig. 10–7).

The total current delivered to a parallel circuit is equal to the sum of the branch currents (Fig. 10–8). This is why a fuse in a household circuit blows out when too many lights or appliances are plugged into its outlets. As each light or appliance is added to the circuit, the current required increases until the total current is greater than the ampere rating of the fuse or circuit breaker. The circuit is then said to be overloaded. It is at this point that the fuse blows or the circuit breaker trips.

Resistance. As branches are added to a parallel circuit, there are more paths through which current can flow. Therefore, the total resistance to the flow of current decreases. The total resistance of a parallel circuit is always less than that of the branch with the lowest resistance (Fig. 10–9).

The total resistance of loads connected in parallel can be found by using one of the following parallel-resistance formulas.

1. If there are two resistances of unequal value,

$$R_t = \frac{R1 \times R2}{R1 + R2}$$

where R_t = total resistance
$R1$ and $R2$ = parallel resistances

Fig. 10–7. Parallel circuits are used for the operation of household appliances, lights, and other products.

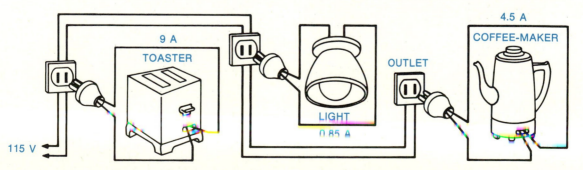

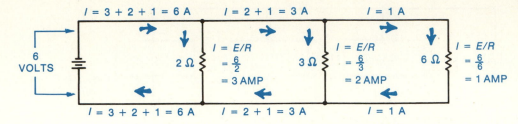

$$I = 3 + 2 + 1 = 6 \text{ A} \qquad I = 2 + 1 = 3 \text{ A} \qquad I = 1 \text{ A}$$

6 VOLTS

$2\,\Omega$

$$I = E/R = \frac{6}{2} = 3 \text{ AMP}$$

$3\,\Omega$

$$I = E/R = \frac{6}{3} = 2 \text{ AMP}$$

$6\,\Omega$

$$I = E/R = \frac{6}{6} = 1 \text{ AMP}$$

$$I = 3 + 2 + 1 = 6 \text{ A} \qquad I = 2 + 1 = 3 \text{ A} \qquad I = 1 \text{ A}$$

Fig. 10–8. The total current in a parallel circuit is equal to the sum of the branch currents.

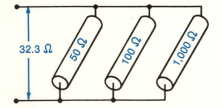

32.3 Ω   50 Ω   100 Ω   1,000 Ω

Fig. 10–9. Three resistors are connected in parallel. The total resistance of the circuit is always less than the resistance of any of its branches.

Fig. 10–10. Diagrams used with parallel-circuit problems

2. If all the resistances are equal in value,

$$R_t = \frac{\text{value of one resistance}}{\text{number of resistances}}$$

3. If the resistances are not all of equal value,

$$R_t = \frac{1}{\dfrac{1}{R1} + \dfrac{1}{R2} + \dfrac{1}{R3}}$$

NOTE: This is a general formula that can be used with any kind of parallel circuit, including the ones in 1 and 2 above.

Problem 1: Two resistors with resistances of 20 and 30 ohms are connected in parallel (Fig. 10–10 at A). Figure the total resistance.

Solution: Formula 1 is the easiest one to use in this case.

$$R_t = \frac{R1 \times R2}{R1 + R2}$$

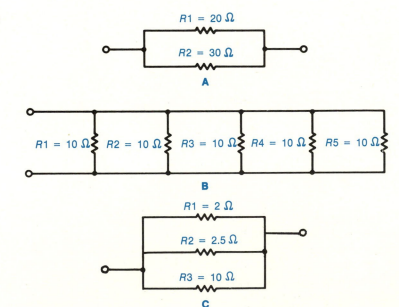

R1 = 20 Ω
R2 = 30 Ω

A

R1 = 10 Ω   R2 = 10 Ω   R3 = 10 Ω   R4 = 10 Ω   R5 = 10 Ω

B

R1 = 2 Ω
R2 = 2.5 Ω
R3 = 10 Ω

C

$$R_t = \frac{20 \times 30}{20 + 30}$$

$$R_t = \frac{600}{50} = 12 \text{ ohms}$$

Problem 2: Five resistors, each with a resistance of 10 ohms, are connected in parallel (Fig. 10–10 at B). Figure the total resistance.

Solution:

$$R_t = \frac{\text{value of one resistance}}{\text{number of resistances}}$$

$$R_t = \frac{10}{5} = 2 \text{ ohms}$$

Problem 3: Three resistors with resistances of 2, 2.5, and 10 ohms are connected in parallel (Fig. 10–10 at C). Figure the total resistance.

Solution:

$$R_t = \frac{1}{\frac{1}{R1} + \frac{1}{R2} + \frac{1}{R3}}$$

$$R_t = \frac{1}{\frac{1}{2} + \frac{1}{2.5} + \frac{1}{10}}$$

$$R_t = \frac{1}{0.5 + 0.4 + 0.1}$$

$$R_t = \frac{1}{1} = 1 \text{ ohm}$$

CELLS IN SERIES AND IN PARALLEL

Energy sources such as cells and batteries often are connected in series and in parallel. These connections affect the total voltage and the total current.

Series Connection. Cells are connected in series to get a higher voltage than can be obtained from one cell. When cells are connected in series, the total voltage is equal to the sum of the individual voltages (Fig. 10–11 at A). The current-delivering capacity of the combination is equal to that of the cell with the lowest capacity. Batteries are made up of a number of cells that are usually connected in series.

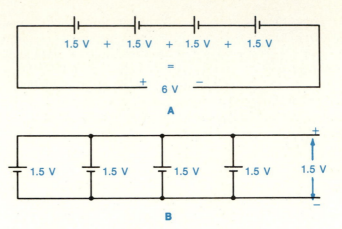

Fig. 10–11. The total voltages obtained when cells are connected in series and in parallel: (A) series connection; (B) parallel connection

Parallel Connection. Cells or batteries are connected in parallel to get more current than can be obtained from one unit. With a parallel connection, the total voltage of the combination is equal to the voltage of each unit (Fig. 10–11 at B). The current-delivering capacity of the combination is equal to the sum of the capacities of all the cells or batteries.

Using a "booster" battery is an interesting and useful example of connecting batteries in parallel. The booster battery

Fig. 10–12. Series-parallel circuit: (A) basic circuit; (B) steps followed in finding the total resistance of a series-parallel circuit; (C) equivalent circuit

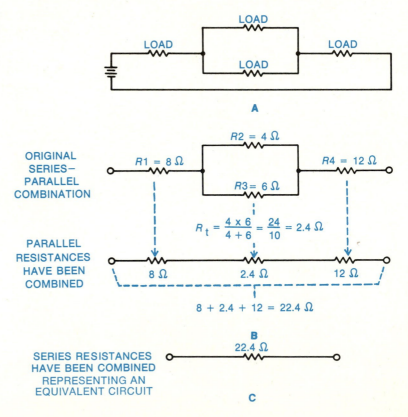

is temporarily connected to an automobile battery that cannot deliver enough current to start a car. The voltage applied to the automobile electrical system is still 12 volts. However, the booster battery can supply enough extra current to turn the cranking motor and start the engine. The plus terminal of the booster battery is connected to the plus terminal of the automobile's battery, and the two negative terminals are also connected.

SERIES-PARALLEL CIRCUIT

A series-parallel circuit has some loads connected in series and some connected in parallel (Fig. 10–12 at A). Such combinations are found in many circuits, including those in radio and television receivers.

To find the total resistance of a series-parallel combination, it is helpful to break down the combination to an equal, but simpler, form (Fig. 10–12 at B). By combining small parts of the circuit and substituting the equal resistance, you can end up with a circuit with one resistance equal to the total resistance of the original circuit (Fig. 10–12 at C).

SELF-TEST

Test your knowledge by writing, on a separate sheet of paper, the word or words that most correctly complete the following statements:

1. The current in all parts of a series circuit is stopped if the circuit is _____ at any point.
2. There is the same amount of _____ in all parts of a series circuit.
3. In a series circuit, the _____ voltage is equal to the _____ of the voltage drops across all the loads in the circuit.
4. The total resistance of a series circuit is equal to the _____ of all the resistances in the circuit.
5. A series network of resistors used to provide different voltages from a single source of energy is called a _____ .
6. In a parallel circuit, the loads are connected across or between the two _____ that lead to the source of energy.

7. All branches of a parallel circuit have the same _____ .
8. The total current in a parallel circuit is equal to the _____ of the _____ currents.
9. The total resistance of a parallel circuit is always _____ than that of the branch with the lowest resistance.
10. The formula for finding the total resistance of two parallel resistances having unequal values is _____ .
11. The formula for finding the total resistance of two or more resistances that are equal in value and connected in parallel is _____ .
12. The general formula for finding the total resistance of two or more resistances not all of equal value and connected in parallel is _____ .
13. When cells are connected in series, the total voltage of the combination is equal to the _____ of the individual voltages.

14. Cells and batteries are connected in parallel to get more _____ than can be obtained from one cell or battery.
15. When cells or batteries with the same voltage are connected in parallel, the total voltage of the combination is equal to the voltage of _____ cell or battery.

FOR REVIEW AND DISCUSSION

1. What is true about the value of current in all parts of a series circuit?
2. State the relationship between the voltage drops and the total applied voltage in a series circuit. Express this with a formula.
3. State the relationship between the total resistance and the individual resistances in a series circuit. Express this with a formula.
4. Describe the voltage and current conditions in a parallel circuit.
5. What is the relationship between the total resistance and the lowest resistance in a parallel circuit?
6. A number of incandescent lamps in a circuit are connected in series. Suddenly, they all go out. Explain the most probable reason for this.
7. A number of incandescent lamps in a circuit are connected in parallel. When one of them burns out, the rest continue to operate normally. Explain the characteristics of a parallel circuit that make this possible.

NOTE: In addition to answering questions 8 through 13, draw diagrams that show the circuits in the questions.

8. Four resistors with the values of 7, 13, 27, and 96 ohms are connected in series. What is their total resistance?
9. Three resistors with the values of 5, 12, and 28 ohms are connected in series with a 90-volt battery. What is the value of current in the circuit?
10. Loads A, B, and C are connected in series.

Loads A and B each have a resistance of 20 ohms. When the circuit is plugged into a 115-volt outlet, 2 amperes of current pass through it. What is the resistance of load C?
11. A parallel circuit has three branches. The loads in these branches each have a resistance of 150 ohms. Figure the total resistance.
12. Two parallel resistors have values of 40 and 60 ohms. What is the total resistance?
13. Three resistors connected in parallel have values of 2, 2.5, and 10 ohms. Figure the total resistance.
14. Six 1.5-volt dry cells are connected in series to form a transistor-radio battery. What is the voltage of the battery?
15. A 12-volt automobile battery has six cells connected in series. What is the voltage of each cell?
16. A series-parallel circuit is shown in Fig. 10–13. Figure the total resistance, the total current, the currents through R2 and R3, and the voltage drop across each resistor in this circuit.

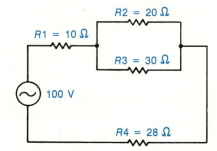

Fig. 10–13. Diagram for question 16

INDIVIDUAL-STUDY ACTIVITY

Obtain a schematic circuit diagram of some electrical device such as a radio, a television set, an electric clothes dryer, and so on. Identify the loads shown in the diagram. Indicate the series, parallel, and series-parallel combinations.

Unit 11   Measuring Electrical Quantities

Electricity cannot be seen. However, electrical quantities such as voltage, current, resistance, and power can be measured with instruments called *meters*. In this unit, you will learn about the most common kinds of meters. These are used in the design, the manufacturing, and the servicing or repairing of all kinds of electrical and electronic products.

HOW A METER WORKS

A typical meter works like a small electric motor. This "motor action" causes the meter pointer to move. The amount of this movement depends on the amount of current that passes through the meter circuit. Although the pointer moves because of current, other electrical quantities, such as voltage and resistance, can be measured. The "face" of the meter can be marked to show the different electrical quantities. The arrangement of parts that causes the pointer to move is called the *meter movement*. Most common meters use a *permanent-magnet moving-coil movement* (Fig. 11–1). The coil is wound around a frame mounted between pivot points so that the coil is free to turn. The pointer is fastened to the coil assembly. A small spring holds the coil so that the pointer is at zero on the meter scale. The ends of the coil are connected to the stationary terminals of the meter.

When current passes through the coil, a small electromagnet is produced. The poles of the electromagnet are repelled by the poles of the permanent magnet. Since the movement

Fig. 11–1. The permanent-magnet moving-coil meter movement (*Triplett Corporation*)

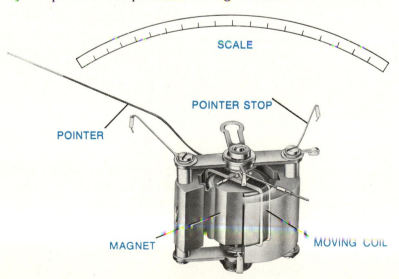

SCALE

POINTER STOP

POINTER

MAGNET

MOVING COIL

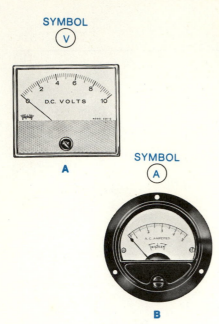

A

B

Fig. 11–2. Panel meters: (A) dc voltmeter; (B) ac ammeter (*Triplett Corporation*)

of the coil and pointer depends on current, the pointer will move. From zero it moves a distance determined by the value of the current. The amount of current needed to move the pointer to the highest reading on the meter scale is called the *full-scale deflection current* of the meter.

VOLTMETERS AND AMMETERS

Voltage is measured with a *voltmeter*. Current is measured with an *ammeter*. Microammeters measure current in millionths, and milliammeters, in thousandths, of an ampere. Very small amounts of current are measured with an instrument called a *galvanometer*.

Voltmeters and ammeters can be built as individual instruments. They can also be part of an instrument called a *multimeter*. Multimeters use a single meter movement but different internal circuits to measure voltage, current, and resistance. *Panel meters* are single-purpose instruments. They usually are made to be mounted on a test or an instrument panel (Fig. 11–2). Panel voltmeters measure either direct or alternating voltage. Panel ammeters measure either direct or alternating current.

OHMMETERS

Electrical resistance is measured with an *ohmmeter*. The meter movement of the ohmmeter is the same as that in voltmeters and ammeters. An ohmmeter can be an individual panel instrument, or it can be a part of a multimeter.

An ohmmeter circuit requires a source of energy such as a cell, a battery, or both. These are contained in the case of the meter.

A *megohmmeter* is used to test materials with very high resistances. Equipment insulation and Bakelite are examples. A high voltage in the megohmmeter is needed to measure these resistances. The source of energy in the megohmmeter, called a *power supply*, provides this high voltage.

MULTIMETER

The most common kind of multimeter is the volt-ohm-milliammeter (VOM) (Fig. 11–3 at A). This instrument can be used as a voltmeter, an ammeter, or an ohmmeter. All you need to do is set its selector switch correctly. Some VOMs are also equipped with a decibel (dB) scale. This is used to measure the sound-output level of audio amplifiers for public address systems.

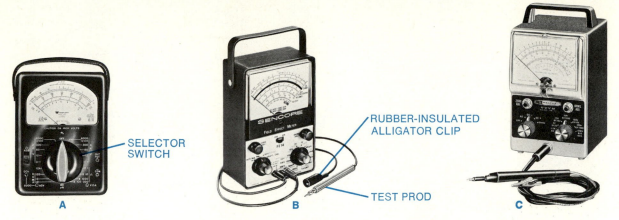

SELECTOR SWITCH

RUBBER-INSULATED ALLIGATOR CLIP

TEST PROD

A B C

Fig. 11–3. Multimeters: (A) volt-ohm-milliammeter (*Triplett Corporation*); (B) field-effect VOM (*Sencore*); (C) vacuum-tube voltmeter (VTVM) (*Heath Company*)

The field-effect volt-ohm-milliammeter (FET VOM) has one or more devices called *field-effect transistors*. These cause the instrument to have a very high input impedance, or resistance to current (Fig. 11–3 at B). As a result, the FET VOM is able to measure voltages without disturbing the circuit. This is very important when low voltages must be measured accurately.

The transistorized voltmeter (TRVM) and vacuum-tube voltmeter (VTVM) are combination voltmeters and ohmmeters (Fig. 11–3 at C). As with the FET VOM, the outstanding feature of these instruments is their high input impedance.

CARE AND SAFE USE OF METERS

All meters are delicate instruments. They must be used with care to keep from damaging them mechanically or electrically. You must choose a voltmeter or an ammeter with the right *range*. This means that the instrument must be able to handle safely the highest voltage or largest current being tested. You should always learn how to operate any meter before trying to use it.

Voltmeters and ammeters are connected to energized circuits or in circuits to which a voltage will be applied. It is very important, therefore, to follow the right safety rules while using them. Voltmeters and ammeters are often connected to a circuit with clips or screw terminals. It is always safer to turn off the power supply before making these connections. The circuit can then be turned on to make the measurement. Also be careful not to touch bare circuit wires when using test prods to connect a meter to a circuit.

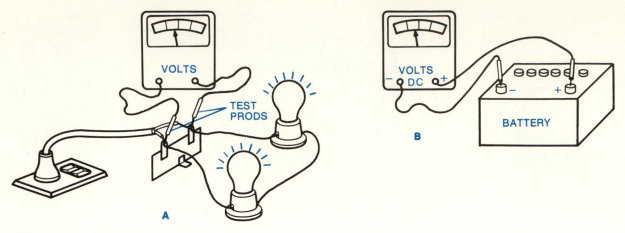

Fig. 11-4. Using the voltmeter

USING THE VOLTMETER

A voltmeter is always connected to the two points of a device or a circuit across which the voltage is to be measured (Fig. 11-4 at A). A dc voltmeter is a *polarized* instrument. This means that it must be connected to a circuit in the correct polarity, or + to + and − to − (Fig. 11-4 at B). If this is not done, the pointer of the meter will move in the wrong direction. This can damage the pointer mechanism. An ac voltmeter can be connected across two points under test without regard to polarity.

USING THE AMMETER

The ammeter is normally connected in series with the conductors and load being tested (Fig. 11-5). It may be permanently damaged if it is connected across an energy source or into a circuit with too much current. A dc ammeter must be

Fig. 11-5. Using the ammeter

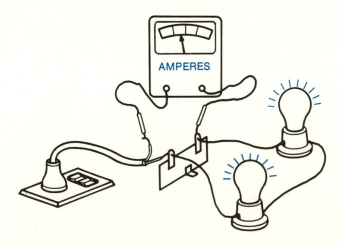

connected into a circuit with the right polarity. If the pointer of a dc ammeter moves in the wrong direction during testing, reverse its connections to the test points. An ac ammeter can be connected into a circuit without regard to polarity.

USING THE OHMMETER

An ohmmeter is connected in parallel to the terminals of the device or circuit to be tested (Fig. 11–6). To get an accurate measurement, an ohmmeter must be "zeroed" before being used. To zero a volt-ohm-milliammeter, bring the test prods together. Next turn the zero adjustment knob until the pointer is at zero on the ohm scale. If you switch the meter to another ohm range, you will have to zero it again.

If an ohmmeter cannot be zeroed, it is usually because of a weak cell or battery in the meter circuit. To keep from ruining an ohmmeter battery, always make sure that the tips of the test prods do not touch, except when zeroing the meter. To prevent accidental shorting of the contacts, turn the meter off if possible. Unplug the test leads when the ohmmeter is not in use. When using an ohmmeter, always be sure that voltage is not being applied to the device or the circuit under test. The meter may be damaged if it is connected across two points between which even a low voltage is present.

Do not let the fingers of both hands touch the tips of the test prods while measuring resistance. If you do so, the ohmmeter will measure the combined resistance of your body and the circuit being tested. Be very careful of this in measuring high values of resistance, when you might not suspect the problem.

READING METER SCALES

Reading the value of an electrical quantity from the position of the pointer on a meter scale is not always easy. To read meter scales quickly and accurately, you must know how they are marked off. In many circuits, small changes in voltage and current will take place, causing the pointer to move only a little. Accurate meter reading means looking and judging carefully.

Panel-meter Scales. Studying the panel-ammeter scales in Fig. 11–7 will help you learn how to read all meter scales. You will notice that the scale of Fig. 11–7 at A is marked off into four main divisions. The values of these are 0.5, 1.0, 1.5, and 2.0 amperes. Each main division is, in turn, divided into 10 subdivisions. The value of each is then equal to 0.5/10, or 0.05, ampere.

OHMMETER SYMBOL

(OHM)

ZERO ADJUSTMENT
POTENTIOMETER KNOB

Fig. 11–6. Using the ohmmeter

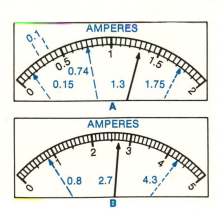

Fig. 11–7. Reading the scales of a panel meter

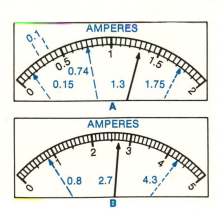

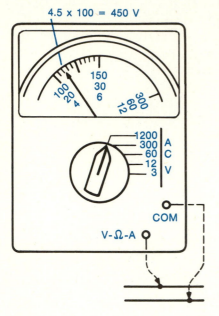

4.5 x 100 = 450 V

Fig. 11–8. Reading an ac VOM scale. The selector switch shows the full-scale deflection.

If the meter pointer is at a position of less than 0.5, the value of the current is equal to the sum of the values of the subdivisions through which the pointer has passed in moving to that position. If the pointer has moved past one of the first three main divisions, the value of the current is equal to the value of that main division plus the sum of the values of the subdivisions to the right of the main division.

When the pointer stops between the two lines of a subdivision, you must estimate the last figure of the reading according to the pointer's position within the subdivision.

In Fig. 11–7 at B, the 0-to-5-ampere scale is marked off into five main divisions. Each of these is divided into 10 subdivisions. The value of each subdivision on this meter scale is then equal to 1/10, or 0.1, ampere.

Volt-ohm-milliammeter Scales. The face of a volt-ohm-milliammeter has a combination of scales. These scales are used to measure a wide range of values in dc voltage and current, ac voltage and current, resistance, and output levels (Fig. 11–3 at A). For this reason, the scale is more complicated than that of a panel meter.

Though several different kinds of volt-ohm-milliammeters are used, all have similar scales and work in similar ways.

Part of a volt-ohm-milliammeter ac scale and range-selector switch is shown in Fig. 11–8. The selector-switch values show the ranges of the meter.

If the range of the voltage being measured is not known, set the selector switch to the highest value, in this case 1200 volts. Once an approximate value for the unknown voltage has been obtained, the selector switch can be set to the position closest to this value but not less than it. For example, if the selector is set at 1,200 and the pointer moves a little to show that the measured voltage is about 10 volts, the selector switch can be set at 12. By doing this, it is possible to get a more accurate reading of the actual voltage, say, 10.8 volts. Note that, although the selector has positions for 3, 12, 60, 300, and 1,200 volts, the face of the meter has only the 12-, 60-, and 300-volt scales. If the 3-volt position on the selector is used, the voltage values read directly on the 300 scale must be divided by 100. With the selector switch set at the 3-volt position and the pointer indicating 153 on the 300-volt scale, the actual voltage being measured is 153/100, or 1.53, volts. Similarly, with the selector on the 1,200-volt position, the voltage shown on the 12-volt scale must be multiplied by 100 to get the actual voltage. Remember that you can always get a reading on a higher position of the selector switch for any voltage less than this value. The lower positions are used merely to get a more accurate reading. You must never set the selector switch to a value less than the actual voltage. If this were done, the pointer might be damaged in swinging past the full-scale deflection. The coil in the meter movement might also burn out.

Fig. 11-9. Reading the ohms scale on a VOM

Measuring Resistance with the Volt-ohm-milliammeter.
The ohms or resistance scale of a typical VOM is read from right to left (Fig. 11-9). The main divisions of this scale are not marked off evenly. Such a meter scale is said to be *non-linear*. While this may be a bit confusing at first, you will find that such a scale is not at all unusual or difficult to read.

Figure 11-10 shows how one type of VOM is used to measure resistance. The ohms-range selector switch of the instrument is set at the ×1,000 position. This means that the resistance-scale reading must be multiplied by 1,000. In Fig. 11-10, the reading of the meter is 14,000 ohms (14 × 1,000). If the range-selector switch were at any other position, the reading of the scale should be multiplied by the number at that position.

The extreme left-hand end of a VOM ohm scale may be marked with an infinity symbol (∞). This means that if the pointer is at this position when the range-selector switch is at the highest ohms range, the resistance being tested is so high that it cannot be measured with the instrument.

CONTINUITY TEST

In addition to measuring resistance, an ohmmeter is used to make continuity tests. These show whether or not there is a continuous electron path from one test point to another. Continuity tests are very useful for checking hidden conductors of coils and other devices in which the complete path cannot be seen. Continuity tests are commonly used for checking switches, transformers, and relays.

When making a continuity test, the ohmmeter (or the VOM) is usually adjusted first to the R × 1 ohms range. It is then connected to the terminals of the device or to the points of the circuit being tested. If the meter pointer does not move from the high resistance end of the ohms scale at any ohms range, a conductor of the device or the circuit being tested is broken or open (Fig. 11-11 at A).

If, on the other hand, the ohmmeter pointer moves to a position that indicates some resistance, there is continuity between the terminals (Fig. 11-11 at B). The amount of resistance will be determined by the characteristics of the device or the circuit being tested. When the resistance is very low, as shown when the meter pointer moves to or near zero

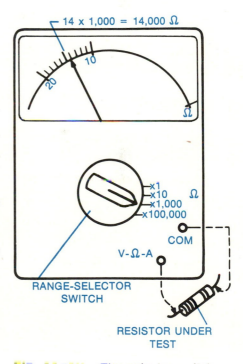

Fig. 11-10. The selector switch on the VOM indicates the multiplier to be used in measuring resistance.

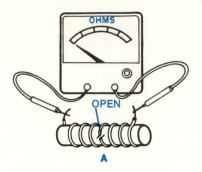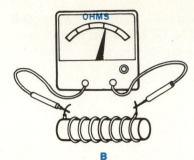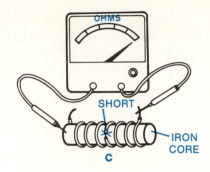

A   **B**   **C**

Fig. 11–11. Making a continuity test with an ohmmeter

on the resistance scale, there is said to be *direct continuity* between the points under test.

A continuity test is also very useful in checking for "shorts" or "grounds." In the test shown in Fig. 11–11 at C, direct continuity exists between one terminal of a coil wound with insulated wire and the core. In this case, the wire insulation has become defective at some point. There is a short between the wire and the core. Since this condition might be dangerous or cause the circuit to work improperly, the coil should be replaced.

DIGITAL VOLT-OHM-MILLIAMMETER

The digital volt-ohm-milliammeter (DVOM) has circuits that give a numerical readout of the voltage, current, and resistance values being tested (Fig. 11–12). This instrument eliminates the need to determine the position of a pointer. It also reduces errors due to parallax, that is, those caused by looking at the pointer at an angle rather than head on.

CLAMP-ON METER

Most meters must be connected directly into a circuit in order to get a reading. The clamp-on meter is designed to avoid this. Clamp-on meters are generally used to measure alternating current and voltage. They are used to measure current by opening the jaws and then closing them over the conductors of the circuit being tested (Fig. 11–13). This makes it easy to measure currents in wires that cannot be disconnected quickly from their terminals.

Fig. 11–12. Digital volt-ohm-milliammeter (*Triplett Corporation*)

LEARNING BY DOING

6. **Measuring Voltage and Current.** Voltage is measured with a voltmeter. Current is measured with an ammeter. The

following procedures will help you gain practical experience in using both ac and dc voltmeters and ammeters.

MATERIALS NEEDED

circuit board, prepunched, 8 × 10 in. (203 × 254 mm)
2 pilot lamps, no. 40
2 pilot lamps, no. 46
2 light bulbs, 60-watt
2 light bulbs, 100-watt
4 miniature lamp sockets with screw base
4 cleat lampholders, medium base
1 attachment plug
5-ft (1.52-m) parallel lamp cord, no. 18 AWG (1.00 mm)
8 round-head machine screws, no. 6-32 × ⅝ in. (M3.5 × 0.6 × 16 mm)
8 hex nuts, no. 6-32 (M3.5)
volt-ohm-milliammeter
2 panel ammeters, ac, 0–1 and 0–2 amperes
6-volt dc power supply or battery

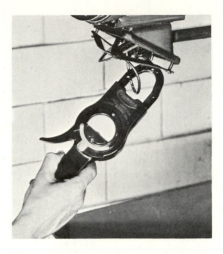

Fig. 11–13. Using a clamp-on meter to measure motor current (*Columbia Electric Manufacturing Company*)

Procedure A (Measuring ac Voltage and Current)

 CAUTION: The following procedures involve working with *live* circuits to which dangerously high voltages are applied. Use extreme care to avoid shock or burns. Make sure that a circuit is unplugged before making or changing any connections. Do not allow your fingers to come into contact with the bare ends of test leads while making any measurement.

1. Connect the circuit shown in Fig. 11–14 at A. Be sure you are standing on dry, insulated flooring. On a sheet of paper, prepare the table shown in the figure.
2. Plug the circuit into a 115-volt receptacle outlet. Be careful not to come into contact with any uninsulated parts of the circuit. *Never place both hands in the circuit.*
3. Measure the voltage across the terminals of each lamp. Record these voltages in the table.
4. Explain why lamps 3 and 4 are not producing a normal amount of light.
5. Unplug the circuit and connect an ammeter into it at point A. Plug in the circuit. Read the value of current at point A. Record this in the table.
6. Repeat step 5 at points B, C, and D of the circuit. Always unplug the circuit first.

Procedure B (Measuring dc Voltage and Current)

1. Connect the circuit shown in Fig. 11–14 at B. On a sheet of paper, prepare the table shown.

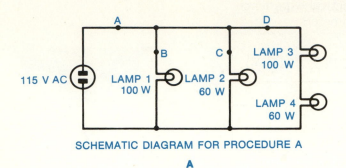

SCHEMATIC DIAGRAM FOR PROCEDURE A

A

Lamp	Voltage across lamp	Point	Current
1		A	
2		B	
3		C	
4		D	

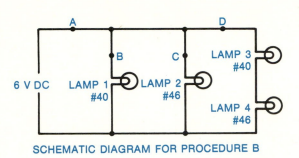

SCHEMATIC DIAGRAM FOR PROCEDURE B

B

Table for Procedure B

Lamp	Voltage across lamp	Point	Current
1		A	
2		B	
3		C	
4		D	

Fig. 11–14. Schematic diagrams for Learning by Doing No. 6, "Measuring Voltage and Current"

2. Measure the voltage across the terminals of each lamp. Record these voltages in the table.
3. Connect an ammeter into the circuit at point A. Read the value of current at this point. Record it in the table.
4. Repeat step 3 at points B, C, and D of the circuit.

7. Verifying Resistance Values with an Ohmmeter. A color code is used with carbon-composition and film resistors. The following procedure will provide practice in verifying this code and using an ohmmeter. See Learning by Doing No. 4, "Using the Resistor Color Code," in Unit 8.

MATERIALS NEEDED

10 different color-coded carbon-composition and/or film resistors
volt-ohm-milliammeter

Procedure

1. On a sheet of paper, prepare Table 11–1 as shown. Write the color code of each resistor in the appropriate space in this table.
2. Write the resistance and the tolerance percentage of each resistor, as indicated by its color code, in columns 3 and 4 of the table.

Table 11-1. Table for Learning by Doing No. 7, "Verifying Resistance Values with an Ohmmeter"

(1) Resistor	(2) Color Code	(3) Resistance	(4) Tolerance, %	(5) Tolerance Range, ohms	(6) Measured Resistance
1					
2					
3					
4					
5					
6					
7					
8					
9					
10					

3. Figure the tolerance range in ohms for each resistor. Record this in the table. For example, a 100-ohm resistor with a tolerance of 10 percent may have an actual resistance 10 percent greater or less than 100 ohms:

> 10 percent of 100 = 10 ohms

Therefore, the allowable range is (100 − 10) = 90 ohms to (100 + 10) = 110 ohms. This range is written as 90–110.

4. Measure and record the actual resistance of each resistor using the VOM.
5. Determine whether or not the actual resistance of each resistor falls within its tolerance range.

8. Verifying Ohm's Law. Although Ohm's law states the exact relationships between voltage, current, and resistance, the law means more when you see these relationships in a circuit.

MATERIALS NEEDED

circuit board, prepunched, 6 × 8 in. (152 × 203 mm)
3 carbon-composition or film resistors: 10 ohms, 20 ohms, and 30 ohms, 1 watt, ±5 percent tolerance
2 no. 6 dry cells or a low-voltage dc power supply with a current rating of at least 0.5 ampere
multimeter (volt-ohm-milliammeter)

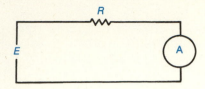

Table for values on schematic diagram

Circuit	E	R	Computed current	Measured current
1	1.5 V	10		
2	1.5 V	20		
3	1.5 V	30		
4	3 V	10		
5	3 V	20		
6	3 V	30		

Fig. 11–15. Schematic diagram and table for Learning by Doing No. 8, "Verifying Ohm's Law"

Procedure

1. On a sheet of paper, prepare a table as in Fig. 11–15. By using Ohm's law, figure the current in each of the circuits, the values of which are given in the table. Record the result for each circuit in the "Computed Current" column of the table.
2. Connect circuit 1. Be sure to note the polarity of the ammeter. The volt-ohm-milliammeter should first be adjusted to a range of at least 400 milliamperes (mA). The range setting can then be lowered as needed.
3. Record the measured current of the circuit in the table.
4. Hook up the rest of the circuits for each combination of E and R in the table. Record the measured current of each.
5. Compare the computed and measured currents for each circuit. What would account for differences between the two values?
6. With the applied voltage at either 1.5 or 3 volts, what effect did doubling the resistance have on the circuit current? Does this follow Ohm's law?
7. With any given resistor (10, 20, or 30 ohms) in the circuit, what effect did increasing the applied voltage from 1.5 to 3 volts have on the current? Does this follow Ohm's law?

9. Additional Experiences with Ohm's Law and the Power Formulas. The following procedure provides practical experience in measuring electrical quantities and making computations involving Ohm's law and the power formulas. It will also show how the resistance of a metal (tungsten) conductor is affected by its temperature.

MATERIALS NEEDED

3 light bulbs, medium base, 40-, 60-, and 100-watt
1 cleat lampholder, medium screw base

3-ft (1-m) parallel lamp cord, no. 18 AWG (1.00 mm)
volt-ohm-milliammeter
ammeter, ac, 0–1 ampere

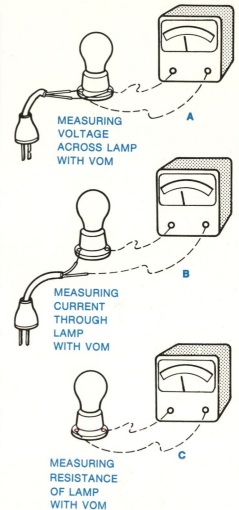

MEASURING
VOLTAGE
ACROSS LAMP
WITH VOM

MEASURING
CURRENT
THROUGH
LAMP
WITH VOM

MEASURING
RESISTANCE
OF LAMP
WITH VOM

Fig. 11–16. Diagrams and table
for Learning by Doing No. 9, "Additional Experiences with Ohm's Law
and the Power Formulas"

Procedure

1. On a sheet of paper, prepare a table like the one in Fig.
   11–16. Use this table to record the results of all measurements and computations.
2. Wire the lamp circuit using the 40-watt lamp and plug it
   into a receptacle outlet (Fig. 11–16 at A). Be careful not
   to touch any uninsulated parts of the circuit. Remember
   also that the lamp will be very hot and can cause bad
   burns if touched.
3. Measure the voltage (E) across the lamp.
4. Figure the operating current of the lamp by using the
   right power formula.
5. Measure the lamp current (Fig. 11–16 at B).
6. Repeat procedures 3, 4, and 5 for the 60- and 100-watt
   lamps.
7. Figure the power used by the 40-watt lamp from the
   power formula, using measured values of E and I. Give
   reasons why this wattage rating may be somewhat different from the 40-watt rating printed on the lamp.
8. Measure the resistance of the lamp filament while it is
   not in operation and record this as the cold resistance in
   the table.
9. Figure the hot, or operating, resistance of the lamp by
   using the right power formula. Which is the higher resistance, cold or hot? Explain the reason for this.

Lamp (watts)	Voltage across lamp (volts)	Computed current (amperes)	Measured current (amperes)	Computed power (watts)	Measured cold resistance (ohms)	Computed hot resistance (ohms)
40						
60						
100						

Test your knowledge by writing, on a separate sheet of paper, the word or words that most correctly complete the following statements:

1. Voltage is measured with a _____. Current is measured with an _____.
2. A _____ measures very small amounts of current.
3. A _____ is a combination of voltmeter, ammeter, and ohmmeter.
4. A voltmeter is always connected to the two _____ of a circuit between which the voltage is to be measured.
5. Dc voltmeters and ammeters are _____ instruments that must be connected to a circuit in the correct _____.
6. The typical ammeter is connected in _____ with one of the conductors of the circuit being tested.
7. An ac voltmeter or ammeter can be connected to a circuit without regard to _____.
8. Electrical resistance is measured with an _____.
9. If an ohmmeter cannot be zeroed, this is often due to a weak _____ or _____ within the meter circuit.
10. An ohmmeter may be seriously damaged if it is connected to two points across which even a low _____ is present.
11. A _____ test is done to find if there is a continuous path for electrons between two points of a circuit or of a device.

FOR REVIEW AND DISCUSSION

1. Name the instruments with which voltage and current are measured.
2. What is a multimeter?
3. What is the advantage of using an FET VOM or a vacuum-tube voltmeter instead of an ordinary VOM?
4. Draw a diagram showing how a voltmeter is connected to circuit points being tested.
5. Draw a diagram showing how a typical ammeter is connected into a circuit.
6. Give a general explanation of how meter scales are read.
7. What condition often makes it impossible to zero an ohmmeter?
8. Briefly describe how an ohmmeter is used for measuring the resistance of a device or of a circuit.
9. What instrument is used to make a continuity test? Why is such a test made?
10. What is a digital VOM?
11. Describe a clamp-on meter.

INDIVIDUAL-STUDY ACTIVITIES

1. Give a demonstration showing how an ohmmeter is used to measure resistance and to make continuity tests. Explain to your class how a continuity test can be used to determine the condition of devices such as switches, incandescent lamps, and coils.
2. Give a demonstration showing how fixed resistors and a potentiometer are tested with an ohmmeter.
3. Give a demonstration describing a typical volt-ohm-milliammeter. Show how it is used to measure voltage and current in a circuit.

Unit 12   Capacitance and Capacitors

Capacitance is the ability of a circuit or a device to "store" electric energy. Although almost all alternating-current circuits have capacitance, it may or may not be desirable. In circuits where a specific value of capacitance is needed, a device known as a *capacitor* may be used (Fig. 12–1). The basic form of a capacitor is two conductors called *plates* separated by insulating material called the *dielectric*.

Capacitors are one of the most common components of electric circuits. They perform several different circuit functions. Since they do not provide a continuous conducting path for electrons, they are used to block a direct current. An alternating current, however, can still flow in the circuit. Capacitors are used with resistors and with coils to filter, or smooth out, a varying direct current. Capacitors, along with resistors, form timing circuits that control other circuits. In radio and television receivers, capacitors are used with coils to form tuning circuits. These let us choose the stations to which we want to listen. Capacitors are also very important components of oscillator circuits. These produce high-frequency alternating voltages.

CAPACITOR ACTION

The action of a simple capacitor is shown in Fig. 12–2. When the plates are connected to a battery, electrons from the plate connected to the positive terminal of the battery move to the battery. This causes the plate to become positively charged. At the same time, an equal number of electrons are repelled to the other plate by the negative terminal of the battery. That plate becomes negatively charged. This produces a voltage across the plates (Fig. 12–2 at A). Because of the dielectric separating the plates, electrons cannot move directly from one plate to the other. Electrons have simply been taken off one plate and put on the other through the circuit that connects them. This flow of electrons in the wires connecting the capacitor to the battery continues until the voltage across the plates is equal to the voltage of the battery. When this happens, the capacitor is fully *charged*. An *electrostatic field* exists between the plates of a charged capacitor. Energy is stored in this field.

When the battery is taken out of the circuit, the plates keep their charges. A voltage continues to exist across them. If the plates are now connected, electrons will flow through the circuit in the opposite direction. They will flow from the nega-

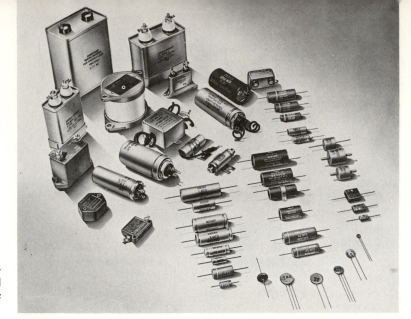

Fig. 12–1. Capacitors are manufactured in many different sizes and shapes. (*Cornell-Dubilier Electric Corporation*)

tively charged plate to the positively charged plate (Fig. 12–2 at B). When the voltage across the plate decreases to zero, the flow of electrons stops. Now the capacitor is fully *discharged*.

Blocking Action. The voltage across the plates of a fully charged capacitor is equal to the voltage of the battery to which it is connected. The capacitor voltage is also of the opposite polarity from the battery voltage. Because of this, a fully charged capacitor blocks current in a circuit (Fig. 12–3). The blocking action of a capacitor is used in many electronic circuits.

Action in an Alternating-current Circuit. When a capacitor is connected to a source of alternating voltage, the polarity of the applied voltage changes each half cycle. As a result, the

Fig. 12–2. Charging and discharging actions of a capacitor: (A) charging; (B) discharging

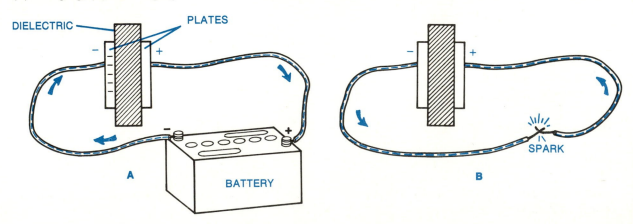

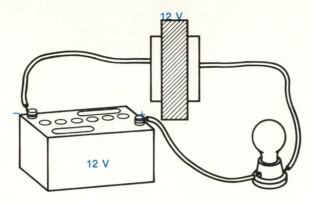

capacitor is alternately charged, discharged, and recharged. It has the opposite polarity during each cycle (Fig. 12-4). However, there is no current flow through the dielectric that separates the plates.

Capacitive Reactance. A capacitor allows the alternating current in a circuit to flow, but it does oppose the current. The opposition of a capacitor to the flow of alternating current is called *capacitive reactance*. The symbol for capacitive reactance is X_c. Its unit of measure is the ohm. The formula for figuring capacitive reactance is

$$X_c = \frac{1}{2\pi FC}$$

where X_c = capacitive reactance in ohms
     F = frequency in hertz
     C = capacitance in farads

Capacitive reactance is inversely proportional to both frequency and capacitance. This means that when the capacitance of a circuit is increased, the capacitive reactance decreases. The capacitive reactance also decreases when the frequency of the current in a circuit increases. These relationships are very important in tuning circuits in which capacitors determine the frequency at which the circuits operate.

Impedance. The total opposition to current in a circuit containing a combination of resistance and capacitive reactance is called *impedance*. It is also measured in ohms. The letter symbol for impedance is Z. Another component of impedance, called *inductive reactance*, is discussed in Unit 15.

CAPACITANCE

The capacitance of a capacitor—its ability to store electric energy—depends mainly on three things: (1) the area of its

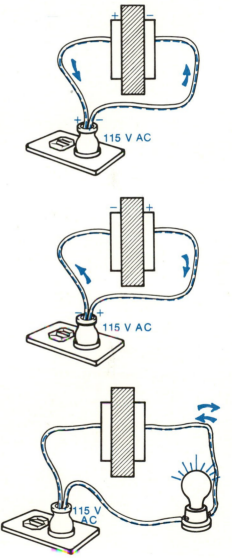

Fig. 12-4. Action of a capacitor in an ac circuit

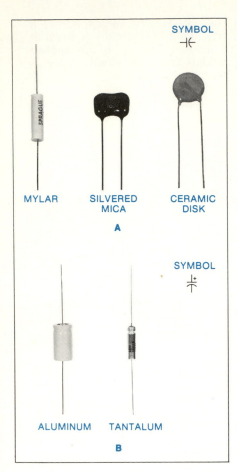

SYMBOL

MYLAR SILVERED CERAMIC
        MICA   DISK

A

SYMBOL

ALUMINUM TANTALUM

B

Fig. 12–5. Fixed capacitors: (A) nonelectrolytic; (B) electrolytic (*Sprague Electric Company*)

plates—the greater the area, the greater the capacitance; (2) the spacing, or distance, between plates—the closer the spacing, the greater the capacitance; and (3) the dielectric material. Energy is stored by a capacitor in the form of an electrostatic field within its dielectric. Different materials have different abilities to maintain an electrostatic field and, thus, to store electric energy. The measure of this ability is known as the *dielectric constant.* Common dielectric materials are air, Mylar, mica, oxide films, and certain ceramic materials. Other capacitor dielectric materials are paper, glass, and polystyrene.

UNIT OF CAPACITANCE

The basic unit of capacitance is the *farad,* abbreviated F. It is named in honor of Michael Faraday. He was an English scientist who lived from 1791 to 1867. A capacitor has a capacitance of one farad when an applied voltage that changes at the rate of one volt per second produces a current of one ampere in the capacitor circuit.

The capacitance of capacitors is most often given in microfarads (μF) or in picofarads (pF). One microfarad is equal to one-millionth of a farad. One picofarad is equal to one-millionth of a microfarad, or one-trillionth of a farad. The prefix *pico–* means 1/1,000,000,000,000 (one-trillionth).

KINDS OF CAPACITORS

Capacitors are made in different ways. They are divided into two general classes: *fixed capacitors* and *variable capacitors.* Fixed capacitors have a specific single value of capacitance. Fixed capacitors are either *nonelectrolytic* or *electrolytic* (Fig. 12–5). This refers to the structure of the dielectric.

Capacitors with a capacitance of not more than one microfarad are generally nonelectrolytic capacitors. The capacitances of electrolytic capacitors usually range from one microfarad to hundreds of thousands of microfarads. Capacitance values and working voltages are usually printed on the body of the capacitor or are shown by means of a color code.

Working Voltage. In addition to capacitance, fixed capacitors are also rated in terms of their working voltage (WVDC). This is given in volts. It is the largest value of direct voltage that can safely be applied to a capacitor. If a higher voltage is applied, the dielectric breaks down and begins to conduct current. When this happens in a nonelectrolytic capacitor, the capacitor is said to be *leaking.* It must be replaced.

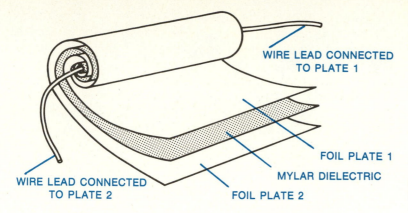

WIRE LEAD CONNECTED
TO PLATE 1

FOIL PLATE 1

MYLAR DIELECTRIC

WIRE LEAD CONNECTED
TO PLATE 2

FOIL PLATE 2

Fig. 12–6. Basic construction of a Mylar capacitor

Mylar Capacitors. A Mylar capacitor is made of thin metal-foil plates and a Mylar dielectric. These are cut into long, narrow strips and rolled together into a compact unit (Fig. 12–6). Capacitance values range from 0.001 to 1 microfarad. Working-voltage ratings are as high as 1,600 volts.

Ceramic Capacitors. Ceramic capacitors have disk-shaped or tubular dielectrics. These are made of ceramic materials such as titanium dioxide and barium titanate. The plates are thin coatings of a silver compound deposited on each side of the dielectric. Capacitances range from 5 picofarads to 0.05 microfarad. Working-voltage ratings exceed 10,000 volts.

Silvered-mica Capacitors. A silvered-mica capacitor is made of a thin sheet of mica (the dielectric) with coatings of a silver compound on each side (the plates). Silvered-mica capacitors have capacitance values that range from 1 to 10,000 picofarads. The working voltage is about 500 volts.

ELECTROLYTIC CAPACITORS

The dielectric of electrolytic capacitors is a thin film of oxide. It is formed by electrochemical action directly on a plate of metal foil. The other plate is a paste electrolyte, either borax or a carbon salt. The very thin dielectric provides a large capacitance within a much smaller volume than would be possible with other materials.

Many electrolytic capacitors have more than one section. Some have two or more individual capacitor units within a single case. Multisection capacitors have a positive lead for each of the capacitor units. They have a single negative lead common to all the capacitor units (Fig. 12–7).

Fig. 12-7. Examples of dual, or two-section, electrolytic capacitors (*Sprague Electric Company*)

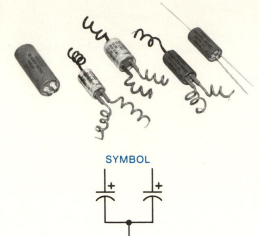

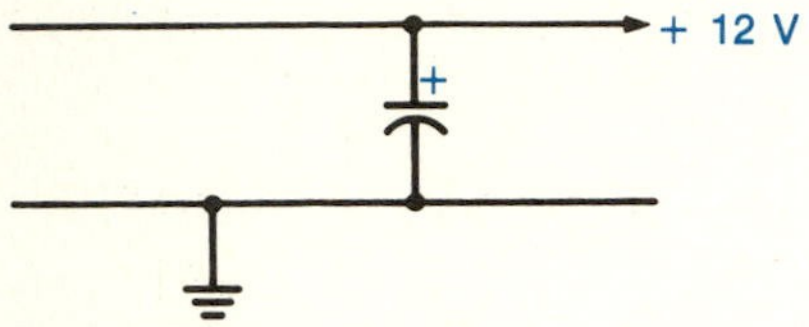

SYMBOL

Fig. 12-8. The polarized electrolytic capacitor must be connected into a circuit according to its polarity markings.

+ 12 V

The typical electrolytic capacitor used in electronic circuits is polarized. It must be connected into a circuit according to the plus and minus markings on the case (Fig. 12-8). Otherwise, the capacitor will overheat. This happens because of excessive leakage of current through the dielectric. Enough gas may be produced to cause the capacitor to explode. Non-electrolytic capacitors are not polarized. They can be connected into a circuit without regard to polarity.

Electrolytic capacitors are used in rectifier circuits to smooth out varying direct currents. They are also often used as blocking capacitors. These block a direct current from a circuit while letting an alternating current pass.

VARIABLE CAPACITORS

Variable capacitors are used primarily to tune circuits. This involves giving the circuits the least opposition to currents with a certain frequency (Fig. 12-9). One kind of variable capacitor, called a *trimmer*, is made of two metal plates separated by a sheet of mica dielectric. The space between the plates can be adjusted with a screw. As the distance between the plates is increased, the value of capacitance decreases.

Another kind of variable capacitor is made of two sets of metal plates separated either by air or by sheets of mica insulation. One set of plates, the *stator assembly*, does not move. It is insulated from the frame of the capacitor, on which it is mounted. The other set of plates, the *rotor assembly*, is connected to the shaft. It can be turned. The rotor plates can move freely in or out between the stator plates. Thus, the capacitance of the capacitor can be adjusted easily from the lowest (plates apart) to the highest (plates together) value.

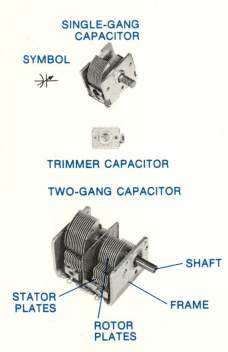

SINGLE-GANG CAPACITOR

SYMBOL

TRIMMER CAPACITOR

TWO-GANG CAPACITOR

SHAFT

STATOR PLATES

FRAME

ROTOR PLATES

Fig. 12-9. Variable capacitors

CAPACITORS IN PARALLEL

Capacitors are often connected in parallel. This is done to get a capacitance value that cannot easily be obtained from one capacitor (Fig. 12–10). The total capacitance of such a combination is equal to the sum of the individual capacitances. The working voltage of the combination is equal to the lowest individual working-voltage rating.

CHARGING RATE OF A CAPACITOR

Unlike a dry cell, a capacitor does not generate a voltage by its own action. However, the voltage developed across the plates of a charged capacitor can be used as a voltage source. One of the important characteristics of a capacitor is that it can be charged. The charge can be stored for a time and then discharged through a load when needed.

The rate at which a capacitor charges and discharges can be controlled by placing resistance in the capacitor circuit. The time needed for the voltage across a charging capacitor to reach 63.2 percent of the applied voltage is known as the *time constant* of the circuit. After five time-constant periods, the voltage across the capacitor will be very near 100 percent of the applied voltage. The time-constant formula is

$$T = R \times C$$

where T = the time constant in seconds
R = the resistance in ohms
C = the capacitance in farads

Capacitor charge and discharge rates can be used to control many circuits, including electronic timers and flashers. The circuit in Fig. 12–11 is a good example. In this circuit, the variable resistor R1 is used to change the time constant. During the first time constant, the capacitor will charge to 63.2 percent of 100 volts, or 63.2 volts. This may be enough voltage across the neon lamp to ionize the neon gas, causing it to conduct current. At low voltages, the neon acts as an insulator, so no current passes through it.

If the 63.2 volts ionizes the lamp, the capacitor will quickly discharge through the lamp. There will be a flash of light. Now the capacitor will again begin to charge. The cycle is repeated. The time between flashes depends on the time constant of the circuit. A smaller value of resistance will let the capacitor charge more quickly. Flashes will occur more often. A larger resistance or capacitance will lengthen the time constant. Flashes will occur less often. This kind of circuit is often called a *relaxation oscillator*.

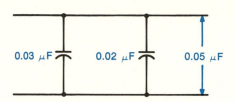

Fig. 12–10. When capacitors are connected in parallel, the total capacitance is equal to the sum of the individual capacitances.

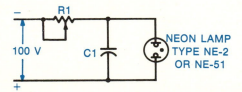

Fig. 12–11. Lamp-flasher circuit

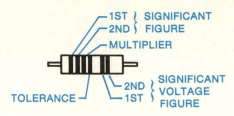

NORMALLY STAMPED FOR VALUE

MOLDED TUBULAR PAPER OR MYLAR

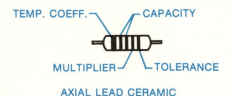

CERAMIC DISK

AXIAL LEAD CERAMIC

Fig. 12–12. How the capacitor color code is used

CAPACITOR COLOR CODES

Capacitance values are often printed directly on the body of capacitors. The values may also be shown by a color code. The numbers represented by colors in this code are the same as those used in the resistor color code. Unless otherwise stated, capacitance values given by the color code are in picofarad units.

How the color code is used with three kinds of nonelectrolytic capacitors is shown in Fig. 12–12. As with the resistor color code, the multiplier color gives the number of zeros that are to be added to the first two numbers of the capacitance value. The temperature-coefficient color in each of the codes shows the degree to which the capacitance will change with an increase of temperature.

CAPACITOR SAFETY

Many capacitors, such as the electrolytic capacitors used in high-voltage rectifier circuits of television receivers, keep their charges for some time after the charging voltages have been removed. There could be dangerous voltages across these capacitors for many hours after their circuits have been turned off. These could give a person a bad electric shock or badly damage test instruments.

To prevent this, all electrolytic capacitors in high-voltage circuits should be discharged before the circuits are handled in any way. This can be done safely by briefly shorting the terminals of the capacitors with an insulated wire. A 10,000-ohm resistor can be connected in series with the wire to reduce sparking at the terminals.

CAPACITOR DEFECTS

Fixed capacitors often become shorted because of a direct contact between their plates. This happens when the dielectric becomes worn or burned through because of age or the application of a voltage higher than the working voltage. A shorted capacitor must be replaced. The capacitor not only may not work properly but also may let too much current pass through other components in its circuit. A capacitor can, but seldom does, become open. This happens when one of its leads becomes disconnected from its plate. Electrolytic capacitors can become defective if the chemical materials inside dry out. Such a capacitor has much less capacitance and should be replaced.

A variable capacitor becomes defective because of shorted plates. This often happens because the plates are bent while the capacitor is being installed or while its circuit is being repaired. A shorted variable capacitor can often be repaired by simply bending the touching plates back into place.

TESTING CAPACITORS

Capacitors can be tested with a capacitor tester (Fig. 12–13). This instrument lets you determine both the condition of the capacitors and their capacitance.

Capacitors can also be tested for shorts with an ohmmeter. The ohmmeter (or volt-ohm-milliammeter) is connected across the terminals of the capacitor being tested. If the capacitor is shorted, the ohmmeter will show continuity or a very low resistance. Leaky or shorted capacitors should be replaced. When testing a capacitor that is wired into a circuit, you should always take it out of, or isolate it from, the circuit (Fig. 12–14). This will keep the ohmmeter from also measuring the resistance of the circuit in parallel with the capacitor.

Good nonelectrolytic capacitors have a very high resistance or no continuity between their terminals. Electrolytic capacitors have a lower resistance between their terminals. This is true even if they are in good condition. The amount of resistance will depend on the capacitance and working-voltage rating. With practice, you can soon learn how to tell if a capacitor is defective by testing it with an ohmmeter. If such a test is indefinite, the capacitor should be tested with a capacitor tester.

An ohmmeter applies a voltage to the component or part of a circuit to which it is connected. To keep from damaging an electrolytic capacitor while testing it with an ohmmeter, make sure that the ohmmeter voltage is not greater than the working-voltage rating of the capacitor. When testing an electrolytic capacitor with an ohmmeter, it is also important to connect the ohmmeter to it in the right polarity, or + to + and − to −.

To test a variable capacitor for shorts, connect an ohmmeter to its terminals. Then slowly turn the shaft as far as it will go in both directions. If the capacitor is shorted at a point, the ohmmeter will show direct continuity at that point.

Fig. 12–13. Capacitor tester (*EICO Electronic Instrument Co., Inc.*)

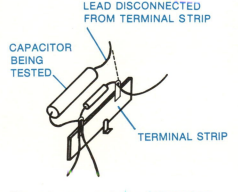

Fig. 12–14. Isolating a capacitor for test purposes

LEARNING BY DOING

10. Capacitor Charge Storage. One of the important properties of a capacitor is its ability to store a charge and then to discharge it through a load at some later time. The following

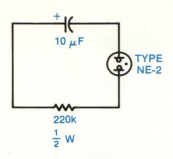

Capacitance (μF)	Lamp remains lighted (seconds)
10	
20	
30	
40	
100	

Fig. 12–15. Schematic diagram and table used for Learning by Doing No. 10, "Capacitor Charge Storage"

procedure will help you observe this action with different capacitors.

MATERIALS NEEDED

circuit board, prepunched, 4 × 6 in. (102 × 152 mm)
4 electrolytic capacitors; 10, 20, 30, and 40 μF, 150 WVDC
1 carbon-composition or film resistor, 220,000 ohms, $\frac{1}{2}$ watt
1 neon lamp, type NE-2
high-voltage dc power supply
stopwatch or timepiece with sweep second hand

Procedure

1. Connect the 10-μF capacitor, the resistor, and the neon lamp as shown in Fig. 12–15. On a sheet of paper, prepare the table shown in the figure.
2. Connect the negative terminal of the power supply to the negative terminal of the capacitor. Adjust the power supply to produce an output voltage of 140 volts. Be careful not to touch any uninsulated parts of the power-supply leads or the circuit.
3. Charge the capacitor by bringing the probe or clip of the positive power-supply lead into contact with the positive terminal of the capacitor for about 10 seconds. Note how long the neon lamp glows after the positive lead is removed from the capacitor. Record this time on the table.

 CAUTION: The capacitor remains charged to a voltage of about 55 volts after the lamp "goes out." To completely discharge the capacitor, short its terminals with a piece of insulated wire.

4. Repeat step 3 with the 20-, 30-, and 40-μF capacitors and with all the capacitors connected in parallel.
5. Look at the data in your table. Explain why the lamp glows for a longer period of time as the capacitance of the circuit is increased.

11.   Making a Relaxation Oscillator. In this interesting circuit, the charge-discharge action of a capacitor causes a neon lamp to flash on and off. When the resistance or the capacitance of the circuit is varied, the time constant is changed. This causes the lamp to flash at a different rate.

circuit board, prepunched, 6 × 8 in. (152 × 203 mm)
1 potentiometer, 1 megohm, ½ watt
1 carbon-composition or film resistor, 1 megohm, ½ watt
2 capacitors, 0.5 μF, 200 WVDC
1 neon lamp, type NE-2
90-volt battery or a dc power supply

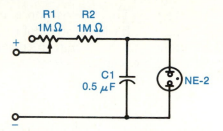

Fig. 12–16. Schematic diagram used for Learning by Doing No. 11, "Making a Relaxation Oscillator"

Procedure

1. Wire the circuit shown in Fig. 12–16.
2. Connect the battery or the power supply to the circuit. If a power supply is used, adjust it to produce an output voltage of 90 volts.
3. Adjust the potentiometer so that the lamp flashes about once per second. The time between flashes can be calculated using the formula $T = RC$. T is time in seconds, R is resistance in ohms, and C is capacitance in farads. This is because 63.2 percent of 90 volts is the approximate ionization voltage of an NE-2 lamp.
4. By using the formula $T = RC$, figure the total resistance of R1 and R2 when the lamp is flashing at the rate of one flash per second.
5. Adjust the potentiometer so that the lamp flashes at a faster speed. Has the total resistance of R1 and R2 been increased or decreased? Explain your answer.
6. Adjust the potentiometer to again get one flash per second. After disconnecting the battery or the power supply from the circuit, connect the other capacitor in parallel with C1. Does this cause the lamp to flash more or fewer times per second? Explain the reason for this.

SELF-TEST

Test your knowledge by writing, on a separate sheet of paper, the word or words that most correctly complete the following statements:

1. Capacitance is the ability of a circuit or a device to store _____ energy.
2. The basic form of a capacitor is two _____ separated by the _____.
3. A capacitor is charged by removing _____ from one of its plates and adding _____ to the other plate.
4. The voltage across the plates of a fully _____ capacitor is equal to the voltage of the source of energy.
5. The energy stored in a charged capacitor is in the form of an _____ field between its plates.
6. The discharge current of a capacitor is in a direction _____ to the direction of the charging current.
7. A capacitor prevents direct current from flowing in a circuit by what is referred to as a _____ action.
8. The opposition of a capacitor to the flow of alternating current in a circuit is called

_____. This opposition is measured in _____. Its letter symbol is _____.

9. The total opposition to current by resistance and capacitive reactance in a circuit is known as _____. This total opposition is measured in _____. Its letter symbol is _____.

10. The measure of the ability of a dielectric material to maintain an electrostatic field between the plates of a capacitor is known as its _____.

11. A capacitor has a capacitance of one farad when an applied voltage that changes at the rate of one volt per _____ produces a current of one _____ in the capacitor circuit.

12. The most common units of capacitance are the _____ and the _____.

13. One _____ is equal to one-millionth of a microfarad.

14. The two general classes of capacitors are _____ capacitors and _____ capacitors.

15. The _____ voltage of a capacitor is the highest dc _____ that can be applied to it without causing the dielectric to break down and conduct current.

16. Three common kinds of fixed nonelectrolytic capacitors are _____ capacitors, _____ capacitors, and _____ capacitors.

17. The typical _____ capacitor is polarized. It must be connected into a circuit with the right _____.

18. An _____ capacitor is used to smooth out the varying dc output of a rectifier circuit.

19. The primary use of variable capacitors is to _____ circuits.

20. When capacitors are connected in parallel, the total capacitance of the combination is equal to the _____ of the individual capacitances.

21. The time needed for the voltage across a charging capacitor to reach 63.2 percent of the applied voltage is known as the _____ of the circuit.

22. For safety, charged high-voltage electrolytic capacitors should be _____ before they are handled in any way.

FOR REVIEW AND DISCUSSION

1. Define capacitance.
2. Describe the basic form of a capacitor.
3. Describe the charging and the discharging action of a capacitor.
4. What is meant by the blocking action of a capacitor?
5. Define capacitive reactance and impedance.
6. What are the three main things that determine the capacitance of a capacitor?
7. What is meant by the dielectric constant of a material?
8. What is meant by the working-voltage rating of a capacitor?
9. Describe the construction of a Mylar capacitor and of a ceramic capacitor.
10. The typical electrolytic capacitor is polarized. What does this mean?
11. Describe the construction of a variable capacitor.
12. What is the effect of connecting capacitors in parallel?
13. Explain the time-constant feature of a resistor-capacitor combination in a circuit.
14. Why is it important to discharge electrolytic capacitors in high-voltage circuits before the circuits are handled in any way?
15. Describe the most common capacitor defects.

INDIVIDUAL-STUDY ACTIVITIES

1. Give a demonstration showing the basic charging action of a capacitor. Explain this to your class.
2. Give a demonstration showing your class how capacitors can be tested with an ohmmeter and with a capacitor tester.

Unit 13 Magnetism

Certain metals and metallic oxides are able to attract other metals (Fig. 13–1). This is called *magnetism*. Materials that have magnetism are called *magnets*. Some magnets are found naturally in metallic ores. Others are manufactured. Magnets that keep their magnetism for a long time are called *permanent magnets*.

Magnetism is produced as a result of electrons spinning on their own axes while rotating about the nuclei of atoms (Fig. 13–2).

In magnetic materials, the atoms within certain of their areas, called *domains*, are aligned. When this happens, most of their electrons spin in the same direction (Fig. 13–3 at A). Magnetization usually results in two *magnetic poles* being formed at the ends of the magnet. These are called the *north pole* and the *south pole*. In the absence of magnetization, the domains within a material are not aligned (Fig. 13–3 at B). The electrons spin in all directions.

ATTRACTION AND REPULSION

The law of magnetic attraction and repulsion states that *unlike* magnetic poles attract each other and *like* magnetic poles repel each other (Fig. 13–4). The force of the attraction or repulsion depends on the strength of the magnets and the distance between them.

As the strengths of magnets increase, the force of attraction or repulsion between them also increases. The force of attraction or repulsion decreases as the distance between the poles of the two magnets increases. If the distance between the unlike poles of two magnets is doubled, the force of attraction between the magnets decreases to one-fourth the original value. If the distance between the poles is reduced by one-half, the force of attraction between the magnets is increased four times. There is another way of stating these relationships. It can be said that the force of attraction or repulsion between the poles of two magnets is inversely proportional to the square of the distance between them.

MAGNETIC FIELD

The energy of a permanent magnet is in the form of a *magnetic field*. This field surrounds the magnet. The magnetic field is made of invisible *lines of force*, or *flux*. Outside the

Fig. 13–1. Metal tools held in place by a magnetic tool holder (*Master Magnetics, Inc.*)

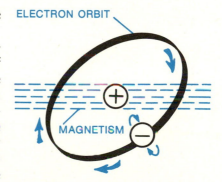

Fig. 13–2. An electron spinning on its own axis while revolving about the nucleus of an atom produces a magnetic field.

111

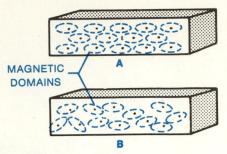

Fig. 13–3. Magnetic domains:
(A) aligned; (B) not aligned

magnet, the lines of force run from the north pole to the south pole (Fig. 13–5). Magnetic lines of force form an unbroken loop. They do not cross one another. The strength of the magnetic field is directly related to the number of domains that have been aligned within the magnet during magnetization. When all the domains have been aligned, the magnetic field is as strong as possible. The magnet is then said to be *saturated.*

MAGNETIC AND NONMAGNETIC MATERIALS

A *magnetic material* is one that is attracted to a magnet and can be made into a magnet. Among these materials are steel and the metallic elements iron, nickel, and cobalt. These are commonly called *ferromagnetic elements. Ferro* refers to iron and to other materials and alloys that have certain properties similar to those of iron. Permanent magnets were originally natural magnets made of an iron ore called *magnetite,* or *lodestone* (Fig. 13–6). During the twelfth century, it was learned that a lodestone needle, when suspended so that it was free to turn, would point to the Earth's magnetic poles. Thus, it could serve as a magnetic compass. Lodestone is no longer used commercially as a magnetic material.

The magnetic materials most often used for making modern permanent magnets are metallic alloys and compounds. These are usually combinations of iron oxide and certain other elements. These materials have a high *retentivity,* the ability to keep a magnetic domain alignment for a long time. They also have high *permeability,* the ability to conduct magnetic lines of force. The magnetic permeability of a material is measured by comparing its ability to conduct magnetic lines of force to the ability of air to do so. Air, a vacuum, or other nonmagnetic substances are assumed to have a permeability of one.

Metals that are not attracted by a magnet and that cannot be made into a magnet are *nonmagnetic materials.* Copper, aluminum, gold, silver, and lead are nonmagnetic. Most non-

Fig. 13–4. The law of magnetic attraction and repulsion

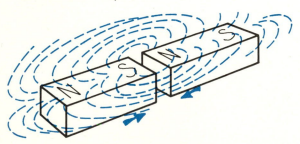

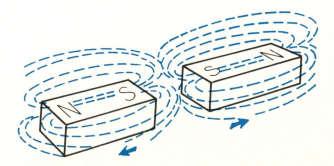

metallic materials such as cloth, paper, porcelain, plastic, and rubber are also nonmagnetic.

Alloys. One of the most common magnetic alloys is alnico. This alloy is made of aluminum, nickel, cobalt, and iron. Permanent alnico magnets come in different shapes and grades. They are used in such products as motors, generators, loudspeakers, microphones, and meters (Fig. 13-7). Other good magnetic alloys are Permalloy (nickel and iron, or cobalt, nickel, and iron), Supermalloy (nickel, iron, molybdenum, and manganese), and platinum-cobalt.

Ceramic Materials. Some hard ceramic magnetic materials are called *ferrites*. They are made by first grinding a combination of iron oxide and an element such as barium into a fine powder. The powder mixture is then pressed into the desired shape and baked at a high temperature. This produces a magnetic material that is very efficient and, unlike alloy materials, has a high electrical resistance.

Ceramic magnets are formed into many shapes (Fig. 13-8). Permanent magnets of this kind are used as latches on refrigerator doors. Such a magnet is in the form of a plastic strip filled with barium ferrite and bent to fit the shape of the door.

Fig. 13-5. The magnetic field surrounding a bar-shaped magnet

Fig. 13-8. Ceramic permanent magnets can be made in many shapes and sizes. (*Allen-Bradley Company*)

Fig. 13-6. Magnetite, commonly known as lodestone, is a natural magnetic material.

BAR

KEEPER

HORSESHOE

DISK

Fig. 13-7. Alnico permanent magnets. The soft-iron *keeper* placed across the poles of the horseshoe magnet helps the magnet keep its full strength for a longer period of time.

HOW MAGNETS ARE MADE

In order to make a magnet, energy in the form of a magnetic field must be applied to some magnetic material. This is called *magnetic induction.*

An example of magnetic induction is shown in Fig. 13-9. A nail made of steel is placed parallel to and near a permanent magnet. Since steel is more permeable than air, the force lines from the permanent magnet pass easily through the nail. This causes some of the domains in the steel to become aligned. Thus, the magnetic induction process produces magnetic poles at the ends of the nail. Note that the magnetic polarity of the nail is opposite to the polarity of the permanent magnet. The nail thus becomes a magnet and is able to attract objects made of magnetic materials.

Commercially, permanent magnets are made with a *magnetizer.* One kind of magnetizer uses electric coils as the source of the magnetization energy (Fig. 13-10). To use the magnetizer, the object to be magnetized is placed over the pole pieces formed by two ends of the metal cores over which the coils are wound. When the switch is turned on, a direct current passes through the coils. This current produces a strong magnetic field that magnetizes the object quickly. The magnetism produced by the current is called *electromagnetism.* (See Unit 14, "Electromagnetism.")

ELIMINATING MAGNETISM

Although magnetism is useful in many ways, sometimes objects must be demagnetized, or have their magnetism removed. Wristwatches made of magnetic materials, for example, will not keep correct time after becoming magnetized. If metal-cutting tools such as drills and reamers become magnetized, they will attract metal chips and filings. This causes the tools to become dull quickly.

Magnetism can be removed from an object by using a *demagnetizer.* In its simplest form, this is a coil of insulated wire through which alternating current passes. The object to be demagnetized is put within the coil. Then the object is slowly removed from the coil. The effect of the varying magnetic field that results is to disturb the alignment of the domains and thus eliminate the magnetism from the object (Fig. 13-11).

A permanent magnet can also be demagnetized or greatly weakened by heating. It can also be demagnetized by striking it sharply with a metal object such as a hammer. In both cases, the domain alignment within the magnet is severely disturbed. Demagnetization is sometimes called *degaussing.*

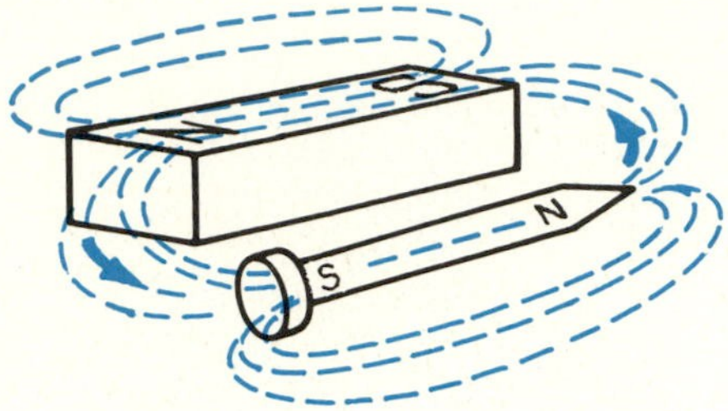

Fig. 13-9. Magnetic induction

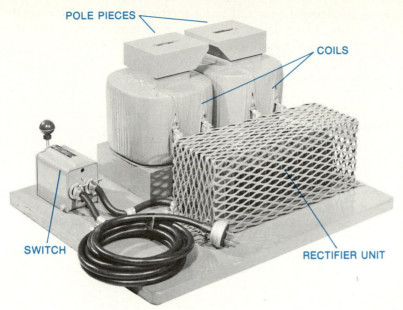

Fig. 13–10. An electromagnetic magnetizer

POLE PIECES

COILS

SWITCH

RECTIFIER UNIT

Fig. 13–11. Using a demagnetizer to remove the magnetism from a metal tool-holding device (*R. B. Annis Company*)

MAGNETIC COMPASS

A *magnetic compass* is a device that is used to find directions. In the compass, a small permanent magnet, the needle, is mounted on a pivot point so that it is free to turn. The poles of the magnet are attracted by the magnetic poles of the Earth. Because of this, the needle of the compass points in a general north-south direction (Fig. 13–12).

The magnetic poles of the Earth are actually about 1,500 miles (2414 km) from the geographic poles. Therefore, a

Fig. 13–12. The needle of a magnetic compass is attracted by the magnetic poles of the Earth. The north pole of a magnet is so named because it points to the Earth's north magnetic pole.

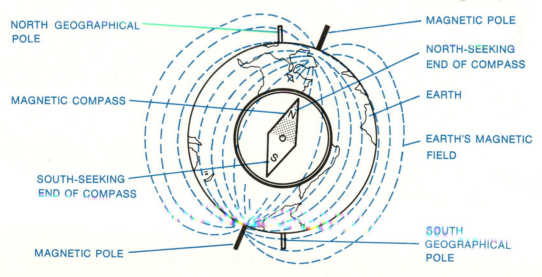

NORTH GEOGRAPHICAL POLE

MAGNETIC POLE

NORTH-SEEKING END OF COMPASS

MAGNETIC COMPASS

EARTH

EARTH'S MAGNETIC FIELD

SOUTH-SEEKING END OF COMPASS

MAGNETIC POLE

SOUTH GEOGRAPHICAL POLE

compass needle does not point to the true geographic north and south. The difference between a compass reading and the true direction is known as the *angle of declination*. This angle must, of course, be considered in finding true geographic directions with a magnetic compass.

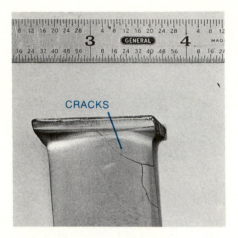

Fig. 13–13. Cracks in the turbine blade of a jet aircraft engine discovered by magnetic-particle inspection (*Magnaflux Corporation*)

MAGNETIC-PARTICLE INSPECTION

Magnetic-particle inspection is a process that uses magnetism to find defects on the surfaces of metal parts. In this process, the object is magnetized and then coated with an oil or other liquid that contains fine particles of iron. The edges of cracks and scratches become magnetic poles to which the particles are attracted. This causes the particles to form a pattern that can easily be seen (Fig. 13–13). Tests of this kind are called *nondestructive testing*.

FUTURE USES OF MAGNETISM

Scientists and engineers are constantly thinking of new and different ways in which to use magnetism. The possibility of using the magnetic energy high above the surface of the Earth is now being studied. Many scientists believe that magnetism may someday be used to counteract the force of gravity. These ideas and the study of other applications of magnetism in the field of electronics are creating a new scientific frontier.

LEARNING BY DOING

12. "Seeing" Magnetic Fields. The magnetic field that surrounds a magnet cannot, of course, actually be seen. However, a "picture" of it can be obtained in a simple and interesting way.

MATERIALS NEEDED

2 pieces of wood, $\frac{3}{8} \times 1 \times 10$ in. ($15 \times 25 \times 254$ mm)
2 pieces of wood, $\frac{3}{8} \times 1 \times 6$ in. ($15 \times 25 \times 152$ mm)
bar, horseshoe, and disk-shaped permanent magnets
iron filings
typing paper, $8\frac{1}{2} \times 11$ in. (216×279 mm)
*photographic print paper, 5×7 in. (127×178 mm) or
    $8\frac{1}{2} \times 11$ in. (216×279 mm)*
photoflood lamp, with socket and cord

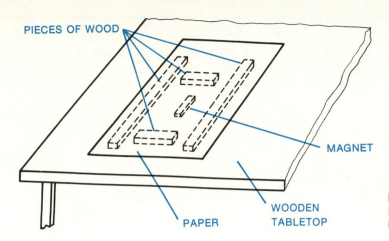

PIECES OF WOOD

MAGNET

PAPER

WOODEN TABLETOP

Fig. 13–14. Assembly used for Learning by Doing No. 12, "Seeing Magnetic Fields"

Procedure

1. Assemble the bar magnet, the paper-support pieces of wood, and a sheet of typing paper on a wooden table as shown in Fig. 13–14.
2. Sprinkle a thin layer of iron filings over the paper directly above all parts of the magnet.
3. Tap the center and the edges of the paper lightly with a pencil. The filings will align themselves with the magnetic lines of force.
4. Describe this magnetic field.
5. Repeat this procedure with the other magnets.
6. A permanent "picture" of a magnetic field can be made by using a sheet of print paper instead of typing paper. After a magnetic field has formed on the print paper, place the paper under the photoflood lamp and leave it there for about 2 minutes. Then fix the print by using a print machine or developer.

Magnetizing and Demagnetizing. The plans and procedures for making an electromagnetism experimenter that can be used as a magnetizer and a demagnetizer are given in Unit 15.

117

Test your knowledge by writing, on a separate sheet of paper, the word or words that most correctly complete the following statements:

1. Magnetism is produced as electrons _____ on their own axes while rotating about the nuclei of atoms.
2. In a permanent magnet, atoms in certain areas called _____ are aligned.
3. Unlike magnetic poles _____ each other, while like magnetic poles _____ each other.
4. The force of attraction or repulsion between the poles of two magnets is inversely proportional to the _____ of the distance between them.
5. A magnetic field is made of invisible lines of _____, the direction of which is from the _____ pole to the _____ pole of a magnet.
6. A magnet is said to be _____ when all its domains have been aligned.
7. A magnetic material is one that is _____ to a magnet and can be _____ into a magnet.
8. The first permanent magnets were natural magnets made of an iron ore called _____, or _____.
9. Three metallic elements that are highly magnetic are _____, _____, and _____.
10. A material has a high _____ when it is able to keep its magnetic-domain alignment for a long time.
11. The _____ of a material is its ability to conduct magnetic lines of force, as compared to air's ability.
12. Permanent magnets are used in products such as _____, _____, _____, and _____.
13. Ceramic magnetic materials are also called _____.
14. Magnetizing an object by putting it in the magnetic field of a permanent magnet is called _____.
15. Permanent magnets are made commercially with a _____.
16. Demagnetization is sometimes called _____.
17. Magnetic-particle inspection is used to find _____ on the surfaces of metals.

FOR REVIEW AND DISCUSSION

1. Explain magnetization in a permanent magnet.
2. State the law of magnetic attraction and repulsion.
3. Describe the magnetic field around a permanent magnet.
4. What is a magnetic material?
5. Define magnetic retentivity and permeability.
6. Name three magnetic and three nonmagnetic materials.
7. Give five uses of permanent magnets made of an alloy such as alnico.
8. Describe the makeup of a ceramic magnetic material.
9. Explain magnetic induction.
10. Describe the making of permanent magnets with a magnetizer.
11. Why must a magnetizer be operated with direct current to get the best results?
12. Give two examples of the undesirable effects of magnetism.
13. State three ways in which magnetism can be removed or greatly weakened.
14. Tell how a magnetic compass works.
15. For what purpose is magnetic-particle inspection used? Describe this process.

INDIVIDUAL-STUDY ACTIVITIES

1. Prepare a written or an oral report telling about the discovery of magnetism, the uses of permanent magnets, and the development of highly efficient magnetic materials.
2. Prepare a written or an oral report about the Earth's magnetism.

Unit 14 Electromagnetism

As you have learned, the magnetism in a permanent magnet is produced by the spinning of electrons on their own axes as they rotate about the nuclei of atoms. Magnetism is also produced as free electrons move through a conductor in the form of a current. This important relationship between electricity and magnetism is known as *electromagnetism,* or the magnetic effect of current. It is used in the operation of many different kinds of circuits, products, and devices. Some of these are discussed in this unit. Others are discussed in later units of this textbook.

ELECTROMAGNETIC FIELD

An *electromagnetic field* is the magnetic field produced by current. It is in the form of circles around the conductor. Its strength depends on the amount of current passing through the conductor. The larger the amount of current, the stronger the field. The direction of the magnetic field depends on the direction of the current (Fig. 14–1). If the current is a steady direct current, the magnetic field is constant in polarity and in strength. An alternating current produces a magnetic field that reverses in polarity and changes in strength. This is called a *moving magnetic field.*

SOLENOID

To concentrate the magnetic field produced by current flowing in a wire, the wire is wound into a coil. When this is done, the magnetic fields around the turns of the coil are added together. This increases the magnetic strength of the coil. A coil wound in this manner is called a *solenoid* (Fig. 14–2). A solenoid has magnetic poles and a magnetic field that have the same properties as those of a permanent magnet.

Fig. 14–1. The magnetic field produced by current in a conductor

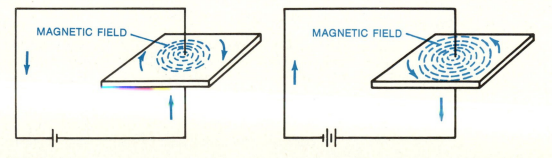

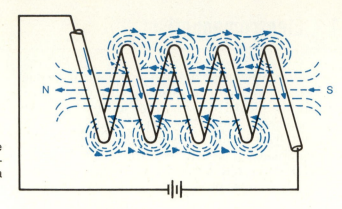

Fig. 14–2. By winding wire in the form of a coil, or solenoid, the magnetic field can be concentrated in a small area.

If the solenoid is energized with direct current, the polarity of its magnetic poles remains fixed. If the solenoid is energized with alternating current, its magnetic polarity reverses with each reversal of the direction of the current.

Solenoid Control. A solenoid coil is usually made by winding many turns of wire around a hollow paper, cardboard, or plastic form. The strong magnetic field created by the coil can be made to produce useful motion.

An example of a solenoid is shown in Fig. 14–3 at A. In this assembly, a plunger, or *armature* (movable piece), made of iron or steel is placed partly inside the coil. When the coil is energized, the armature is drawn further into it. With the right coupling, the movement of the armature can be used to control things such as valves, switches, and clutch mechanisms. A typical example of this kind of control is found in automatic washing machines, where water valves are operated by solenoid assemblies (Fig. 14–3 at B).

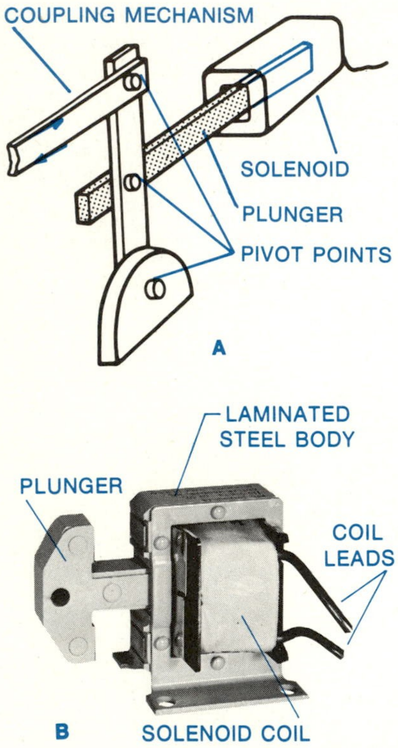

Door-chime solenoid. The door chime is another common use of the solenoid. In this device, an iron or steel armature with a rubber or plastic tip is put inside the coil (Fig. 14–4 at A). When the coil is energized, the armature is attracted upward into it. This causes the tip to strike the chime bar and produce sound (Fig. 14–4 at B).

IRON-CORE ELECTROMAGNET

An electromagnet can be made more powerful by winding magnet wire around an iron core. This is because iron is more *permeable* than air. That means that iron can conduct magnetic lines of force more easily. As a result, many more lines of force can pass between the poles of the electromagnet, producing a stronger magnetic field (Fig. 14–5). If soft iron is used as the core, very little *residual*, or leftover, *magnetism*

COUPLING MECHANISM

SOLENOID

PLUNGER

PIVOT POINTS

A

LAMINATED STEEL BODY

PLUNGER

COIL LEADS

B   SOLENOID COIL

Fig. 14–3. Solenoid-control mechanism: (A) basic construction and operation; (B) typical solenoid assembly (*Photo courtesy of Cutler-Hammer*)

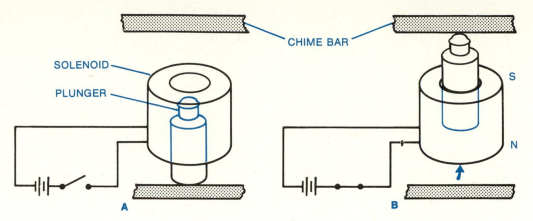

Fig. 14–4. Operation of a door chime

stays after the electromagnet has been deenergized, or turned off. This is because soft iron has a low *retentivity*, or little ability to stay magnetized.

The strength of an electromagnet is given by a unit called *ampere-turns*. This is the product of the coil current times the number of turns of wire with which the coil is wound. The ampere-turns unit is useful for comparing the strengths of electromagnets that operate at the same voltage.

Electromagnets of different sizes and strengths are used in a large number of devices. These include doorbells, buzzers, horns, relays, circuit breakers, and electric clutches. As you will learn in later units of this book, electromagnetism or a combination of electromagnets and permanent magnets is used in many electronic products. Electromagnets are, for example, used in transformers, generators, loudspeakers, microphones, phonograph cartridges, and audio equipment for recording and playback.

BELLS AND BUZZERS

Electric doorbells, buzzers, and some kinds of horns are an interesting group of *vibrating devices*. In these devices, electromagnetism is used to produce a rapid back-and-forth motion of an armature. The basic construction and operation of a typical doorbell are shown in Fig. 14–6.

Before the doorbell button is pressed, the contact strip is held against the contact point by the tension of the spring (Fig. 14–6 at A). When the button is pressed, the circuit is completed and current flows through the coil. The electromagnet then attracts the armature to the position shown in Fig. 14–6 at B. This causes the armature to move away from the contact point. The circuit is now turned off, and the electromagnet is deenergized. The spring is then able to pull the

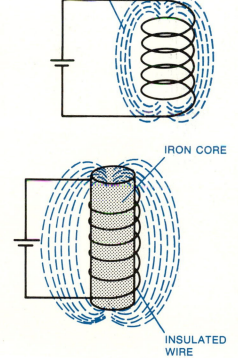

Fig. 14–5. The use of a soft-iron core increases the strength of an electromagnet.

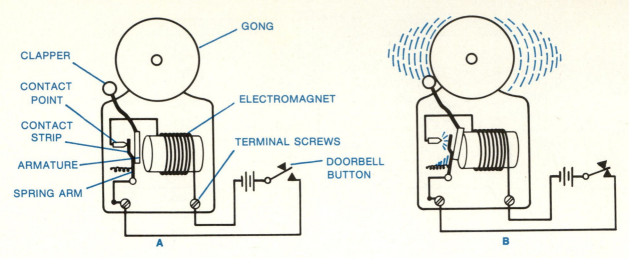

Fig. 14-6. Basic construction and operation of a doorbell

armature back to the contact point. This action is repeated very rapidly as long as the button is pressed. The *clapper* attached to the armature strikes the gong each time the armature is pulled toward the electromagnet. This produces a continuous ringing sound. In a buzzer, the sounds are produced as the armature strikes against one end of the electromagnet core.

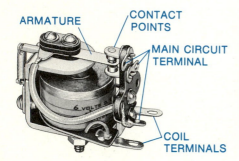

Fig. 14-7. Single-pole–double-throw relay

THE RELAY

A *relay* is a magnetically operated switch that makes or breaks one or more of the contacts between its terminals (Fig. 14–7). As with mechanical switches, the action of relays is described by the number of lines (poles) that are controlled and the number of contacts (throws) each pole can make. The relay in Fig. 14–7 controls one line (single-pole) and can touch either of two contacts (double-throw).

The basic operation of a single-pole–single-throw relay is shown in Fig. 14–8 at A. When the switch in the relay circuit is closed, the electromagnet is energized. It thus attracts the armature to the fixed contact point. There is now continuity between terminals 1 and 2 and the lamp is turned on. When the relay circuit switch is opened, the relay coil is deenergized. This lets the spring pull the armature from the fixed contact point. The circuit connected to terminals 1 and 2 is thus turned off. Other relay switching arrangements are shown in Fig. 14–8 at B.

A relay can control a large load current at a high voltage by means of a small relay-energizing current at a low voltage. An example of this is shown by the circuit of Fig. 14–9. Here the relay is controlled by the intensity of the light striking the

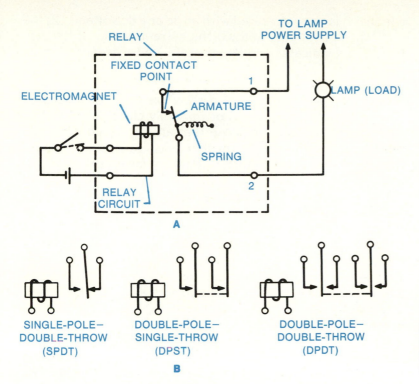

SINGLE-POLE–
DOUBLE-THROW
(SPDT)

DOUBLE-POLE–
SINGLE-THROW
(DPST)

DOUBLE-POLE–
DOUBLE-THROW
(DPDT)

B

Fig. 14–8. Relay operation: (A) single-pole–single-throw relay; (B) switching action of other common relays

photoconductive cell. The current in the control circuit, which consists of the cell, the relay coil, and the battery, is quite small. However, since the motor circuit that is controlled by the relay has a different source of energy, the current in this circuit can be and is quite large.

Rating. The general-purpose power relay is rated in terms of (1) the operating-voltage rating of the relay coil and

Fig. 14–9. A sensitive relay used in a light-activated control circuit

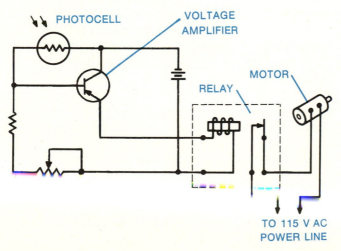

whether it is to be operated with an ac or a dc voltage, (2) the resistance of its coil, and (3) the current rating of its contacts. This indicates the largest safe load current the relay can control.

Contacts. The contacts of a relay are often described as being normally open (N.O.) or normally closed (N.C.). Normally open contacts are those that are separated when the relay is not energized. Normally closed contacts are those that are closed, or in contact, when the relay is not energized.

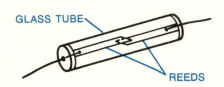

Fig. 14–10. Magnetic-reed relay

The Magnetic-reed Relay. The switching assembly of a typical magnetic-reed relay, or reed switch, is made of ferromagnetic reeds enclosed in a sealed glass tube (Fig. 14–10). In a complete relay assembly, the tube is placed in a coil. When the coil is energized, the reeds come together as a result of magnetic attraction. This kind of relay is very sensitive, which means that it can be operated with a very small amount of current.

In another kind of magnetic-reed relay, the make and break action of the reeds is controlled with a permanent magnet that is brought near to or removed from the reed enclosure. When used in this way, the reed relay is sometimes called a magnetic *proximity switch*.

MAGNETIC CIRCUIT BREAKER

The magnetic *circuit breaker* is a device that protects a circuit from too much current. In one kind of circuit breaker, the coil of an electromagnet and two contact points are connected in series with one wire of a circuit (Fig. 14–11).

When the current is more than the ampere rating, or the trip size, of the circuit breaker, the electromagnet becomes strong enough to attract the armature. This moves and latches the contact point in the open position, and the circuit is broken. After being reset mechanically, the circuit breaker is once again ready to protect the circuit.

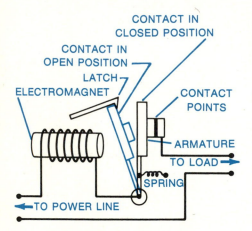

Fig. 14–11. Basic construction and operation of a magnetic circuit breaker

LEARNING BY DOING

13. The Solenoid in Action. The solenoid described below is an interesting device. It shows how the mechanical motion of a plunger can be produced with an electromagnet. The same basic idea is used to operate many kinds of door chimes, advertising-sign displays, relays, and switches.

*2 pieces of wood, ³⁄₄ × 1¹⁄₂ × 4 in.
   (19 × 38 × 102 mm)
4-in. (102-mm) length of plastic or heavy-paper tube,
   ⁵⁄₁₆-in. (8-mm) inside diameter
300-ft (91.4-m) PE magnet wire, no. 22 (0.63 mm)
2 Fahnestock clips, 1 in. (25 mm)
2 round-head wood screws, no. 5 × ¹⁄₂ in.
2 solder lugs, no. 8
piece of round soft-iron rod, ¹⁄₄ × 4¹⁄₂ in.
   (6.4 × 114.3 mm)
small amount of wood glue or plastic cement
5-ft. (1.5-m) hookup wire
6-volt dc power supply or battery
bell transformer, 10-volt secondary equipped with primary leads and attachment plug*

Procedure

1. Assemble the first eight items on the list of materials as shown in Fig. 14–12.
2. Connect the battery to the clips.
3. Put the iron rod in the tube. You will notice the rather strong pulling action of the electromagnet after the rod is about halfway into the tube.

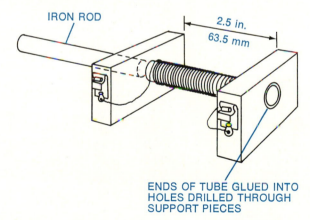

IRON ROD

2.5 in.
63.5 mm

ENDS OF TUBE GLUED INTO
HOLES DRILLED THROUGH
SUPPORT PIECES

Fig. 14–12. Using the principle of electromagnetism in a solenoid

4. Disconnect the battery. Put the solenoid in an upright position with its bottom edge about 2³⁄₄ in. (70 mm) above a table. Drop the rod through the tube and connect the battery to the circuit. The plunger will now move upward very rapidly.
5. Connect the transformer to the solenoid and put the rod in the tube. Hold a metal object against one end of the

rod. The blade of a knife or the tip of a screwdriver will do. If held lightly, the object will "chatter" against the end of the rod. What causes this?

6. Explain the operation of the solenoid and plunger.

SELF-TEST

Test your knowledge by writing, on a separate sheet of paper, the word or words that most correctly complete the following statements:

1. The electromagnetic field produced by current is in the form of _____ around the conductor.
2. A direct current produces a magnetic field that is _____ in polarity and strength.
3. An alternating current produces a magnetic field that reverses in _____ and changes in _____.
4. When current passes through a coil of wire, the magnetic fields around the turns of the coil are _____ together to produce the total electromagnetic field.
5. If a solenoid is energized with alternating current, its magnetic polarity reverses with each reversal of the _____ of the current.
6. A movable iron piece put in a control solenoid is called an _____.
7. An electromagnet that is more powerful than a solenoid can be made by winding insulated wire around an _____.
8. The magnetism remaining in the core of an electromagnet after its coil has been deenergized is called _____ magnetism.
9. Electromagnets are used in devices such as _____, _____, _____, _____, and _____.
10. A relay is a _____ operated _____.
11. The general-purpose power relay is rated in terms of _____, _____, and _____.
12. Normally open relay contacts are those that are separated when the relay is not _____.

FOR REVIEW AND DISCUSSION

1. Define electromagnetism.
2. What are the differences between the magnetic fields that are produced by direct current and by alternating current?
3. What is a solenoid?
4. Why does the magnetic polarity of a solenoid periodically reverse when it is operated with alternating current?
5. Describe a practical solenoid. Tell how it can be used as a control device.
6. Explain the operation of a door-chime solenoid.
7. What is the effect of adding an iron core to an electromagnetic coil? Explain the reason for this.
8. Define residual magnetism.
9. Tell how the strength of a given iron-core electromagnet can be increased without adding turns of wire to its coil.
10. Explain why adding turns of wire to an electromagnet coil beyond a certain point will cause the strength of the electromagnet to decrease.
11. Describe the construction of a typical doorbell. Explain its operation.
12. What is a relay? How is the switching action of a relay described?
13. Explain the operation of a simple relay.
14. What can a relay be used to control?

INDIVIDUAL-STUDY ACTIVITIES

1. Prepare a written or an oral report on the uses of electromagnetism and electromagnetic devices.
2. Give a demonstration showing electromagnetic devices such as solenoids, buzzers, or relays. Describe these to your class and explain the operation of each.

Unit 15   Electromagnetic Induction and Inductance

If a magnet is moved near a wire, its magnetic field cuts across the wire and produces a voltage across the ends of the wire. If the ends of the wire are connected to form a closed circuit, the voltage causes current to flow through the circuit (Fig. 15–1 at A). A voltage that is produced in this way is called an *induced voltage.* A current produced by an induced voltage is called an *induced current.* This important relationship between electricity and magnetism is called *electromagnetic induction.* When the magnet is moved in the opposite direction, the polarity of the induced voltage changes and the direction of the current in the circuit also changes (Fig. 15–1 at B).

Motion is needed to produce a voltage by electromagnetic induction. The magnetic field must move past the wire, or the wire must move past the magnetic field. If both the magnetic field and the wire are stationary, no voltage can be induced across the ends of the wire.

One of the most important uses of electromagnetic induction is in *alternators* and other kinds of *generators.* These machines supply the electric energy that is distributed to our homes by electric power companies. Electromagnetic induction is also used to operate many other kinds of machines and devices. These include motors, transformers, microphones, phonograph cartridges, and tape-player recording and playback heads.

MOVING MAGNETIC FIELD

As you have learned, electromagnetic induction takes place only when there is relative motion between a magnetic field and a conductor. In a generator, this motion is produced by turning its *rotor* assembly or its *field-coil* assembly with an engine of some kind. In a device such as a transformer that does not have any moving parts, a moving magnetic field is produced as the result of changes in the value of the current.

Alternating Current. The magnetic field produced by an alternating current is shown in Fig. 15–2 at A. When the current is increasing, the magnetic field moves outward from the conductor. As the current decreases, the magnetic field moves toward the conductor. A moving magnetic field such as this can produce electromagnetic induction.

Direct Current. A steady direct current does not produce a moving magnetic field. For this reason, such a current cannot

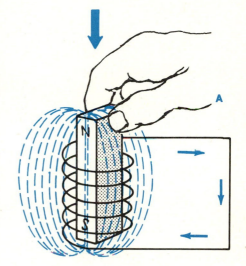

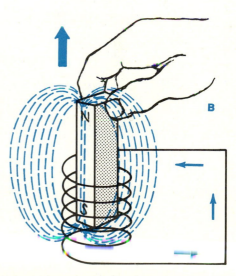

Fig. 15–1. Electromagnetic induction

127

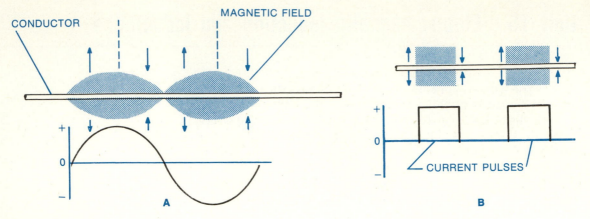

Fig. 15–2. Moving magnetic fields: (A) field produced by an alternating current; (B) field produced by a pulsating direct current

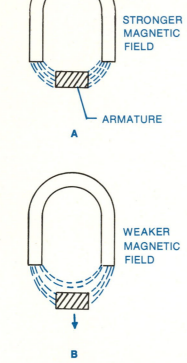

Fig. 15–3. Moving magnetic field produced by changing the position of a soft-iron armature within a magnetic air gap

produce electromagnetic induction. When a steady direct current must be used with a device such as a transformer, the current must be broken up into *pulses*. A pulsating direct current does produce a moving magnetic field, which can cause electromagnetic induction (Fig. 15–2 at B).

Magnetic Air Gap. The magnetic field of a stationary permanent magnet can be made to vary in strength. This is done by changing the permeability of its *air gap*, or the space between its poles. The air gap is changed by moving a soft-iron armature within it.

As the armature moves farther into the air gap, the permeability of the gap increases. This allows a stronger magnetic field to exist between the poles of the magnet (Fig. 15–3 at A). When the armature moves away from the air gap, the permeability of the gap decreases. When this occurs, the strength of the magnetic field also decreases (Fig. 15–3 at B). This method of changing the strength of a magnetic field is used in a *magnetic phonograph cartridge.*

TRANSFORMER

The word *transform* means "to change." A *transformer* is used to change the value of voltage or current in an electrical system (Fig. 15–4). If it reduces a voltage, it is called a *step-down transformer.* If it increases a voltage, it is called a *step-up transformer.* Some transformers do not change the value of voltage. These are called *isolation transformers.* Isolation transformers are used when the electrical equipment must not be grounded through the power line.

Transformers are used to change the value of voltage in many kinds of circuits. For example, they are used to change the power line voltage of 120 volts to a voltage needed to operate rectifier circuits. These circuits are used to change alternating current into direct current in products such as radios, television receivers, and music systems. The ignition coil in an automobile ignition system is a step-up transformer that supplies a high voltage to the spark plugs. Both step-up and step-down transformers are used in systems that distribute electric energy from power plants to buildings.

The operation of a simple transformer is shown in Fig. 15–5. Here one winding, or coil, of the transformer is connected to the source of energy. This is called the *primary winding*. The other is electrically insulated from the primary winding and is connected to the load. This is called the *secondary winding*.

The alternating current in the primary winding produces a moving magnetic field. This induces a voltage in the secondary winding. As a result, energy is transferred from the primary winding to the secondary winding. This is an example of what is known as *mutual induction*.

Polarity of Induced Voltage. The polarity of the voltage produced by electromagnetic induction depends on the direction in which magnetic lines of force cut across a conductor. Since a transformer is operated with alternating current or with a pulsating direct current, the magnetic field around the primary winding expands and collapses as it cuts across the secondary winding. This means that the magnetic field cuts across the secondary winding first in one direction, then in the other. Therefore, the voltage induced across the secondary winding of any transformer is an alternating voltage.

Transformer Cores. The transformer in Fig. 15–5 is an *air-core transformer*. The coils of transformers operated either with alternating currents of less than about 20,000 hertz or with pulsating direct currents are wound around *iron cores*. Despite their names, these cores are usually made of thin sheets of steel to which a small amount of silicon is added. These sheets are called *laminations*. Two kinds of cores and

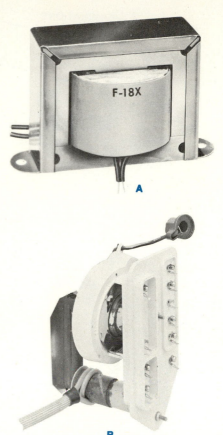

Fig. 15–4. Small transformers: (A) audio transformer; (B) television high-voltage flyback transformer (*Triad-Utrad Distributor Division of Litton Precision Products Inc.*)

Fig. 15–5. Operation of a simple transformer

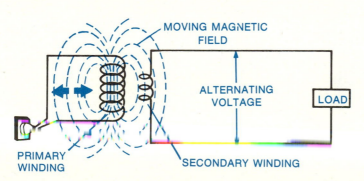

MOVING MAGNETIC FIELD

ALTERNATING VOLTAGE

LOAD

PRIMARY WINDING

SECONDARY WINDING

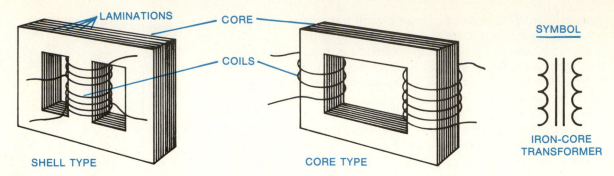

LAMINATIONS   CORE   SYMBOL

COILS

IRON-CORE
TRANSFORMER

SHELL TYPE   CORE TYPE

Fig. 15–6. Types of cores used with iron-core transformers

windings commonly used with iron-core transformers are shown in Fig. 15–6.

The iron core allows a stronger magnetic field to exist in a transformer. This, in turn, permits a greater amount of energy to be transferred from the primary to the secondary winding.

A laminated, as opposed to solid, core is used to lessen the amount of energy wasted in the form of heat. Some of this heat is produced by *eddy currents.* These are currents induced in the core of a transformer. Another energy loss in the transformer core is known as a *hysteresis loss.* This is the heat produced within a core as its magnetic domains try to align themselves with the changes in the polarity of a magnetic field produced by an alternating current.

Value of Induced Voltage. The value of the voltage induced across the secondary winding of a transformer depends on the number of turns of wire it has as compared with the number of primary-winding turns. This is called the *turns ratio* of the transformer. If the secondary winding has half as many turns as the primary winding, the voltage will be stepped down to one-half the primary-winding voltage (Fig. 15–7 at A). If the secondary winding contains twice as many turns as the primary winding, the voltage will be stepped up to twice the primary-winding voltage (Fig. 15–7 at B).

Fig. 15–7. The voltage induced across the secondary winding of a transformer depends on the winding turns ratio of the transformer. (A) Step-down transformer; (B) step-up transformer.

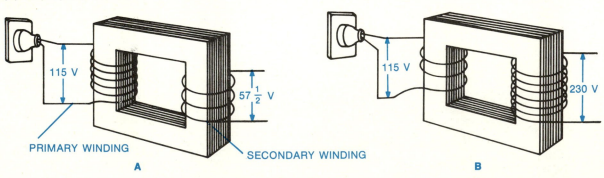

115 V   57 $\frac{1}{2}$ V   115 V   230 V

PRIMARY WINDING   SECONDARY WINDING

A   B

POWER TRANSFORMERS

A transformer that supplies the voltages needed in electronic equipment is often called a *power transformer* (Fig. 15–8). A power transformer used in a rectifier circuit is often called a *rectifier transformer.*

Some power transformers have more than one secondary winding. Each of these windings is electrically insulated from the other. This makes it possible to obtain different values of output voltages (Fig. 15–9). A color code is often used to identify the different windings. Transformers of this kind are commonly found in older radios, amplifiers, and television sets that have electron tubes instead of transistors.

RELATIONSHIP BETWEEN VOLTAGE AND CURRENT

Although a power transformer can step up or step down voltage, it cannot deliver a greater amount of energy to a load from its secondary winding than is supplied to the primary winding. In fact, the energy delivered by the secondary winding is always less than that supplied to the primary winding. This is because of power losses within the core and the windings.

As an example, suppose that the primary winding of a transformer operating at 115 volts has a current of 2 amperes. If the secondary winding has 10 times as many turns of wire as the primary, the voltage across the secondary winding will be stepped up to 115×10, or 1,150 volts (Fig. 15–10).

Since $P = EI$, the wattage of the primary winding is equal to 115×2, or 230 watts. Disregarding power losses within the transformer, the wattage of the secondary winding will also be 230 watts. The largest current that the secondary winding can deliver to a load is then equal to

$$I = \frac{P}{E}$$

$$= \frac{230}{1,150} = 0.2 \text{ ampere}$$

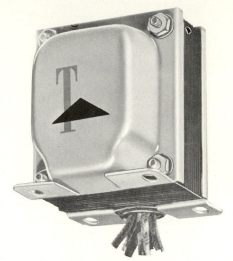

Fig. 15–8. Power transformer (*Triad-Utrad Distributor Division of Litton Precision Products Inc.*)

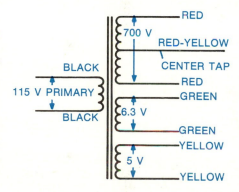

Fig. 15–9. Schematic of a power transformer showing the color code used to identify the various windings

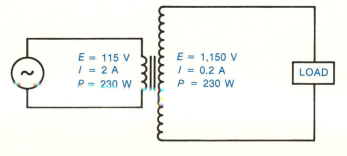

E = 115 V
I = 2 A
P = 230 W

E = 1,150 V
I = 0.2 A
P = 230 W

LOAD

Fig. 15–10. Relationship between voltage and current in a transformer

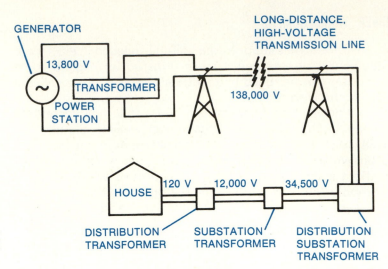

Fig. 15–11. Simplified representation of a power transmission and distribution system

Therefore, as the voltage is increased 10 times, the current in the secondary winding is decreased 10 times, or from 2 amperes to 0.2 ampere.

Likewise, if a transformer steps down voltage, the largest current that can be delivered to a load by the secondary winding will be greater than the current in the primary winding. This condition must exist if the wattage ($P = EI$) of the secondary winding is to equal the wattage of the primary winding.

POWER-TRANSMISSION LINE

The relationships between input and output voltages and currents in transformers are exemplified by a *transmission line.* Such a line is used to deliver electric energy from a power generating plant to homes and other buildings (Fig. 15–11). Here, the alternating voltage produced by the generator is stepped up by a transformer at the power plant. Because of this very high voltage, it is possible to deliver the required amount of energy from the plant to the users along the long-distance transmission line with a low current. This low current, in turn, lessens the power losses within the transmission line. It also permits the use of smaller conductors.

The substation transformers used with this power-transmission system step down the high voltage of the transmission line to much lower voltages. These voltages are then further stepped down by distribution transformers. Because of this stepping down of voltages, the total current available to the users is many times greater than the current present in

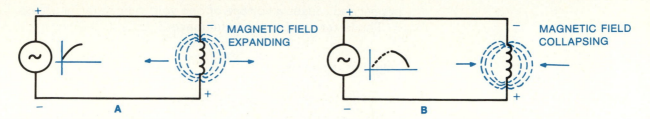

Fig. 15–12. Self-inductance

the transmission line. However, the total power is never greater than that produced originally by the generator. Actually, because of line losses, the available power is much less than that generated.

SELF-INDUCTANCE

When a current that is changing in value passes through a single coil, a moving magnetic field is produced around it. This action induces a voltage across the coil itself by means of electromagnetic induction. This is called *self-inductance,* or simply *inductance.* The voltage generated by self-inductance is always of a polarity that opposes any change in the value of the current through the coil.

Consider the circuits shown in Fig. 15–12. When the alternating voltage applied to these circuits is increasing in value during the first quarter of its cycle, the current through the coil also increases. This produces an expanding magnetic field around the coil (Fig. 15–12 at A). A voltage is now induced across the coil of a polarity that tends to oppose the change (increase) of the current through the coil.

When the applied voltage begins to decrease during the second quarter of its cycle, the magnetic field around the coil starts to collapse (Fig. 15–12 at B). As a result, the polarity of the voltage induced across the coil reverses. This voltage is now of a polarity that once again tends to oppose the change (decrease) of the current through the coil.

As the current moves through the second half of its cycle, the polarities of the applied voltage and the induced voltages reverse. However, the induced voltage continues to be of a polarity that tends to oppose any change in the value of the current. Because of this, the induced voltage also opposes the applied voltage that causes the current changes. For this reason, the voltage induced by self-inductance is called a *counter voltage* or a *counter electromotive force* (cemf).

The unit of inductance is the *henry.* It is named after Joseph Henry. He was an American teacher and physicist who lived from 1797 to 1878. A coil has an inductance of one henry when a current that changes at the rate of one ampere per second causes a voltage of one volt to be induced across it. The letter symbol for inductance is L.

INDUCTIVE REACTANCE

Inductive reactance is the opposition to a changing current in a coil that is caused by the inductance of the coil. The unit of inductive reactance is the ohm. Its letter symbol is X_L. The value of inductive reactance depends on the inductance of the coil and the frequency of the current that passes through it. The formula for inductive reactance is

$$X_L \text{ (in ohms)} = 2\,fL$$

where f = the frequency of the current in hertz
L = the inductance of a coil in henrys

This formula shows that inductive reactance is directly proportional to both frequency and inductance. As the frequency increases, the inductive reactance will increase. The inductive reactance will also increase if the inductance of a circuit is increased.

When inductive reactance and resistance are both present in a circuit, their total opposition to current is called *impedance*. The unit of impedance is the ohm. Its letter symbol is Z.

INDUCTORS

Inductors are coils that are wound so as to have a certain amount of inductance. The inductance of an inductor depends mainly on four things: (1) the number of turns of wire with which it is wound, (2) the cross-sectional area of the coil, (3) the permeability of its core materials, and (4) the length of the coil.

Iron-core inductors are sometimes called *choke coils* or *reactors*. They often are used to *filter*, or smooth, the output current from a rectifier circuit (Fig. 15–13). Inductors are also used with capacitors in radio and television tuning circuits and in oscillator circuits.

TESTING TRANSFORMERS AND INDUCTORS

The most common defects of transformers are these: (1) burned-out windings caused by too much current, (2) windings grounded or shorted to the core, (3) shorts between adjacent turns of wire, and (4) shorts between the primary and secondary windings. The first three of these defects also occur in inductors.

SYMBOL

Fig. 15–13. Choke coils (*Essex International, Inc.*)

Continuity Tests. The different windings of a transformer can be identified easily by making *continuity tests*. A continuity test also can be made to see if there are any open windings or to see if there are any shorts between windings or between a winding and the core.

A short between the turns of a transformer or an inductor winding usually cannot be located with a continuity test. Such a short most often will cause a winding to become overheated even when it is operating with a normal amount of current.

Continuity tests of a transformer or an inductor should be made only after the device has been disconnected from its circuit. If this is not done, other components in the circuit may interfere with the tests.

Voltage Tests. After testing for continuity, the secondary winding should be given a voltage test. To do this, it is necessary to apply voltage to the primary winding. This should be done only after checking to see that bare ends of secondary-winding leads are not in contact with each other. Where very high secondary voltages are expected, a reduced primary voltage may be used and the voltage ratio determined. For example, a transformer with a 120-volt primary and a 600-volt secondary winding has a voltage ratio of 120:600, or 1:5. If 10 volts is applied to the primary, the secondary should read about 50 volts.

CAUTION: Power transformers often have one very high voltage winding. This winding is a severe shock hazard. Extreme care must be taken when applying the rated primary voltage. Never hold both winding leads at the same time when making tests.

After you have found the voltage of each secondary winding, label the leads for future reference.

LEARNING BY DOING

14. Electromagnetic Induction. The generation of voltage by electromagnetic induction can be seen in a very interesting way by the following procedure. This will help you better understand what electromagnetic induction is and how it is used for the operation of transformers and generators.

MATERIALS NEEDED

1 piece of wood, $\frac{3}{4} \times 4 \times 6$ in. (19 × 102 × 152 mm)
1 piece of wood dowel rod, $\frac{1}{2} \times 1\frac{1}{4}$ in.
  (12.7 × 32 mm)
plastic disk, 3 in. (76 mm) in diameter and at least
  $\frac{1}{16}$ in. (1.6 mm) thick
300 ft (91.4 m) PE magnet wire, no. 25 (0.45 mm)
2 Fahnestock clips, 1 in. (25.4 mm)
2 round-head wood screws, no. $5 \times \frac{1}{2}$ in.
2 solder lugs, no. 8
1 common-head nail, 1 or $\frac{3}{4}$ in. (25 or 19 mm)
bar or rod permanent magnet
galvanometer with center-zero dial

Procedure

1. Assemble the materials as shown in Fig. 15–14. You will find it convenient to wind the coil before mounting the clips on the base.
2. Connect the galvanometer to the clips.
3. Hold a pole of the magnet over the center of the coil. Do not move either the magnet or the coil. Is a voltage now being induced across the coil? Why?
4. Move the north pole of the magnet over the disk from the center outward and notice in which direction the meter pointer moves. Now reverse the magnet and move its south pole from the center outward. Has the polarity of the voltage induced across the coil reversed? Why?
5. Slowly move a pole of the magnet over the disk from the center of the coil outward and back to the center. As this is done, you will notice that the meter pointer moves from zero toward one end of the scale, back to zero, then to the other end of the scale, and finally back to zero. What kind of voltage has been generated across the coil as a result of this action?
6. Very slowly move a pole of the magnet from the center of the coil outward and back again several times. Note the highest reading of the meter pointer. Now move the mag-

Fig. 15–14. Assembly used for Learning by Doing No. 14, "Electromagnetic Induction"

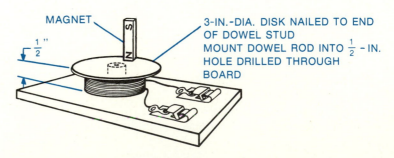

MAGNET
$\frac{1}{2}$"
3-IN.-DIA. DISK NAILED TO END OF DOWEL STUD
MOUNT DOWEL ROD INTO $\frac{1}{2}$ - IN. HOLE DRILLED THROUGH BOARD

net over the coil faster. What effect does the faster movement have on the value of the voltage induced across the coil?

7. Hold a pole of the magnet a little above the center of the coil. Move the coil back and forth. Does this cause a voltage to be induced across the coil? This proves that electromagnetic induction occurs when a magnetic field moves past a coil or when a coil is moved within a stationary magnetic field.

15. Making an Electromagnetism Experimenter. This assembly can be used as an electromagnet, a magnetizer, a demagnetizer, and for doing several very interesting experiments with electromagnetism. It can be operated at 115 volts ac or with dc voltages of up to 6 volts.

MATERIALS NEEDED

1 piece of wood, $\frac{3}{4} \times 7\frac{1}{2} \times 10\frac{1}{2}$ in.
    $(19 \times 191 \times 267$ mm)
2 round, plastic or Masonite disks, $\frac{1}{8} \times 4$ in.
    $(3.18 \times 102$ mm) in diameter
enough sheet metal (transformer iron) to form a laminated core $1\frac{1}{2} \times 1\frac{1}{2} \times 4$ in. $(38 \times 38 \times 102$ mm).
    The laminations from a large discarded transformer
    are excellent for this purpose.
1 piece of band iron, $\frac{1}{8} \times 1\frac{1}{2} \times 7$ in.
    $(3.18 \times 38 \times 178$ mm)
1 round, soft-iron rod, 1×4 in. $(25.4 \times 102$ mm)
1 round, soft-iron rod, 1×10 in. $(25.4 \times 254$ mm)
1 aluminum ring or washer, $1\frac{1}{2}$ or 2 in. $(38$ or 50 mm)
    in diameter
1 round, sheet-copper or sheet-aluminum disk, $2\frac{1}{2}$ in.
    $(63.5$ mm) in diameter
350 ft $(107$ m) PE magnet wire, no. 14 $(1.60$ mm)
18 ft $(5.5$ m) PE magnet wire, no. 18 $(1.00$ mm)
10 in. $(254$ mm) bare wire or PE magnet wire, no. 14
    $(1.60$ mm)
5 ft $(1.5$ m) two-conductor service cord, no. 16
    $(1.25$ mm)
2 ft $(0.6$ m) parallel lamp cord, no. 18 $(1.00$ mm)
1 pilot lamp, no. 40 or no. 46
1 miniature lamp socket, screw base
1 heavy-duty attachment plug
1 15-ampere surface toggle switch, spst, with mounting
    screws
2 wire nuts, medium size
5 insulated staples
3 flat-head machine screws, no. $\frac{1}{4}$–20 $\times$ 1 in. $(M6 \times 1)$
3 round-head wood screws, no. $8 \times \frac{3}{4}$ in.

cambric cloth, medium thick
insulating varnish
plastic electrical tape
low-voltage, battery-eliminator dc power supply

Procedure A (Basic Construction)

1. Assemble the electromagnet coil core and the coil form as shown in Fig. 15–15 at A.
2. Tightly wind the coil in layers with the no. 14 (1.6-mm) PE magnet wire (Fig. 15–15 at B). Insulate every third layer of the wire with a single layer of cambric cloth. After the coil is wound, apply a heavy coating of insulating varnish to its outside surfaces. Let it dry thoroughly.
3. Assemble materials and connect the circuit shown in Fig. 15–15 at C.

Procedure B (The Electromagnet)

1. Plug the experimenter into a 115-volt ac receptacle outlet and turn the switch on. By holding an object made of a magnetic material near the coil-core pole piece, you will find that a strong electromagnetic field exists in the area near this pole piece.

 SAFETY
The coil will overheat if operated continuously for an extended time. When this occurs, the circuit should be turned off and the coil allowed to cool.

Procedure C (Magnetizing and Demagnetizing)

1. To use the experimenter as a magnetizer, connect it to the power supply adjusted to a dc voltage of 6 volts. Put the object to be magnetized across its pole pieces. One of the pole pieces is formed by placing the 4-in. (102-mm) iron rod over the band iron in an upright position (Fig. 15–15 at D). The rod can be moved as needed for the right spacing. Turn the switch on and, after a few seconds, turn it off. The object should now be magnetized. Explain this process of magnetization.
2. To use the experimenter as a demagnetizer, plug it into a 115-volt ac receptacle outlet and turn the switch on. Hold the object to be demagnetized a little above the coil-core pole piece. Then slowly move it away from the pole piece for a distance of about 18 in. (457 mm). Repeat if necessary. Explain this process of demagnetization.

Procedure D (Transformer Action)

1. Assemble the coil and lamp shown in Fig. 15–15 at E.

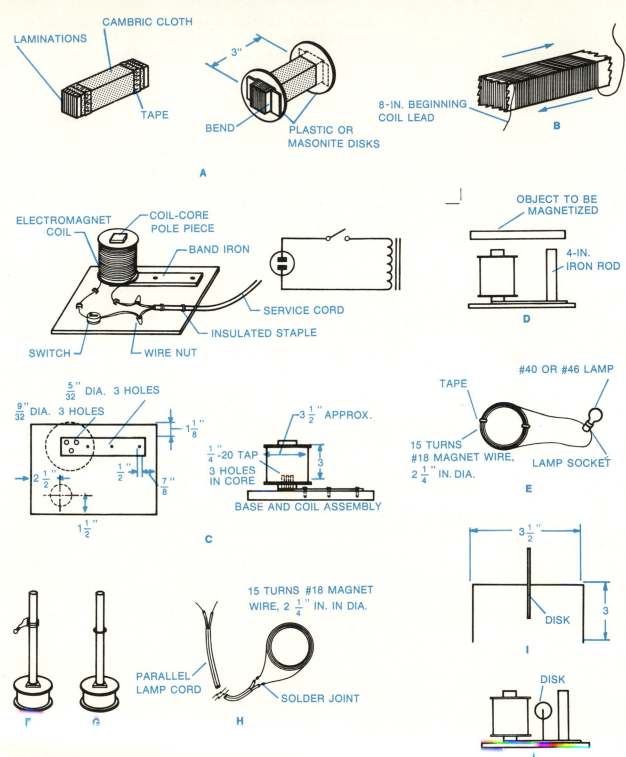

Fig. 15–15. Assemblies used for Learning by Doing No. 15, "Making an Electromagnetism Experimenter"

139

2. Place the 10-in. (254-mm) iron rod on the coil-core pole piece in an upright position. Plug the experimenter into a 115-volt ac receptacle outlet. Turn the switch on.

3. Slowly move the coil down over the iron rod. Do not let it touch the rod (Fig. 15–15 at F). As you do this, the lamp will begin to burn more and more brightly. Explain how energy is transferred from the electromagnet coil to the lamp circuit.

4. You will notice that the iron rod becomes quite warm after the experimenter is operated in this way for a short time. This is an example of induction heating. Induction heating is described in Unit 33, "Producing Heat."

Procedure E (The Jumping Ring and Coil)

1. Place the 10-in. (254-mm) iron rod on the coil-core pole piece in an upright position. Plug the experimenter into a 115-volt ac receptacle outlet and turn the switch on.

2. Put the aluminum ring down over the end of the rod (Fig. 15–15 at G). Explain why the ring remains suspended above the coil-core pole piece. Turn the switch off.

3. Drop the ring over the iron rod and onto the coil-core pole piece. Put one hand over the end of the rod and turn the switch on. What happens? What causes this?

4. Form the coil shown in Fig. 15–15 at H.

5. Put the coil over the iron rod. With the switch on, slowly make and break contact between the ends of the lamp cord connected to it. Why does the coil jump upward when the ends of the lamp cord are in contact? Why does the coil remain stationary when the ends are not touching?

Procedure F (Motor Action)

1. Drill a $\frac{1}{16}$-in. (1.6-mm) hole through the center of the copper or aluminum disk. Pass the piece of no. 14 (1.6-mm) wire through the hole. Bend the wire as shown in Fig. 15–15 at I.

2. Put the legs of the disk-support wire into small holes drilled into the wood base of the experimenter. The disk should be midway between the coil and the 4-in. (102-mm) iron rod (Fig. 15–15 at J).

3. Plug the experimenter into a 115-volt ac receptacle outlet. Turn the switch on.

4. Spin the disk with your fingers. It will continue to rotate. This is a simple induction motor. Induction motors are described in Unit 32, "Electric Motors."

16. The Effects of Self-inductance. The following procedure will let you see the effects of self-inductance in dc and in ac circuits. This will help you understand an important property of electromagnetism that is used in many kinds of circuits.

MATERIALS NEEDED

1 piece of wood, ³⁄₄ × 4 × 6 in. (19 × 102 × 152 mm)
2 pieces of plastic or Masonite, ¹⁄₈ × 1³⁄₄ × 2 in.
   (3.18 × 44 × 50 mm)
1 plastic, heavy-paper, copper, or aluminum tube,
   1³⁄₄ in. (44.5 mm) long, ⁹⁄₁₆-in. (14.3-mm) inside
   diameter
1 round, soft-iron rod, ¹⁄₂ × 4 in. (12.7 × 102 mm)
100 ft (30.5 m) PE magnet wire, no. 21 (0.71 mm)
1 pilot lamp, no. 1493
1 double-contact candelabra-lamp bayonet socket with
   mounting bracket
2 Fahnestock clips, 1 in. (25.4 mm)
3 round-head wood screws, no. 5 × ¹⁄₂ in.
low-voltage, battery-eliminator dc power supply
ac power supply or transformer having a 6.3-volt output
   with a current rating of at least 2.5 amperes

Procedure

1. Assemble the coil form shown in Fig. 15–16. If a copper or an aluminum coil-core tube is used, it should be covered with a layer of plastic electrical tape before the coil is wound. A copper or aluminum tube can be made from sheet metal cut to size and formed around a wood dowel rod ¹⁄₂ in. (12.7 mm) in diameter.
2. Wind the coil in layers with the no. 21 (0.71-mm) magnet wire.
3. Cement the coil assembly to the wood base. Mount other components.
4. Wire the circuit.
5. Connect the circuit to the 6.3-volt ac power supply or to the secondary winding of the transformer.
6. Slowly put the iron rod into the coil. What effect does this have on the amount of light produced by the lamp? Explain the reason for this.
7. Disconnect the ac power supply or the transformer from the circuit. Connect the circuit to the dc power supply adjusted to produce a voltage of about 6.3 volts.
8. Again, slowly put the iron rod into the coil. Explain why the lamp now continues to operate normally.

ENDS OF TUBE CEMENTED INTO HOLES DRILLED THROUGH PLASTIC OR MASONITE SUPPORT PIECES

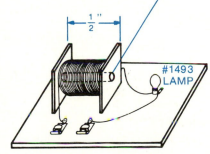

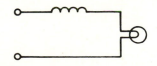

Fig. 15–16. Assembly used for Learning by Doing No. 16, "The Effects of Self-inductance"

Test your knowledge by writing, on a separate sheet of paper, the word or words that most correctly complete the following statements:

1. If a magnet is moved near a wire, a _____ is produced across the ends of the wire.
2. _____ is needed to produce a voltage by electromagnetic induction.
3. The magnetic field produced by an alternating current can be thought of as a _____ magnetic field.
4. If a transformer is operated with direct current, the current must be broken up into _____.
5. The transformer is used to _____ the value of a voltage or current.
6. If a transformer reduces voltage, it is called a _____ transformer. If it increases voltage, it is called a _____ transformer.
7. The winding of a transformer that is connected to a source of energy is called the _____ winding.
8. A transformer is an example of _____ induction.
9. The voltage induced across the secondary winding of any transformer is an _____ voltage.
10. The thin sheets of steel from which transformer cores are made are called _____.
11. A laminated transformer core reduces the amount of energy that is wasted in the form of _____.
12. The voltage induced across a coil by its own moving magnetic field is an example of _____ -inductance.
13. The total opposition to current presented by inductive reactance is called _____.
14. Coils that are wound so as to have a certain amount of inductance are called _____.
15. Iron-core inductors used to filter the output current from rectifier circuits are called _____ coils.

FOR REVIEW AND DISCUSSION

1. Define induced voltage and electromagnetic induction.
2. What condition must exist between a conductor and a magnetic field in order to produce electromagnetic induction?
3. How can the polarity of the voltage generated by electromagnetic induction be reversed?
4. Describe the moving magnetic field produced by an alternating current.
5. How can a moving magnetic field be produced by direct current?
6. Tell how a magnetic air gap can be used to produce a moving magnetic field.
7. Explain the construction and the operation of a simple transformer.
8. What is meant by a step-up transformer? A step-down transformer?
9. Name and define the unit of inductance.
10. Define inductive reactance and write its symbol.
11. For what purpose are choke-coil inductors used?
12. How are transformers tested?
13. What transformer or inductor defects cannot be detected by making continuity tests?

INDIVIDUAL-STUDY ACTIVITIES

1. Give a demonstration showing how adding an iron core increases the strength of an electromagnet. Explain why.
2. Obtain a variety of transformers. Explain the operation of each and give the purpose for which it is used.
3. Talk or write to a representative of the electrical company that serves your area and ask him or her for information about how transformers are used in their distribution system. Report orally or in writing about what you have learned.
4. Give a demonstration of a continuity test for a transformer.
5. Give a demonstration of a voltage test for a transformer.

Energy Sources

Unit 16   Chemical Cells and Batteries

A *voltaic chemical cell* is a combination of materials used to change chemical energy into electric energy in the form of voltage. The words *cell* and *battery* are often used to mean the same thing. However, this is not technically correct. A *cell* is a single unit. A *battery* is formed when two or more cells are connected together in series or in parallel (Fig. 16–1).

BASIC ACTION OF A CELL

A chemical cell is made up of two *electrodes*, or *plates*, in contact with a substance in which there are many ions. As you learned in Unit 1, an ion is a charged atom. A substance that has many ions is called an *electrolyte*. Water solutions made with acids, bases, or salts are electrolytes. Salt water, for example, is an electrolyte. It is also a good conductor of electricity.

The chemical actions that cause a combination of substances to produce a voltage are complicated. Studying how a very simple cell works, however, will help you understand how chemical cells in general work.

In the cell, the electrolyte ionizes to form positive and negative ions (Fig. 16–2 at A). At the same time, chemical action also causes the atoms within one of the electrodes to ionize. Because of this, electrons are deposited on the electrode. Positive ions from the electrode pass into the electrolyte. This creates a negative charge on this electrode. It leaves the area near it positively charged (Fig. 16–2 at B).

Some of the positive ions produced because of the ionization of the electrolyte are then repelled to the other elec-

SYMBOL

POSITIVE
TERMINAL

Fig. 16–1. A battery consists of two or more cells connected together. (*Union Carbide Corporation*)

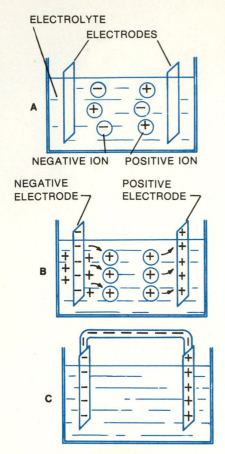

ELECTROLYTE
ELECTRODES

A

NEGATIVE ION   POSITIVE ION

NEGATIVE
ELECTRODE

POSITIVE
ELECTRODE

B

C

Fig. 16–2. Basic action of a chemical voltaic cell

Fig. 16–3. Battery charger and the charging circuit (*Schauer Manufacturing Corporation*)

trode. At this electrode, these ions combine with electrons. Since this removes electrons from the electrode, it becomes positively charged. Because the chemical action has caused the electrodes to have opposite charges, there is now a voltage between them.

If a wire is connected between the electrodes of the cell, excess electrons from the negative electrode will pass through the wire and into the positive electrode (Fig. 16–2 at C). This current will continue until the materials in the cell become chemically inactive.

The electrolyte of a cell may be a liquid or a paste. If it is a liquid, the cell is often called a *wet cell*. Cells in which the electrolyte is a paste are called *dry cells*.

PRIMARY AND SECONDARY CELLS

Primary cells are those that cannot be *recharged*. This means they cannot be returned to good condition after their output voltage drops too low.

Secondary cells can be recharged. During recharging, the chemicals that provide electric energy are restored to their original condition. This is done by passing direct current through a cell in a direction opposite to the direction of the current that the cell delivers to a circuit.

A cell (or battery) is recharged by connecting it to a battery charger in "like-to-like" polarity as shown in Fig. 16–3. The charger is a rectifier circuit that can produce a variable output voltage. Many battery chargers have a voltmeter and an ammeter that show the charging voltage and current.

VOLTAGE AND CURRENT OF A CELL

The *voltage rating* of a cell is the voltage that the cell produces when it is not connected to a circuit. This is called the

open-circuit voltage. The value of the open-circuit voltage depends on the materials from which the cell is made.

The *capacity* of a cell is its ability to deliver a given amount of current to a circuit. This depends on two things: the amount and condition of the electrolyte and the size of the electrodes. A large cell can usually deliver more current for a longer period of time than a smaller cell with the same kinds of electrodes and electrolyte.

In order to get a higher voltage, cells are connected in series. To get greater current, cells are connected in parallel. To review these connections, see Unit 10, "Series and Parallel Circuits."

SHELF LIFE

All dry cells, even when not in use, will lose energy to some extent. The *shelf life* of a cell is that period of time during which the cell can be stored without losing more than about 10 percent of its original capacity. A cell loses capacity because its electrolyte dries out and chemical actions change the materials within it. Since heat speeds both of these processes, the shelf life can be made longer by keeping a cell in a cool place.

CARBON-ZINC CELLS AND BATTERIES

The *carbon-zinc cell* is one of the oldest and most widely used primary dry cells. Its electrolyte is a paste of ammonium chloride and zinc chloride dissolved in water. These and other chemicals are contained in the *bobbin* of the cell (Fig. 16–4).

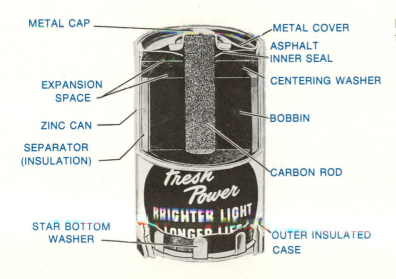

METAL CAP

METAL COVER

ASPHALT INNER SEAL

EXPANSION SPACE

CENTERING WASHER

ZINC CAN

BOBBIN

SEPARATOR (INSULATION)

CARBON ROD

STAR BOTTOM WASHER

OUTER INSULATED CASE

Fresh Power

BRIGHTER LIGHT LONGER LIFE

Fig. 16–4. Principal parts of a typical carbon-zinc dry cell

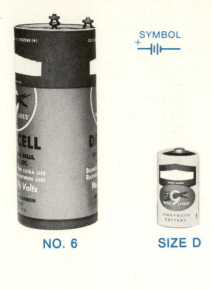

NO. 6   SIZE D

SIZE C   SIZE   SIZE   SIZE
   AA   AAA   N

Fig. 16–5. Common sizes of carbon-zinc cells

Fig. 16–6. Construction of dry-cell batteries: (A) cylindrical-cell battery containing fifteen 1½-volt cells connected in series to produce a total battery voltage of 22½ volts; (B) flat-cell battery containing forty-five 1½-volt cells connected in series to produce a total battery voltage of 67½ volts

The negative electrode is a zinc can. The positive electrode is a mixture of a black mineral called manganese dioxide and powdered carbon. The carbon is used to decrease the resistance of the electrode. A solid carbon rod passes through the center of the mixture. The rod provides a good electrical contact between the positive electrode and the positive terminal of the cell.

Polarization. During the operation of a carbon-zinc cell, the carbon rod becomes coated with hydrogen gas by a process known as *polarization*. This gas is removed from the cell by the manganese dioxide, which in this case serves as a *depolarizing agent*. As a result, the cell is able to give much better service.

Sizes and Voltage. Carbon-zinc cells, which are available in several sizes, have an open-circuit voltage of from 1.5 to 1.6 volts (Fig. 16–5). There are a number of different kinds of carbon-zinc batteries. The most common have voltages of 3, 4.5, 6, 9, 13.5, 22.5 and 45 volts. In some batteries, the cells are cylindrical (Fig. 16–6 at A). In others, they are flat (Fig. 16–6 at B).

Operating Efficiency. Ordinary carbon-zinc cells and batteries provide the most efficient service when they are used for short periods of time at relatively low currents. This lets the cells and batteries stay polarized.

ALKALINE CELLS

The *secondary*, or rechargeable, *alkaline cell* was a major advance in portable energy sources. In this dry cell, the elec-

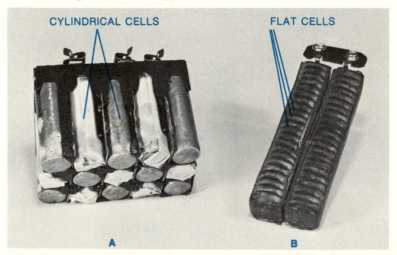

CYLINDRICAL CELLS   FLAT CELLS

A   B

trolyte is potassium hydroxide. The negative electrode is zinc. The positive electrode is manganese dioxide (Fig. 16–7).

Secondary alkaline cells have an open-circuit voltage of 1.5 volts. They commonly come in AA, C, and D sizes. They can be used as direct replacements for carbon-zinc cells of the same sizes. The most common secondary alkaline batteries have voltages of 4.5, 7.5, 13.5, and 15 volts. Both the cells and batteries come in a fully charged condition. They should be allowed to discharge before being recharged. For best efficiency, recharging should be done as the manufacturer recommends.

Primary alkaline cells are similar in makeup to the rechargeable kinds and have the same open-circuit voltage. These dry cells last longer than carbon-zinc cells of the same size when used in the same way. Both the primary and secondary alkaline cells (and batteries) give good service when used with high-current loads. For this reason, they are widely used as sources of energy in radios, television sets, tape recorders, cameras, and cordless appliances.

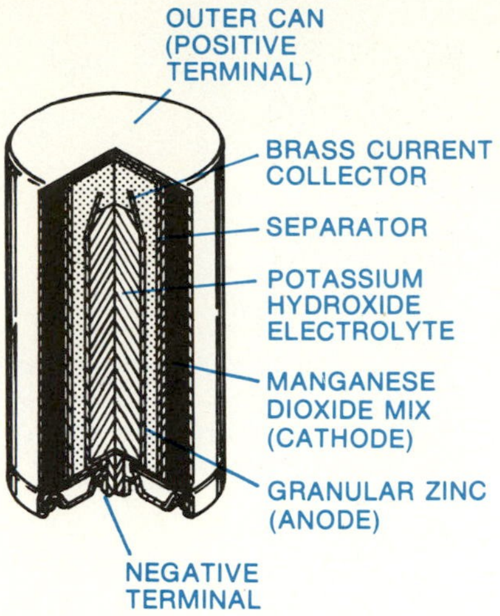

Fig. 16–7. Principal parts of a rechargeable alkaline cell (*Radio Corporation of America*)

MERCURY CELLS

In the typical *mercury cell*, the electrolyte is a paste of potassium hydroxide and zinc oxide. The negative electrode is a compound of zinc and mercury. The positive electrode is mercuric oxide. Mercury cells are primary cells. They have an open-circuit voltage of 1.35 or 1.4 volts, depending on the electrolyte mixture. Mercury batteries have a variety of voltage ratings.

Mercury cells and batteries have a very good shelf life and are extremely rugged. They have the ability to provide a rather constant output voltage under different load conditions. Thus, they are used in all kinds of products, including electric watches, hearing aids, test instruments, and alarm systems. They come in a wide range of shapes and sizes (Fig. 16–8).

NICKEL-CADMIUM CELL

The *nickel-cadmium cell* was originally developed in Europe as a secondary wet cell for use in automobiles. In the secondary nickel-cadmium dry cell, the electrolyte is potassium hydroxide. The negative electrode is nickel hydroxide. The positive electrode is cadmium oxide.

The open-circuit voltage of nickel-cadmium dry cells is 1.25 volts. These cells come in several sizes, including the common AA, C, and D cells and the flat button shapes (Fig. 16–

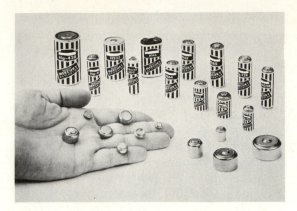

Fig. 16–8. Mercury cells (*Gould Inc., Burgess Battery Division*)

9). The most common nickel-cadmium batteries have voltages of 6, 9.6, or 12 volts.

Nickel-cadmium cells are rugged. They give good service under extreme conditions of shock, vibration, and temperature. They are used in many different kinds of products.

TESTING DRY CELLS AND BATTERIES

The condition of a dry cell or battery can be checked by measuring its voltage when connected to a load. If the value of the voltage is less than 80 percent of the open-circuit voltage, the cell or battery should be replaced. The voltage test must be made with the load connected and the current flowing. This is because the load current will produce a voltage drop across the internal resistance of the cell or battery. If a dry cell or battery is not in good condition, its internal resistance is high. This is due to the drying out of the electrolyte. Moreover, a relatively large internal voltage drop occurs. Therefore, the voltage of the terminals is greatly reduced. If

Fig. 16–9. Nickel-cadmium cells (*Gould Inc., Burgess Battery Division*)

the cell or battery is tested under an open-circuit condition, there is only a very small internal voltage drop. As a result, the voltage across the terminals of the cell or the battery may be close to the rated voltage. This would be true even if the battery were not in good condition.

LEAD-ACID CELL AND BATTERY

The lead-acid secondary storage battery is mainly used in the automobile. In a fully charged *lead-acid cell,* the electrolyte is a solution of water and sulfuric acid. About 27 percent of the total volume is acid. The active material on the positive (brown-colored) plates is *lead peroxide.* The active material on the negative (gray-colored) plates is pure *lead* in a spongelike form.

The lead-acid cell has an open-circuit voltage of a little more than 2 volts. In the typical automobile battery, six cells are connected in series to produce a total voltage of 12 volts (Fig. 16–10).

Chemical Action. As a lead-acid cell discharges, some of the acid within the electrolyte leaves the electrolyte. The acid combines with the active material on the plates (Fig. 16–11). This chemical action changes the material on both plates to *lead sulfate.* When the cell is being charged, the reverse action takes place. Now the acid that was absorbed by the plates is returned to the electrolyte. As a result, the active materials on the plates are changed back into the original lead peroxide and lead.

The Hydrometer. The *hydrometer* (Fig. 16–12) is a device used to measure the specific gravity of the electrolyte in a

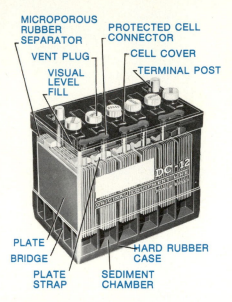

Fig. 16–10. Principal parts of an automobile lead-acid battery (*Delco-Remy Division of General Motors Corporation*)

Fig. 16–11. Chemical actions of a lead-acid cell: (A) discharging; (B) charging

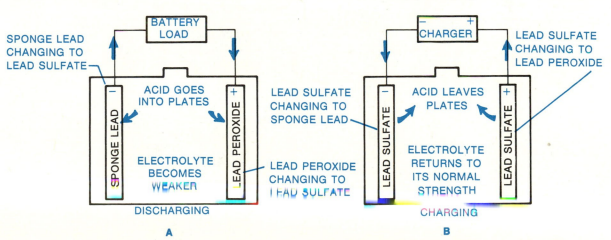

149

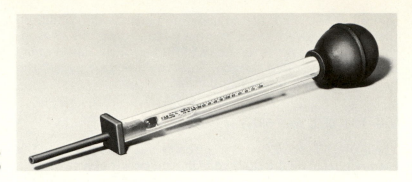

Fig. 16–12. Hydrometer used to test battery acid (*ESB Brands, Inc.*)

lead-acid cell. The *specific gravity* of a liquid is its weight as compared to that of an equal volume of pure water. Since sulfuric acid is heavier than water, the specific gravity of the electrolyte in a lead-acid cell decreases as the cell discharges. Therefore, by measuring the specific gravity of the electrolyte, the charge condition of lead-acid cells can be determined.

A sample of the electrolyte is obtained by squeezing the rubber bulb of the hydrometer while the end of the hydrometer is in the electrolyte. When the bulb is released slowly, the electrolyte will be drawn into the glass tube. The float in the hydrometer will then rise within the electrolyte to a level that depends on the specific gravity of the electrolyte. After the float has settled to its level, the specific gravity is read from the markings on the float (Fig. 16–13). Each cell of the battery must be tested. Since the cells are connected in series, any cell that is defective will cause the entire battery to be defective.

 CAUTION: The sulfuric acid used in a battery is a highly corrosive chemical. It can cause severe burns. When using a hydrometer, do not allow any of the electrolyte to touch your skin or clothing. In case of contact, immediately wash the affected area with large quantities of soap and water.

The specific gravity of the electrolyte within a fully charged automobile battery is about 1.280. As the cell discharges, the specific gravity gradually decreases to about 1.130. At that point, the cell is considered to be fully discharged.

Rating. The capacity rating of a lead-acid battery is usually given in ampere-hours (Ah) for a specific discharge period. The most commonly used rating system is based on the Society of Automotive Engineers (SAE) 20-hour rate. This means that a new battery rated at 100 ampere-hours should deliver 5

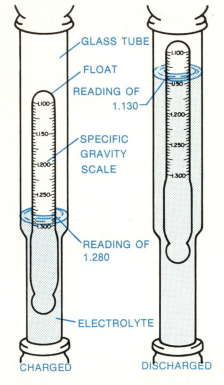

GLASS TUBE

FLOAT

READING OF 1.130

SPECIFIC GRAVITY SCALE

READING OF 1.280

ELECTROLYTE

CHARGED   DISCHARGED

Fig. 16–13. Reading of a hydrometer scale when a battery is fully charged and when it is discharged

amperes of current continuously for 20 hours and maintain at least 1.75 volts per cell.

Dry Charge. Most lead-acid batteries are shipped from the manufacturer as dry-charge batteries. The plates of a dry-charge battery are in the condition of a fully charged battery. This means that one plate is lead and the other is lead peroxide. To prepare the battery for use, sulfuric acid of the proper specific gravity is added to each cell. The battery is then ready to be used.

BATTERY MAINTENANCE

Although the modern automobile battery is rugged, it must be used and maintained in the right way. The following suggestions will be helpful.

Battery cells must be kept properly filled at all times (Fig. 16–14). Terminals should be clean and free of corrosion. Battery-cable clamps should be connected solidly. The specific gravity of the electrolyte should be checked often.

A battery may show a low charge because of (1) faulty wiring in the electrical system of the automobile, which is causing the battery to be discharged, (2) a defect in the generator or regulator, or (3) a loose or greasy alternator (fan) belt. Do not overcharge a lead-acid battery. This will weaken the electrolyte and may cause serious damage to the cell plates through overheating.

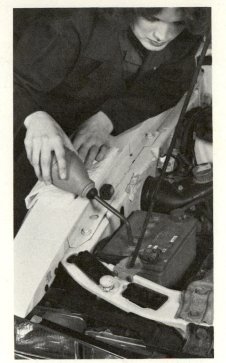

Fig. 16–14. Adding water to a lead-acid battery (*Ford Motor Company*)

SAFETY
Hydrogen, an explosive gas, is released from a battery while it is being charged. For this reason, a battery should be charged in a well-ventilated area, away from open flames or sparks.

If a battery is to be stored, it should first be fully charged. The plates of a stored battery that is partly or fully discharged soon become coated with a sulfate compound. This compound hardens and causes the plate material to become chemically inactive. A partly or fully discharged battery will also freeze at a temperature of about 20°F (6.7°C).

SELF-TEST

Test your knowledge by writing, on a separate sheet of paper, the word or words that most correctly complete the following statements:

1. A cell that changes chemical energy into electric energy is called a _____ cell.
2. A _____ is formed when two or more cells are connected.
3. A chemical cell is made up of two _____ in contact with an _____.
4. A cell in which the electrolyte is a liquid is often called a _____ cell. A cell in which the electrolyte is a paste is called a _____ cell.
5. Cells that cannot be recharged are called _____ cells. Cells that can be recharged are called _____ cells.
6. A cell or a battery is recharged by passing current through it in a direction _____ to the direction of its discharge current.
7. When charging a cell or a battery, its positive terminal is connected to the _____ terminal of the battery charger. Its negative terminal is connected to the _____ terminal of the charger.
8. The open-circuit voltage of a cell depends on the _____ from which it is made.
9. The amount of current that a cell can deliver to a load depends on the amount and condition of its _____ and the size of its _____.
10. The _____ of a cell is the period of time during which it can be stored without a significant loss of capacity.
11. The open-circuit voltage of carbon-zinc cells ranges from _____ to _____ volts.
12. The open-circuit voltage of alkaline cells is _____ volts.
13. Alkaline cells are both _____ and _____.
14. Mercury cells have an open-circuit voltage of _____ or _____ volts.
15. Nickel-cadmium dry cells have an open-circuit voltage of _____ volts.

FOR REVIEW AND DISCUSSION

1. What is the difference between a cell and a battery?
2. Describe the construction of a simple chemical cell.
3. Explain the basic action by means of which a chemical cell produces voltage.
4. Define primary and secondary cells.
5. State the correct way of connecting a battery charger to a battery.
6. On what does the open-circuit voltage of a cell depend?
7. What two factors determine the current-delivering capacity of a cell?
8. Define the shelf life of a cell or a battery.
9. Give several common sizes of carbon-zinc cells and identify their voltages.
10. Under what operating conditions will carbon-zinc cells and batteries provide the most efficient service?
11. Name the two kinds of alkaline cells. What advantages do these cells provide as compared to carbon-zinc cells?
12. State two desirable characteristics of mercury cells.
13. What is the outstanding feature of nickel-cadmium dry cells?
14. How are dry cells and batteries tested with a voltmeter?
15. Describe the basic construction of an automobile-battery lead-acid cell. What is the open-circuit voltage of such a cell?
16. How is a battery-test hydrometer used?
17. What is meant by a dry-charge automobile battery?

INDIVIDUAL-STUDY ACTIVITIES

1. Show your class a number of different kinds and sizes of dry cells. Describe these and tell the characteristics of each.
2. Give a demonstration showing how a dry cell or battery is tested with a voltmeter.
3. Prepare a written or an oral report telling about the proper care and use of a common lead-acid automobile battery.

Unit 17 Generators

The word *generate* means to "produce." An *electric generator* is a machine that produces a voltage by means of electromagnetic induction. This is done by rotating coils of wire through a magnetic field or by rotating a magnetic field past coils of wire. The modern generator is the result of work done with electromagnetic induction by Michael Faraday and Joseph Henry in the early 1800s. Now more than 95 percent of the world's electric energy is supplied by generators.

BASIC GENERATOR

A continuous alternating voltage can be produced by rotating a coil of wire between the poles of a permanent magnet (Fig. 17–1). This is a simple generator. The coil is called the *armature*. The ends of the armature coil are connected to *slip rings*. These are insulated from each other and from the armature shaft on which they are mounted. The stationary brushes press against the slip rings. They make it possible to connect the rotating armature to an external circuit. The armature must be driven by mechanical force. Thus, a generator can be defined as a machine that changes mechanical energy into electric energy.

GENERATOR ACTION

The value of the voltage induced by generator action at any instant of time depends on three things. These are (1) the flux density of the magnetic field through which a conductor is moving (the greater the flux density, the greater the induced voltage); (2) the velocity of the conductor motion (the induced voltage increases as the velocity of the conductor in-

Fig. 17–1. A simple one-coil ac generator

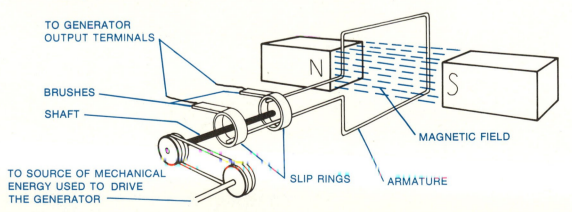

TO GENERATOR OUTPUT TERMINALS

BRUSHES

SHAFT

TO SOURCE OF MECHANICAL ENERGY USED TO DRIVE THE GENERATOR

SLIP RINGS

ARMATURE

MAGNETIC FIELD

N   S

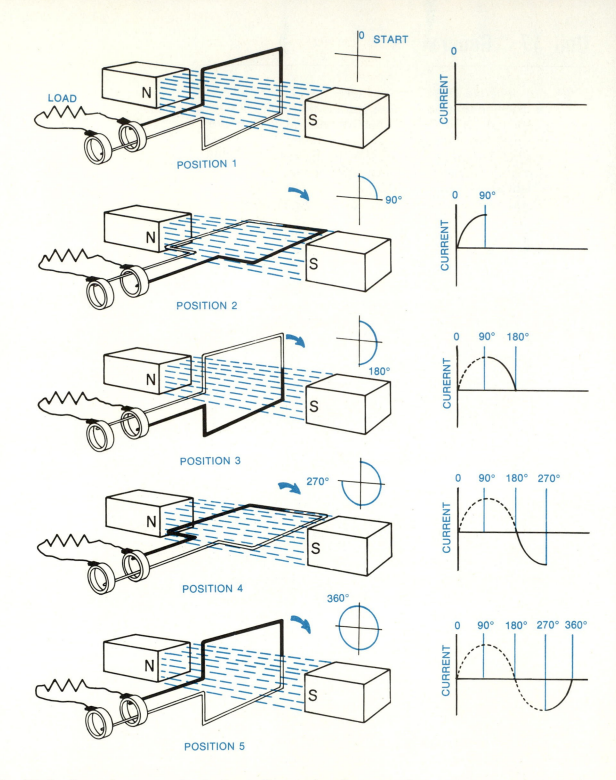

Fig. 17-2. Generating one cycle of voltage with a single-coil ac generator

creases); and (3) the angle at which a conductor cuts across flux lines (the greatest voltage is induced when the conductor cuts across flux lines at a 90° angle).

The action of a single coil in producing a complete cycle of alternating voltage is shown in Fig. 17–2. If a load is connected across the terminals, an alternating current will flow through the circuit.

In Fig. 17–2, with the coil in position 1, no voltage is being induced because the armature is not cutting across any flux lines. No current flows through the load circuit.

As the armature moves from position 1 to position 2, it cuts across more and more flux lines. Thus, the voltage increases from zero to its highest value in one direction. This increase in voltage results in a similar increase in current. This is shown by the first quarter of the sine wave. At position 2, the coil is cutting flux lines at a 90° angle. This produces the highest voltage.

In moving from position 2 to position 3, the armature cuts across fewer flux lines at sharper angles, but in the same direction. As a result, the voltage decreases from its highest value to zero. During this time, the current also decreases to zero. This is shown by the second quarter of the sine wave.

As the armature continues to rotate to position 4, each of its sides cuts across the magnetic field in the opposite direction. This changes the polarity of the voltage and the direction of the current. Once again, the voltage and current increase from zero to their highest values during the third quarter of the sine wave.

From position 4 to position 5, the armature rotates back to its starting point. During this time, the voltage and current decrease from their highest values to zero. This completes the cycle.

DIRECT-CURRENT GENERATOR

In a direct-current generator, the ends of the armature coil or coils are connected to a *commutator*. This device is needed to produce a direct current. In its basic form, a commutator is a ringlike device made up of metal pieces called *segments*. The segments are insulated from each other and from the shaft on which they are mounted.

The operation of a simple dc generator is shown in Fig. 17–3. In Fig. 17–3 at A, the armature coil is cutting across the magnetic field. This motion produces a voltage that causes current to move through the load circuit in the direction shown by the arrows. In this position of the coil, commutator segment 1 is in contact with brush 1. Commutator segment 2 is in contact with brush 2.

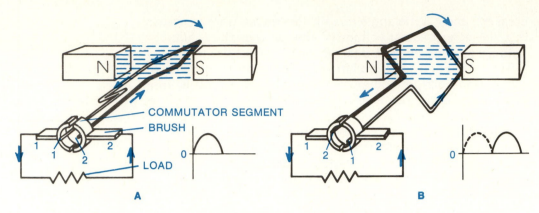

Fig. 17–3. Basic operation of a dc generator

Fig. 17–4. Main parts of a commercial dc generator (*Westinghouse Electric Corporation*)

Commutating Action. As the armature rotates a half turn in a clockwise direction, the contacts between the commutator segments and the brushes are reversed (Fig. 17–3 at B). Now segment 1 is in contact with brush 2. Segment 2 is in contact with brush 1. Because of this *commutating action*, the side of the armature coil in contact with either of the brushes is always cutting across the magnetic field in the same direction. Therefore, brushes 1 and 2 have a constant polarity. A direct voltage is applied to the external load circuit.

The Practical Generator. The main parts of a commercial dc generator are shown in Fig. 17–4. In this generator, the armature is made up of coils of magnet wire. These are placed in the slots of the laminated armature core. The ends of the coils are connected to the commutator segments. The field

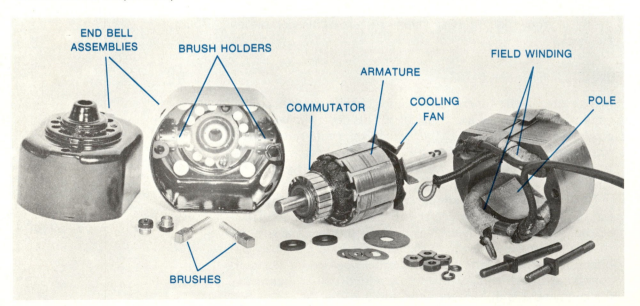

windings are electromagnets. They produce the magnetic field needed for the operation of the generator.

The direct current used to energize the field windings of a generator is called the *excitation current*. In a dc generator, this current is obtained from the output of the generator (Fig. 17–5).

Most small dc generators are driven by alternating-current motors or by gasoline engines. They are used in trains, in welders, for charging batteries, and for the operation of field telephone equipment.

ALTERNATING-CURRENT GENERATORS

The rotating part of large ac generators is called the *rotor*. It is turned by steam turbines, hydro- (water-driven) turbines, or diesel engines. These generators produce the electric energy used in our homes and in industry. Small ac generators are usually driven by gasoline engines (Fig. 17–6). Such generators are commonly used to provide emergency power. Alternating-current generators are also called *alternators*.

Excitation. In the small ac generator, the excitation current needed to energize the field windings is often obtained from a battery or from the output of the generator itself. The excitation current must be direct, but the output current is alternating. Thus, the output current must first pass through a rectifier and be changed to direct current. In some alternators, the rectifier circuit is located within the generator housing itself. In a large ac generator, the excitation current is obtained from a dc *exciter generator*. This is either mounted on the shaft of the main generator or is located nearby.

Rotating-armature Generator. In small ac generators, it is usually the armature that is the rotating part, or *rotor*. The rotor turns within the magnetic field produced by stationary field windings, called the *stator*. The rotor has a collector or slip rings that are in contact with carbon brushes (Fig. 17–7).

Rotating-field Generator. In a rotating-field ac generator, the armature is stationary. It is made up of windings placed in the slots of the frame assembly (Fig. 17–8 at A). The field windings are wound around pole pieces on the rotor assembly. They are connected to slip rings (Fig. 17–8 at B). The excitation current is passed to the field windings by means of carbon brushes in contact with the slip rings. Generators of this kind are used in most large power generating plants.

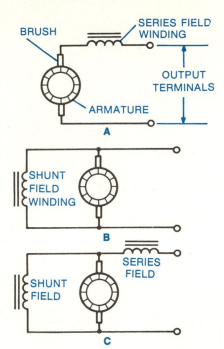

Fig. 17–5. Schematic diagrams of common dc generators: (A) series generator with field windings connected in series with the armature; (B) shunt generator with field windings connected in parallel with the armature; (C) compound generator containing both series and shunt field windings

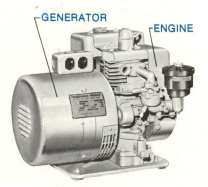

Fig. 17–6. A gasoline-engine driven ac generator (*Fairbanks Morse and Company*)

157

Fig. 17–7. Slip rings mounted on the armature assembly of a rotating-armature ac generator (*Generac Corporation*)

Frequency. The frequency of the alternating current produced by a generator depends on the speed of the rotor (either the armature or the field windings) and on the number of magnetic poles formed by the field windings. The power-company generators in most parts of the United States produce a frequency of 60 hertz. Special-purpose generators may have a higher or lower frequency.

Voltage. The output voltage of a generator depends mostly on the speed of the rotor, the number of armature coils, and the strength of the magnetic field produced by the field windings. Power-station generators usually have an output voltage of 10,000 volts or more.

MAGNETO

A *magneto* is an ac generator in which the magnetic field is produced by one or more permanent magnets rather than by electromagnets. In some magnetos, the permanent magnets are in the rotor assembly (Fig. 17–9).

A flywheel magneto is commonly used with small gasoline engines. One or more permanent magnets of this generator are mounted on the flywheel assembly. As the flywheel turns, a magnetic field cuts across a stationary coil. This is the ignition coil, and it is a step-up transformer (Fig. 17–10). The voltage induced across the secondary winding of the ignition coil is applied to the spark plug.

THREE-PHASE GENERATOR

Fig. 17–8. Principal parts of a rotating-field generator: (A) armature (stator); (B) field (rotor)

A generator with a single set of windings and one pair of slip rings will produce a single wave of voltage. This is known as a *single-phase system.*

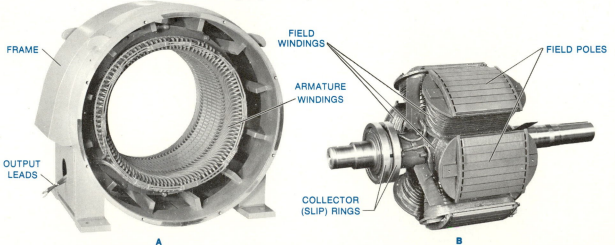

FRAME

FIELD WINDINGS

ARMATURE WINDINGS

FIELD POLES

OUTPUT LEADS

COLLECTOR (SLIP) RINGS

A

B

A *three-phase generator* has three separate sets of windings. One end of each winding is connected to a slip ring (Fig. 17–11 at A). In such a generator, each complete turn of the rotor produces three separate voltages (Fig. 17–11 at B). These are applied to a load by means of a three-conductor power line. A three-phase power system is usually protected with circuit breakers. A circuit breaker is connected in series with each of the conductors. A typical automobile alternator is an example of a three-phase generator.

A three-phase power system delivers a more steady supply of electric energy to a load. Because of this, three-phase systems are used for heavy-duty equipment that operates at a voltage of 208 volts or more. Such equipment includes large motors, welders, and heating units.

Almost all power companies in this country use three-phase generators and three-phase power-distribution lines. Single-phase loads that operate at a voltage of about 120

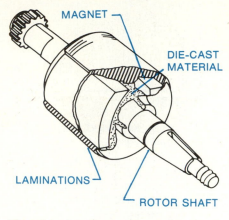

Fig. 17–9. Rotor assembly of a typical magneto showing the bar-shaped permanent magnets embedded in the rotor (*Fairbanks Morse Engine Accessories Operation, Colt Industries*)

Fig. 17–10. Ignition circuit of a typical single-cylinder gasoline engine using a magneto

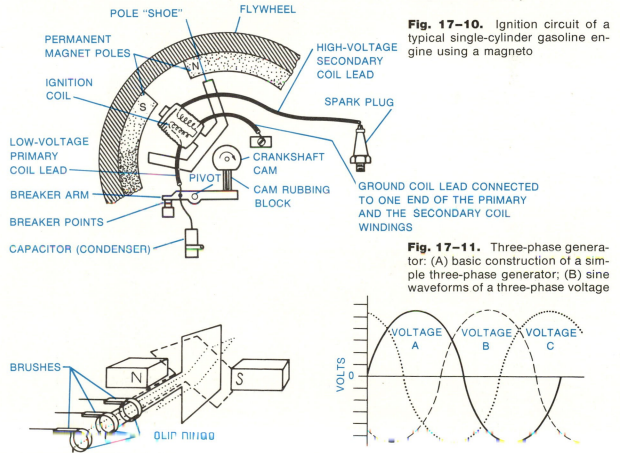

Fig. 17–11. Three-phase generator: (A) basic construction of a simple three-phase generator; (B) sine waveforms of a three-phase voltage

volts are connected to one of the three conductors of the power line and to a fourth conductor, called a *neutral wire*. Ordinary loads in homes and other buildings are so connected.

SELF-TEST

Test your knowledge by writing, on a separate sheet of paper, the word or words that most correctly complete the following statements:

1. An electric generator produces a voltage by means of _____.
2. A generator can be operated by rotating coils of wire through a _____ or by rotating a _____ past coils of wire.
3. A generator can be defined as a machine that converts _____ energy into _____ energy.
4. In its basic form, a commutator is a ring-like device made up of metal pieces called _____.
5. A commutator is used in a _____ generator.
6. Electromagnets that produce the magnetic field needed for the operation of a generator are called _____.
7. The direct current used to energize the field windings of a generator is called the _____ current.
8. Most dc generators are driven by _____ or by _____.
9. Large ac generators are driven by sources of mechanical energy such as _____, _____, and _____.
10. Alternating-current generators are also called _____.
11. A rotating armature in an ac generator is equipped with _____ or _____ rings that are in contact with _____.
12. The frequency of the alternating current produced by the generator depends on the speed of the _____ and on the number of magnetic _____ formed by the field windings.
13. A three-phase generator has three separate sets of _____.

FOR REVIEW AND DISCUSSION

1. Define an electric generator.
2. Describe the construction of and explain the operation of a simple alternating-current generator.
3. What three things determine the value of the voltage induced by generator action?
4. What is the purpose of a commutator in a dc generator?
5. Explain commutating action.
6. Name the main parts of a practical dc generator.
7. What is meant by the excitation current of a generator?
8. How is the excitation current in a dc generator obtained?
9. Name three means by which the rotors of large ac generators are turned.
10. Are large ac generators usually the rotating-field kind?
11. What two things determine the frequency of an ac generator?
12. What is a magneto?
13. Explain the operation of the flywheel magneto commonly used on small gasoline engines.
14. Describe three-phase voltage.
15. What is the advantage of three-phase power as compared to single-phase power?

INDIVIDUAL-STUDY ACTIVITIES

1. Prepare a written or an oral report on the development of large commercial generators.
2. Demonstrate the operation of a magneto. Describe its construction and explain how it is able to produce voltage.
3. Prepare a written or an oral report telling about three-phase power systems and the purposes for which they are used.

Unit 18 Other Sources of Electric Energy

Energy sources are an important part of technology. Technology involves using scientific ideas in making machines that can do work for us. Countries such as the United States can maintain their social and technical standards only by using large amounts of energy. However, this use of energy creates several problems. These include: (1) the *depletion* (using up) of our fuel supplies, (2) the pollution of our environment, (3) a dependence on foreign countries for our fuel supplies, and (4) economic difficulties caused by the need to spend large amounts of money on fuels imported from other countries.

Electric energy is one of the cleanest forms of energy. It is easy to transport. If ways can be found to produce large quantities of electric energy at a reasonable cost, our society will be able to maintain its high living standards. At the same time, damage to the environment must be reduced. Consider an electric automobile powered by rechargeable batteries. Such a vehicle would be clean and quiet and would probably have few maintenance problems. More information about the electric automobile is given in Unit 35, "The Automobile Electrical System."

Today, the production of most electric energy begins with heat energy. Fossil fuels, such as coal, oil, and natural gas, are burned to produce heat. This heat is used to heat water and change the water into steam. The steam is then used to power turbines that drive large generators.

Heat is also produced by atomic fission in nuclear power plants. The operation of these and other common kinds of power plants is discussed in Unit 44, "The Electric Power Industry."

The world's supplies of coal, oil, natural gas, and fissionable materials are limited. Thus, other sources of energy must be used. Some of these are sunlight, wind, ocean heat, wave motion, and *organic* (animal or plant) materials. These sources are particularly useful for producing electricity.

SUNLIGHT

The use of sunlight as a source of energy to generate electric power has many advantages. Sunlight is inexhaustible (cannot be used up) and is available everywhere. The use of sunlight should do little damage to the environment and produce

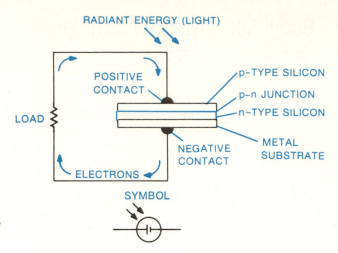

RADIANT ENERGY (LIGHT)

POSITIVE CONTACT

p-TYPE SILICON

p-n JUNCTION

n-TYPE SILICON

LOAD

NEGATIVE CONTACT

METAL SUBSTRATE

ELECTRONS

SYMBOL

Fig. 18–1. Cross-sectional view of a silicon solar cell

few wastes. *Photovoltaic*, or *solar, cells* generate electricity directly from sunlight using no fuel. Changing sunlight into electricity may help solve the long-term energy problems of the United States.

Solar Cells. These cells usually are made up of a layer of *silicon*, a common nonmetallic element, that coats a metal base known as the substrate. The silicon is treated chemically to make what are called *p-type* and *n-type* forms. The

Fig. 18–2. Experimental water purifier powered by solar energy (*U.S. Department of Defense*)

BATTERIES

SOLAR PANELS

area where these two forms touch each other is called a *p-n junction*.

A cross section of a silicon solar cell is shown in Fig. 18–1. The top layer of p-type silicon is very thin. Light energy will pass through it and reach the p-n junction. The light energy gives energy to the electrons in the p-type layer. This causes these electrons to move across the p-n junction and into the n-type layer. The n-type layer then becomes negatively charged. A voltage is produced across the layers on each side of the p-n junction.

The amount of voltage produced by a single solar cell is quite small. For this reason, many solar cells are often connected to form a *solar panel*. One very important use of solar panels is to provide energy for charging batteries and operating circuits in space vehicles.

Figure 18–2 shows a solar-powered water purifier. This experimental system uses several different solar cell designs mounted on panels. The solar cells produce 11 kilowatts when the sun is at its highest point, generally around noon each day. Energy is also stored in 20 automotive lead-acid batteries. These are used during cloudy days and when more power is needed to start motors.

WIND

Energy from the wind has been used in the United States for many years. The *windmill* is a machine that converts wind energy into useful work. In the rural United States, windmills were often used to pump water. Only recently have windmills been seriously considered as another means of generating electricity.

Figure 18–3 shows some modern designs for windmills that will produce electricity. The blades of a windmill are sometimes shaped like airplane propellers. When the wind presses against the blades, it causes them to turn. To produce electricity, a generator is connected to the shaft on which the blade turns. A rudderlike device steers the blades toward the wind. Figure 18–4 shows a large modern wind generator that is in operation in Clayton, New Mexico. This wind generator can produce 200 kilowatts of electricity. Feeding directly into the local utility system, this is about enough power to electrify 60 homes. At a wind speed of 8 miles per hour (12.9 km/h), the machine will start generating power. At 18 miles per hour (29.0 km/h), it produces its maximum output of 200 kilowatts. From 18 to 35 miles per hour (29.0 to 56.3 km/h), the blades are adjusted in pitch. This keeps the output steady at 200 kilowatts. At winds above 35 miles per hour (56.3 km/h), the machine shuts itself off.

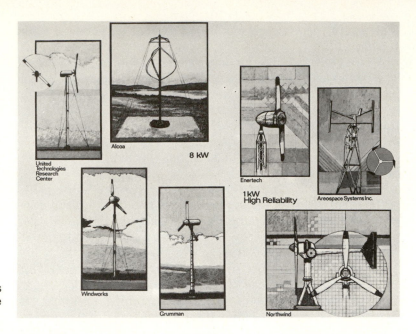

Fig. 18-3. Experimental designs of small windmills for limited use (*Rockwell International*)

Fig. 18-4. The first large wind generator to be used in the United States in several decades (*National Aeronautics and Space Administration*)

OCEAN HEAT

Another possible way to generate electricity is to use the *thermal* (heat) energy of the ocean. This involves using the ocean water to collect and store energy from the sun. A system for doing this may take the form of a large floating structure anchored to the ocean floor (Fig. 18-5). This system makes use of the difference in temperature between the heated upper layers of water and the colder layers below. The warmer water can change a liquid such as ammonia into a gas. Gas pressure can be used to drive a turbine. An electric generator can be connected to the turbine. The gas, after passing through the turbine, can be changed back into a liquid by the colder water. This can go on and on. This system works somewhat like the steam electric plant described in Unit 44. However, an ocean thermal energy conversion system would operate at much lower temperatures and pressures. Very large amounts of ocean water would be used. Power would be transmitted to land by underwater cables where they would connect with regular power lines. Large numbers of such systems may be built. They would probably be located close to major population centers.

GEOTHERMAL

Scientists are working to find other natural sources of heat energy. One possible source, now little used, is *geothermal*

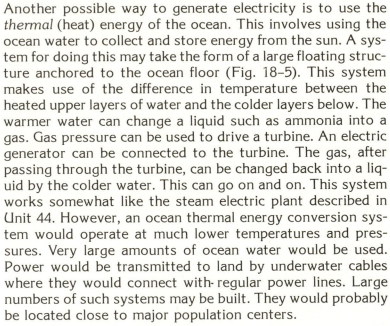

164

energy, the natural heat of the Earth. This vast source could be tapped by drilling holes deep into the Earth. These holes release steam, which could be piped to turbines that drive large electric generators.

WAVE MOTION

Another source of energy being considered is the use of the motion of ocean waves. One way of doing this uses an anchored *buoy* (float). You may have seen buoys marking the entrances to channels and harbors. A buoy for using wave motion to produce electricity has two parts. The lower section is the part that floats. The upper part is shaped like a ball with its bottom open. Waves cause water to move up and down in this upper part. This forces air in and out through a tube containing a rotor with windmill-like blades. The air moves the blades of the rotor, which are designed to always turn in the same direction. Connected to the shaft of the rotor is an electric generator. Such a system could be an inexpensive way of generating electricity for countries with a seacoast.

ORGANIC MATERIALS

Organic materials that can be changed into usable fuels are another possible source of energy. Such materials include agricultural wastes from both crops and animals and sewage. These can be changed into methane gas, alcohol, and oil. Some crops can be grown specifically to be made into fuels. These fuels can replace some of the oil or gasoline used by the engines that turn electric generators.

HIGH-ALTITUDE TRANSMISSION OF POWER

Figure 18–6 shows an artist's idea of one kind of satellite power station, a Space-Based Power Conversion System. This system is being studied by the National Aeronautics and Space Administration (NASA). Such a system can be placed in a fixed *orbit* (path) around the Earth. There, it can collect pollution-free energy from the sun. This energy can be transmitted to the Earth in the form of *microwaves*. These are ultra-high frequency radio waves. The receiving station on Earth can then change the microwaves to electricity. Located high above the Earth, this system would not be affected by clouds or night. Thus, it could operate continuously.

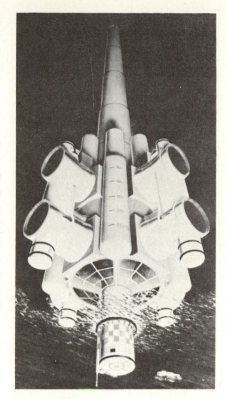

Fig. 18–5. Artist's concept of a floating structure for an ocean thermal energy conversion system (*Lockheed Missiles and Space Co., Inc.*)

165

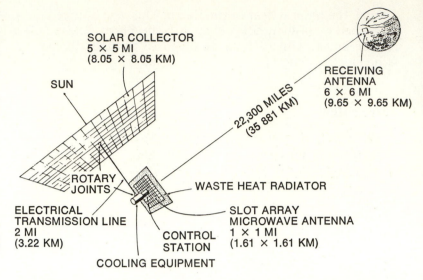

SOLAR COLLECTOR
5 × 5 MI
(8.05 × 8.05 KM)

SUN

RECEIVING
ANTENNA
6 × 6 MI
(9.65 × 9.65 KM)

22,300 MILES
(35 881 KM)

ROTARY
JOINTS

WASTE HEAT RADIATOR

ELECTRICAL
TRANSMISSION LINE
2 MI
(3.22 KM)

SLOT ARRAY
MICROWAVE ANTENNA
1 × 1 MI
(1.61 × 1.61 KM)

CONTROL
STATION

COOLING EQUIPMENT

Fig. 18–6. Artist's concept of a space-based power conversion system (*National Aeronautics and Space Administration*)

FUEL CELLS

In a *fuel cell,* chemical reactions between oxygen and a fuel cause chemical energy to be changed directly into electric energy. The fuels most commonly used in such cells are hydrogen and methane.

The basic construction of a hydrogen-oxygen fuel cell is shown in Fig. 18–7. In this cell, hydrogen in either liquid or gas form is supplied to the cathode, or negative electrode. The hydrogen then spreads through the electrolyte. This causes electrons to be released from the electrolyte. These electrons are deposited on the cathode, causing it to become negatively charged. The electrons then move through the load to the anode, or positive electrode.

Fig. 18–7. Basic operation of a fuel cell

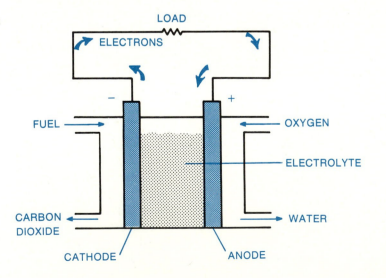

LOAD

ELECTRONS

−   +

FUEL →   ← OXYGEN

ELECTROLYTE

CARBON
DIOXIDE ←   → WATER

CATHODE   ANODE

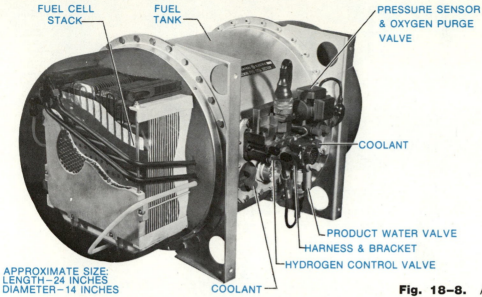

FUEL CELL STACK
FUEL TANK
PRESSURE SENSOR & OXYGEN PURGE VALVE
COOLANT
PRODUCT WATER VALVE
HARNESS & BRACKET
HYDROGEN CONTROL VALVE
COOLANT

APPROXIMATE SIZE:
LENGTH—24 INCHES
DIAMETER—14 INCHES

Fig. 18–8. A space-vehicle fuel cell (*General Electric*)

At present, fuel cells are most commonly used in space vehicles (Fig. 18–8). In the future, larger fuel cells may be used to supply large amounts of electric energy for ordinary use.

MAGNETOHYDRODYNAMIC (MHD) CONVERSION

Another device that could generate large amounts of electricity is the *magnetohydrodynamic* (MHD) *converter*.

MHD converters are based on two principles. The first is that gases can be ionized by high temperatures and thus made into good conductors of electricity. The second is the principle of electromagnetic induction.

Figure 18–9 shows the basic elements of the MHD converter. Gas is heated to a temperature of about 5000°F

Fig. 18–9. Basic elements of a magnetohydrodynamic (MHD) converter

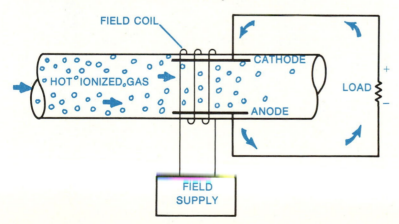

FIELD COIL
CATHODE
HOT IONIZED GAS
LOAD
ANODE
FIELD SUPPLY

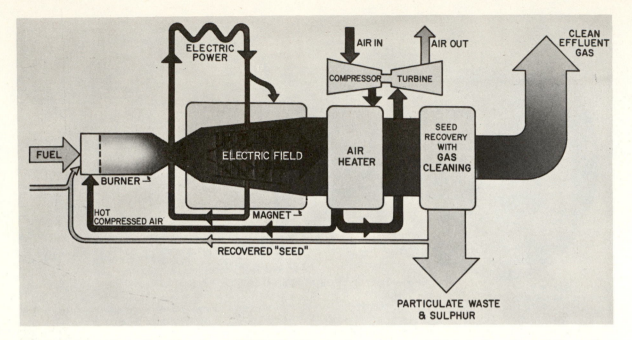

Fig. 18–10. Flow diagram of a magnetohydrodynamic (MHD) electric power converter (*Avco Everett Research Laboratory*)

(2760°C). This ionizes the gas, which is then forced through a strong magnetic field. The electrical force that results makes contact with electrodes at the top and bottom of the gas stream. Current flows in the external circuit. The gas moves to a condenser unit and then back to the heat source to complete the cycle. The ionized gas thus takes the place of metal conductors.

MHD converters have few mechanical moving parts. New materials will allow them to withstand the tremendous heat and pressures needed. Someday it may be possible to design and build MHD converters with outputs of more than 1 million watts. These will be smaller, lighter, more efficient, and more reliable than equivalent turbine-generator combinations. Figure 18–10 is a flow diagram that shows the operation of an MHD converter.

THERMIONIC ENERGY CONVERSION

Another system for changing heat directly into electricity is the *thermionic energy converter*. It consists of two electrodes in a vacuum. The *emitter* electrode is heated to produce free electrons. The *collector* electrode, at a much lower temperature, receives the electrons released by the emitter.

A diagram of a *coaxial thermionic energy converter* is shown in Fig. 18–11. The heat source in the center of the converter will heat the emitter. The heated emitter will give off electrons. The electrons will then move to the collector and on to the external load.

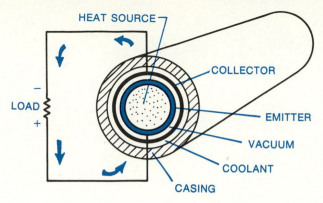

HEAT SOURCE

COLLECTOR

EMITTER

VACUUM

COOLANT

CASING

LOAD

Fig. 18–11. Coaxial thermionic energy converter

An alternating current can be produced directly in the thermionic converter. This is done by applying a small *modulating* (varying) *signal* to a *grid electrode* located between the emitter and the collector. Large thermionic converters could someday replace ordinary turbine-driven generators. (A general discussion of modulating signals can be found in Unit 40; the effect of grid electrodes is discussed in Unit 19.)

SELF-TEST

Test your knowledge by writing, on a separate sheet of paper, the word or words that most correctly complete the following statements:

1. _____ is an inexhaustible source of energy.
2. Devices that change sunlight directly into electricity are called _____ .
3. The wind can be converted to electric energy by a _____ .
4. Natural heat from the Earth is called _____ energy.
5. A satellite power station placed in a fixed orbit around Earth would transmit energy to Earth in the form of _____ .
6. A device that converts chemical fuel energy directly into electric energy is the _____ .
7. A device that changes heat energy into electric energy by using a hot ionized gas is the _____ converter.
8. Applying a modulating signal to a grid electrode in a thermionic energy converter can produce _____ .

FOR REVIEW AND DISCUSSION

1. List some of the problems created by a high rate of energy use.
2. List some of the advantages of an electric automobile.
3. Identify some sources of energy that may provide alternatives to today's fuels.
4. Discuss the advantages of sunlight as a source of energy.
5. Explain how a satellite power station might change sunlight into useful energy.

INDIVIDUAL-STUDY ACTIVITIES

1. Write a report on the world's present supplies of coal, oil, and gas. Discuss the immediate and long-term problems with these fuels.
2. Calculate the roof area of your school. Estimate how many silicon solar cells, each having an area of 1 square inch (645 square millimeters), could be placed on the roof. Assume that each silicon solar cell develops 0.2 volt at 10 milliamperes in bright sunlight. Calculate the power output for the entire sun battery.

4 Electronics—Theory and Devices

Unit 19   Electron Tubes

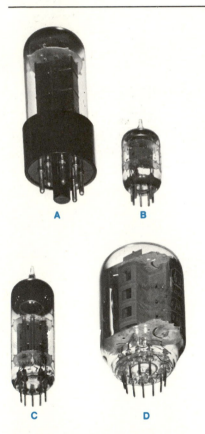

Fig. 19–1. Common types of electron tubes: (A) octal base—the center (insulated) post properly orients the metal pins in the tube socket; (B) 7-pin miniature; (C) 9-pin miniature; (D) 12-pin (duo-decar) base.

Electron tubes have been largely replaced by solid-state devices. However, electron tubes are still found in some electronic products. Solid-state devices include semiconductor diodes and transistors. They are discussed in Unit 20. This unit is about the construction of typical electron tubes and their operation as rectifiers and amplifiers. Special kinds of electron tubes used in oscilloscopes and television cameras and receivers are discussed in later units of this book.

CONSTRUCTION AND BASIC OPERATION OF TUBES

Electron tubes consist of electrodes contained within a glass envelope from which most of the air has been removed. Early tubes had a high vacuum. For this reason, present-day tubes are still often called *vacuum tubes*. The electrodes of a tube are connected internally to pins that extend from the base (Fig. 19–1).

An electron tube is put into a circuit by being plugged into a tube socket. The tube and the rest of the circuit are connected by wires attached to the lugs of the socket (Fig. 19–2). Some special miniature tubes have wire leads that are soldered directly into the circuit.

One of the electrodes in an electron tube is called a *cathode*. When this electrode is heated, it emits, or gives off, electrons. The cathode may be heated directly or indirectly. The process of electrons being emitted from a heated surface is known as *thermionic emission*.

Cathodes. A directly heated cathode is usually made of a nickel-alloy metal. It is in the form of a filament coated with a

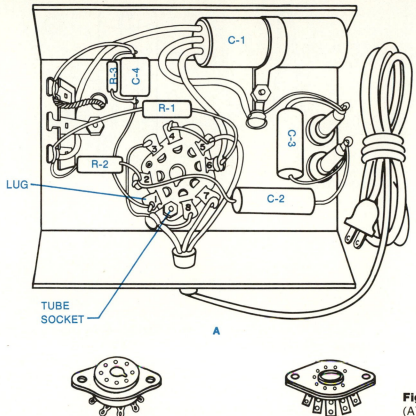

LUG

TUBE
SOCKET

A

B

C

Fig. 19–2. Electron-tube sockets: (A) wired socket; (B) molded socket; (C) wafer socket

material that emits electrons (Fig. 19–3 at A). The filament is connected to a source of voltage. It is heated by the current that passes through it.

An indirectly heated cathode is in the form of a thin metal sleeve. It is also coated with a material that emits electrons when heated. This kind of cathode is heated by a heater wire inside the sleeve but not in contact with the sleeve (Fig. 19–3 at B).

Electrostatic Attraction. The cathode has a negative charge. The electrons emitted from the cathode are attracted to a positively charged tube electrode called the *plate* (Fig. 19–4). This is known as *electrostatic attraction*. When the plate is not positively charged, the electrons will not flow. (Actually, a small number of electrons will reach the plate even if it is not charged. This is called the *Edison effect* in honor of Thomas Edison, who discovered it.)

Heater Glow. The light seen when a tube is on is caused by the glow of the heated filament. The heating process uses most of the energy supplied to an electron tube.

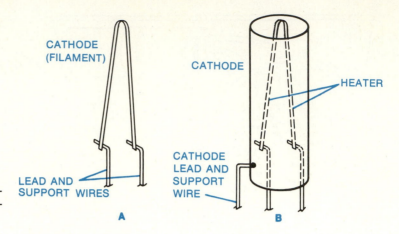

Fig. 19–3. Electron-tube cathodes: (A) directly heated; (B) indirectly heated

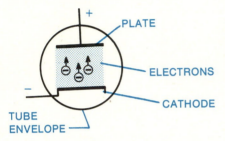

Fig. 19–4. Electrostatic attraction of electrons to the plate, or anode, of an electron tube

Fig. 19–5. The diode tube

DIODE TUBE

A *diode* (*di* for two and *ode* for part) is the simplest kind of electron tube. It has two electrodes, a cathode and a plate (Fig. 19–5).

If an alternating voltage is applied across the plate and cathode, the plate will be positive with respect to the cathode during one-half of each voltage cycle. Electrons will thus flow in the tube from the cathode to the plate only during this half of the cycle (Fig. 19–6). As a result, the tube will emit a pulsating direct current. Thus, a diode tube can act as a *half-wave rectifier*. A *rectifier* is a device or a circuit that changes alternating current into direct current.

The *full-wave-rectifier* electron tube has a cathode and two separate plates. The tube is connected to the ac voltage in

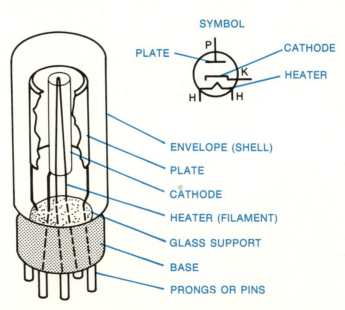

such a way that one of the plates is always positive with respect to the cathode. Because of this, both halves of the cycle produce a current, but always in the same direction.

TRIODE TUBE

In a *triode* electron tube, a third electrode, called the *control grid*, is put between the cathode and the plate. The grid is usually a spiral of fine wire placed around the cathode and insulated from it (Fig. 19–7). The control grid controls the number of electrons that pass from the cathode to the plate. When the grid is positive with respect to the cathode, the grid lets electrons from the cathode reach the plate (Fig. 19–8 at A). When the grid is negative with respect to the cathode, the grid repels electrons, preventing many of them from reaching the plate (Fig. 19–8 at B).

A small change in the value of the voltage applied between the control grid and the cathode of a triode tube can control a large amount of current in the plate circuit of the tube. This action allows the tube to act as an amplifier. An *amplifier* is a device or circuit that increases the value of a voltage or a current.

OTHER TUBES

Diodes and triodes are only two of the several kinds of electron tubes. Others have more electrodes, which are designed to improve electron flow. One such tube is the *pentode.* In

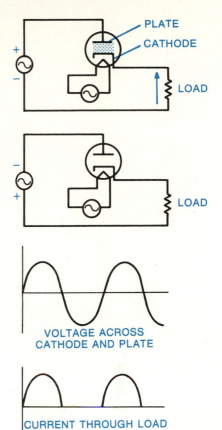

Fig. 19–6. Electrons flow through a diode tube only when the plate is positive with respect to the cathode.

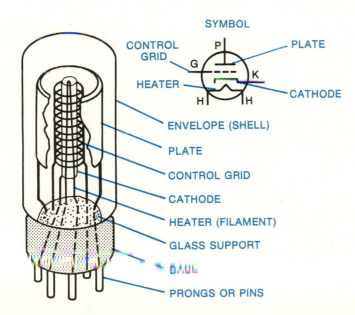

Fig. 19–7. The triode tube

173

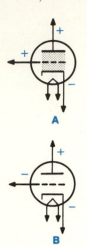

Fig. 19–8. Action of the control grid in a triode

addition to a control grid, it has a *screen grid* and a *suppressor grid.* Some tubes, commonly known as *multifunction* or *multisection tubes,* consist of two or three tubes in one envelope. Such tubes have the electrodes of a double diode and a triode, a triple diode, a double triode, or a triode and a pentode.

BASING DIAGRAMS

A tube basing diagram shows, by means of symbols, the kinds of electrodes in the tube. It also shows the particular base pin to which each electrode is connected (Fig. 19–9 at A). To identify the pins, look at a tube from its base. Then count each pin in a clockwise direction, beginning with the key (Fig. 19–9 at B).

On some tubes, an electrode is connected to a terminal located at the top of the envelope. This is usually the plate or control grid. This is indicated on a basing diagram with a square, as shown in Fig. 19–9 at A. The basing diagrams and other important information about many different kinds of electron tubes are given in publications called *tube manuals.*

TUBE DESIGNATION AND SUBSTITUTION

Electron tubes are identified by a combination of numbers and letters, such as 5U4GB, 12BA6, and 50C5. These are called *type designations.* The first number, for example, 12, is the voltage at which the heater filament is operated. Tubes with the same type designations have the same electrical characteristics even though they may have different brand names. In some circuits, it is possible to substitute one tube for another tube that has a different type-designation number. Tubes that can be used as replacements for another tube are listed in *tube-substitution manuals.*

Fig. 19–9. Identifying electron-tube base pins

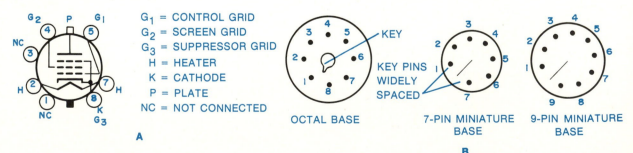

G_1 = CONTROL GRID
G_2 = SCREEN GRID
G_3 = SUPPRESSOR GRID
H = HEATER
K = CATHODE
P = PLATE
NC = NOT CONNECTED

OCTAL BASE   7-PIN MINIATURE BASE   9-PIN MINIATURE BASE

174

GETTER SPOT

A *getter spot* is a dark spot on the inside of an electron-tube envelope. This is caused by the burning of a material placed inside a tube after the tube has been sealed during manufacture. After being sealed, a tube is put into an oven and heated. At a certain temperature, the getter material burns. It thus removes any oxygen that may be in the envelope after air has been pumped from it. It is necessary to remove the oxygen since oxygen would cause the tube electrodes to evaporate more rapidly. The presence of a getter spot *does not* mean that a tube is defective.

ELECTRON-TUBE DEFECTS

The most common defect of tubes is a burned-out filament. This keeps the tube from working. Other common electron-tube defects are discussed below.

Leaks. A leak in a tube is most often caused by a crack in its envelope. Leaks can also be caused by a broken seal between the envelope and the pins. When a leak occurs, air enters the tube. A milky white spot on the inside surface of the envelope indicates that this has happened. A leaky tube must be replaced.

Gas. The cathodes of electron tubes work at very high temperatures. This sometimes causes them to partly vaporize, or turn into gas. This gas disrupts the flow of electrons and affects the working of the tube. Too much gas in a vacuum tube often produces a bluish glow that you can see. When this happens, you should replace the tube.

Shorts. A tube becomes shorted when there is contact between two or more electrodes that should be separated. This condition often causes the electrodes or parts of the circuit to overheat. A shorted tube must be replaced.

TESTING ELECTRON TUBES

Electron tubes can be checked with a tube tester (Fig. 19–10). You can test a particular kind of tube with this instrument by adjusting its controls according to the operating-chart control settings for that tube. When using a tube tester, always be sure that the filament-voltage control is set to the right value for the tube being tested. Too much voltage applied to a filament or heater wire will cause it to burn out.

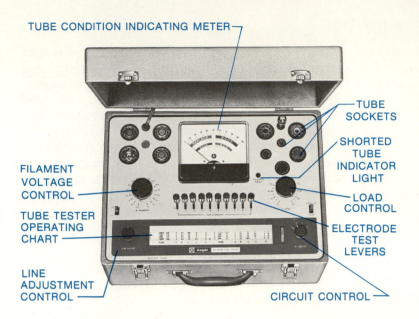

TUBE CONDITION INDICATING METER

TUBE SOCKETS

SHORTED TUBE INDICATOR LIGHT

LOAD CONTROL

ELECTRODE TEST LEVERS

FILAMENT VOLTAGE CONTROL

TUBE TESTER OPERATING CHART

LINE ADJUSTMENT CONTROL

CIRCUIT CONTROL

Fig. 19-10. Tube tester

A continuity test done with an ohmmeter can be used to test for open filaments. You adjust the ohmmeter to a low or medium resistance range and then connect it to the heater pins. A tube manual will help you find the right pin numbers.

SELF-TEST

Test your knowledge by writing, on a separate sheet of paper, the word or words that most correctly complete the following statements:

1. Electrons move from cathode to plate in a tube when the plate is _____ charged with respect to the cathode.
2. The full-wave-rectifier tube contains a cathode and two separate _____.
3. The three electrodes of a triode tube are the _____, the _____, and the _____.
4. The current-control action of a triode tube allows it to act as an _____.
5. A tube _____ diagram shows the kinds of electrodes in a tube and the base _____ to which they are connected.
6. The dark spot on the inside of an electron-tube envelope is called the _____.

FOR REVIEW AND DISCUSSION

1. Describe the basic construction of a typical electron tube.
2. What is the purpose of a cathode?
3. What is meant by thermionic emission?
4. What is electrostatic attraction in an electron tube?
5. Describe the construction of a diode tube. Explain how it works as a rectifier.
6. Explain the function of the control grid in a triode tube.
7. Name and describe four common tube defects.
8. How are electron tubes tested?

INDIVIDUAL-STUDY ACTIVITY

Give a demonstration showing your class several different kinds of electron tubes. Give the name of each tube and explain what its type designation means.

Unit 20　Semiconductors and Diodes

A *semiconductor material* is one that, in its pure state and at normal room temperature, is neither a good conductor nor a good insulator. However, pure semiconductor materials are seldom used in electronics. Instead, small amounts of other substances are added by a process called *doping.* Thereby the resistance of semiconductor materials can be made much lower than that of the pure materials.

The most widely used semiconductor materials are the elements germanium and silicon. They are used in solid-state diodes and transistors. A *solid-state diode* is one that, unlike the electron tube, is made as a solid unit. It does not have a glass envelope or a filament (Fig. 20–1).

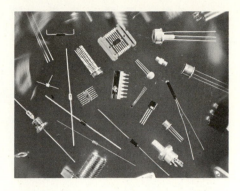

Fig. 20–1. Semiconductor solid-state devices (*Texas Instruments Incorporated*)

SEMICONDUCTOR STRUCTURE

The atoms of germanium and silicon have four *valence electrons,* those in the outermost shell (Fig. 20–2). In these materials, the atoms are arranged in what is called a *crystal-lattice structure.* Each atom shares each of its valence electrons with nearby atoms. This forms what are called *electron-pair,* or *covalent, bonds* between the atoms (Fig. 20–3).

Semiconductor Materials: n-type. One impurity with which a pure semiconductor material, such as silicon, is often doped is arsenic. Arsenic has five valence electrons. Four of the five valence electrons of each arsenic atom form electron-pair bonds with the valence electrons of four silicon atoms. One valence electron of each arsenic atom is then left unattached. It becomes a free electron (Fig. 20–4). Since the silicon then contains free electrons, or *negatively* charged particles, it is called an *n-type semiconductor material.* The arsenic donated free electrons to the silicon. Because of this, arsenic is called a *donor impurity.*

Semiconductor Materials: p-type. Pure silicon is often doped with an impurity, such as the element indium, that has three valence electrons. In this case, the valence electrons of each indium atom form electron-pair bonds with the valence electrons of three silicon atoms. This leaves one valence electron of a nearby silicon atom without an electron-pair bond. As a result, there is a lack of one electron in the crystal-lattice structure of the silicon. This lack is called a *hole* (Fig. 20–5 at A).

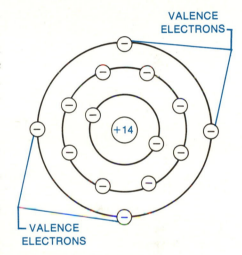

Fig. 20–2. Representation of a silicon atom

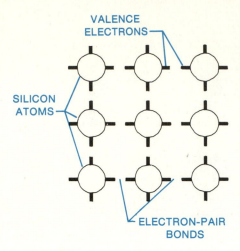

Fig. 20–3. Electron-pair, or covalent, bonds in the crystal-lattice structure of pure silicon

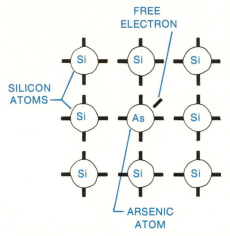

Fig. 20–4. Free, or excess, electron resulting from the addition of arsenic to silicon

If a voltage is applied to the silicon, an electron from an electron-pair bond can gain enough energy to break the bond and move into the hole. This creates a new hole. This hole is then ready to accept another electron that has broken its electron-pair bond. As this continues, the hole is said to *move* through the silicon, as shown in Fig. 20–5 at B.

Since a hole represents the lack of an electron, it can be considered a *positively* charged current carrier. Therefore, a semiconductor material containing many holes is called a *p-type material*. Indium is called an *acceptor impurity*. This is because each atom of indium added to silicon or germanium can accept an electron from an electron-pair bond.

THE P-N-JUNCTION DIODE

The basic solid-state diode is a combination of p-type and n-type semiconductor materials. It is made as a single unit through a chemical process. The section where the two types of materials meet is known as a *p-n junction* (Fig. 20–6). The diodes are enclosed in protective cases of various shapes with lugs or leads for making the needed circuit connections (Fig. 20–7 at A). The p-type material in a diode is called the *anode*. The n-type material is called the *cathode*. The anode and cathode terminals of a diode are identified in several different ways, as shown in Fig. 20–7 at B.

RECTIFICATION

Most power-company generating stations that supply electric energy to homes, businesses, and factories produce alternating current. One of the main reasons for this is that alternating-current energy can be sent over long distances very efficiently. Many processes and devices, however, can be operated only with direct current. For example, transistors, electron tubes, and certain control devices need direct current. If circuits containing these devices are to be operated from ordinary outlets, they must use rectifier circuits. *Rectifier circuits* change alternating current to direct current. Diodes are used as rectifiers in battery chargers and in other power-supply units that provide direct current. They also are used in rectifier circuits in radios, television receivers, and other electronic and electrical products.

HOW A DIODE WORKS

The common solid-state diode rectifier is made of silicon. It works by acting as a gate. A *gate* lets current pass through it in one direction, but not in the other. The polarity of the

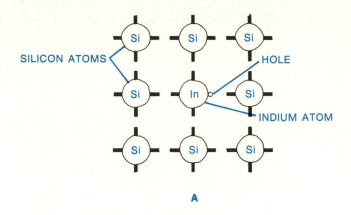

A

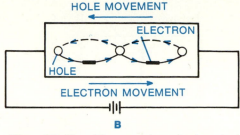

B

Fig. 20–5. The semiconductor hole: (A) hole created when indium is added to silicon; (B) movement, or drift, of holes and electrons through a p-type material

voltage applied to a diode determines whether or not the diode will conduct current. The two polarities of an applied voltage are known as *forward bias* and *reverse bias*.

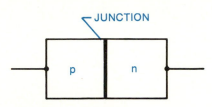

Fig. 20–6. Basic p-n semiconductor junction

Forward Bias. A diode is forward-biased when the positive terminal of a voltage source, such as a battery, is connected to its anode and the negative terminal is connected to its cathode (Fig. 20–8 at A). Under this condition, the positive terminal of the battery repels holes within the p-type material toward the p-n junction. At the same time, free electrons within the n-type material are repelled toward the junction by

Fig. 20–7. Silicon rectifier diodes: (A) typical diodes, (B) methods used to identify the cathode lead or terminal of small-size diodes (*Photo courtesy of International Rectifier*)

A

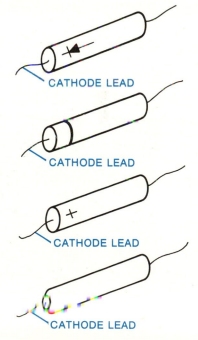

B

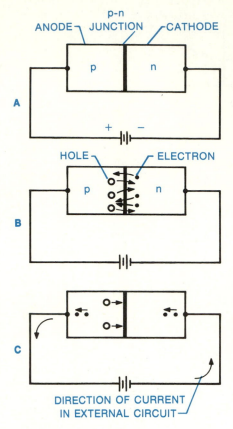

Fig. 20–8. Diode circuit with forward bias

Fig. 20–9. Diode circuit with reverse bias

the negative terminal of the battery. When holes and electrons reach the junction, some of them break through it (Fig. 20–8 at B). Then holes combine with electrons within the n-type material. Electrons combine with holes within the p-type material.

Each time that a hole combines with an electron or an electron combines with a hole near the junction, an electron from an electron-pair bond within the p-type material (anode) breaks its bond. This electron then enters the positive terminal of the battery. At the same time, an electron from the negative terminal of the battery moves into the n-type material (cathode) of the diode (Fig. 20–8 at C). This produces a negative-to-positive flow of electrons in the external circuit to which the diode is connected.

Reverse Bias. A diode is reverse-biased when its anode is connected to the negative terminal of the battery and its cathode is connected to the positive terminal of the battery (Fig. 20–9 at A). In this condition, holes within the p-type material are attracted toward the negative terminal of the battery and away from the p-n junction. At the same time, free electrons within the n-type material are attracted toward the positive terminal of the battery and away from the junction (Fig. 20–9 at B). This action causes the diode to present a high resistance to current. Under this condition, no current will flow in the external circuit.

HALF-WAVE-RECTIFIER CIRCUIT

When a diode is connected to a source of alternating voltage, it is alternately forward-biased and then reverse-biased during each cycle. Therefore, when a single diode is used in a rectifier circuit, current passes through the circuit load during only one half of each cycle of the input voltage (Fig. 20–10). For this reason, the circuit is called a *half-wave rectifier.*

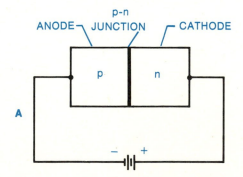

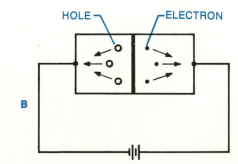

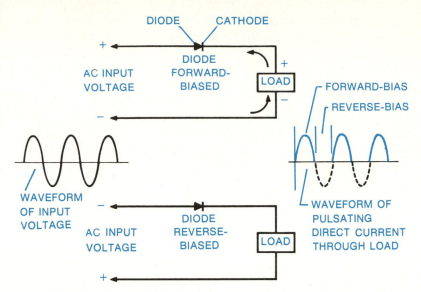

Fig. 20–10. Half-wave-rectifier circuit

DIODE   CATHODE

AC INPUT VOLTAGE

DIODE FORWARD-BIASED

LOAD

FORWARD-BIAS

REVERSE-BIAS

WAVEFORM OF INPUT VOLTAGE

AC INPUT VOLTAGE

DIODE REVERSE-BIASED

LOAD

WAVEFORM OF PULSATING DIRECT CURRENT THROUGH LOAD

Output. The output of a half-wave-rectifier circuit is a pulsating direct current. Such a current can be used in some circuits. However, it produces a loud hum in radios, television sets, and amplifiers. To eliminate or reduce this hum, the pulsating direct current is *filtered,* or smoothed out.

Filtering. The basic filtering component of rectifier circuits is an *electrolytic capacitor* (Fig. 20–11). The action of such a capacitor, when used in a half-wave-rectifier circuit, is shown in Fig. 20–12.

During the time that the diode is forward-biased, voltage builds up across the output circuit. The capacitor becomes charged. As the voltage decreases, the capacitor begins to discharge through the resistor R1. When the diode is reverse-biased, it ceases to conduct. Meanwhile, the capacitor keeps discharging through R1. When the diode is forward-

Fig. 20–11. Electrolytic capacitor used in the rectifier circuit of a timing assembly (*Gibbs Manufacturing and Research Corporation*)

ELECTROLYTIC CAPACITOR

CAPACITOR CHARGING   CAPACITOR DISCHARGING

CR1

+

R1

Fig. 20–12. Filtering action of an electrolytic capacitor in a half-wave-rectifier circuit

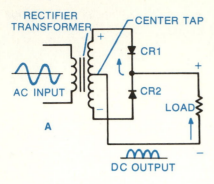

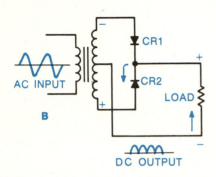

Fig. 20–13. Basic two-diode, full-wave-rectifier circuit: (A) diode CR1 conducting, diode CR2 not conducting; (B) diode CR2 conducting, diode CR1 not conducting

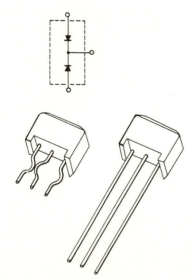

Fig. 20–14. Two-diode rectifier units

biased again, the capacitor, which was not fully discharged, is brought back to full charge. It then begins the discharge cycle again. Thus the output voltage and current are smoothed.

FULL-WAVE-RECTIFIER CIRCUIT

A *full-wave-rectifier* circuit is one that rectifies the entire cycle of an applied voltage. The basic full-wave-rectifier circuit uses two diodes. The action of these diodes during each half cycle of the applied voltage is shown in Fig. 20–13. The diodes may be individual units, or they may both be in a single package (Fig. 20–14).

The rectifier transformer used in this kind of circuit must have a center-tapped secondary winding. The dc output voltage of the rectifier circuit will depend mainly on the voltage across the secondary winding of the transformer.

Full-wave rectifiers have a smoother output voltage than half-wave rectifiers. The reason for this is that the full-wave rectifier produces a pulse of voltage in the output circuit during each half cycle of the applied voltage. After filtering, the load current can be quite smooth.

Another kind of full-wave-rectifier circuit is called a *bridge circuit.* In this circuit, four diodes are used. The action of these diodes during each half cycle of the applied alternating input voltage is shown in Fig. 20–15. Note that the rectifier transformer used in a bridge circuit does not have a center-tapped secondary winding. The diodes may be individual units or be packaged into a single unit (Fig. 20–16).

DIODE RATINGS

Rectifier diodes are usually rated in terms of the largest current that they can safely conduct and their *peak inverse voltage* (PIV). The PIV rating of a diode is the highest reverse-bias voltage that can be applied to the diode. If more than this voltage is applied, the diode may be badly damaged or completely ruined.

TESTING DIODES

When a semiconductor diode becomes defective, it is usually open or shorted. Each of these conditions generally occurs because of overheating produced by too much current passing through the diode. This causes the atomic and molecular structures in the semiconductor material to become seriously disarranged.

Diodes are best tested with a diode tester. Rectifier diodes can also be checked with an ohmmeter.

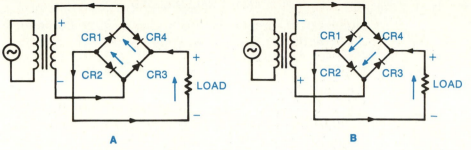

Ohmmeter testing of a diode is based on forward and reverse bias. When the negative terminal of the ohmmeter power supply is connected to the diode's cathode, the diode is forward-biased. Its resistance will be relatively low (Fig. 20–17 at A). If the positive terminal of the ohmmeter is connected to the diode's cathode, the diode is reverse-biased. Its resistance will be relatively high (Fig. 20–17 at B).

To test a diode with an ohmmeter, connect the meter to the diode. Then note the resistance of the diode. Next reverse the connections of the ohmmeter to the diode. If the resistance of the diode is much greater in one direction than in the other, you can assume that the diode is in good condition. If the test shows direct continuity or the same amount of resistance for both connections, the diode is defective and must be replaced.

Fig. 20–16. Four-diode bridge rectifier units (*Varo Semiconductor, Inc.*)

LEARNING BY DOING

17. The Diode as a "Gate." By following the procedure on the next page, you can see a diode work as a one-way current gate. Diodes are widely used in switching and computer circuits because they conduct current in one direction only.

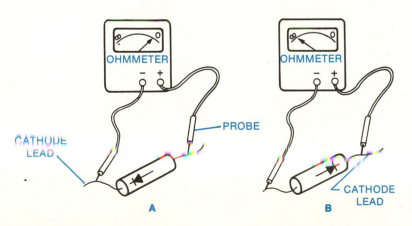

Fig. 20–17. Testing a diode with an ohmmeter

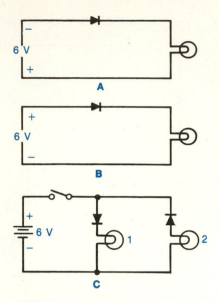

Fig. 20–18. Circuits for Learning by Doing No. 17, "The Diode as a Gate"

MATERIALS NEEDED

circuit board, prepunched, 6 × 8 in. (152 × 203 mm)
2 silicon diodes, 1 ampere, 200 PIV
2 miniature lamps, no. 40
2 miniature lamp sockets, phenolic, screw-base
1 low-voltage, push-button switch, spst
6-volt battery or dc power supply with a current rating of at least ½ ampere

Procedure

1. Connect the circuit shown in Fig. 20–18 at A. Does the lamp light? Give the reason for this.
2. Interchange the connections to the battery or to the power supply (Fig. 20–18 at B). Does the lamp light? Why?
3. Connect the circuit as shown in Fig. 20–18 at C. Do both lamps light? Reverse the battery. How could such a circuit be used to test polarity?

18. The Diode as a Rectifier. By wiring the following circuits, you will better understand how a diode works as a rectifier. You will also be able to see the difference between half-wave- and full-wave-rectifier circuits. Since you must use an oscilloscope, your teacher will have to help you connect and adjust this instrument in the circuit. You may also want to refer to Unit 47, "Electronic Test Instruments," for a discussion of the oscilloscope.

MATERIALS NEEDED

circuit board, prepunched, 6 × 8 in. (152 × 203 mm)
rectifier transformer, 115-volt primary, 12.6-volt secondary with center tap; 1-ampere, full-load secondary current
4 silicon diodes, 1 ampere, 200 PIV
electrolytic capacitor, 100 μF, 50 WVDC
2 carbon-composition or film resistors, 560 ohms, 2 watts
1 switch, spst
6 Fahnestock clips, 1 in. (254 mm), or solderless spring terminals
volt-ohm-milliammeter
oscilloscope

Procedure

1. Wire the rectifier circuits shown in Fig. 20–19 at A. Leave the capacitor switch in the open position.

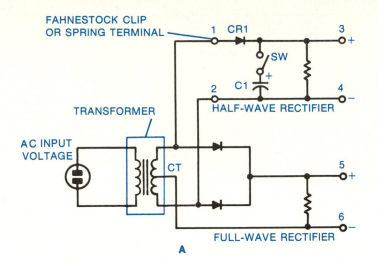

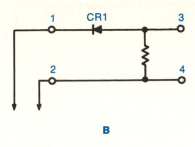

Fig. 20–19. Circuits for Learning by Doing No. 18, "The Diode as a Rectifier"

2. With the help of your teacher, connect the oscilloscope to the ac input of the half-wave-rectifier circuit (points 1 and 2). Observe the waveform.

3. Connect the oscilloscope to the output of the half-wave-rectifier circuit (points 3 and 4). Observe this waveform. Describe the difference between the input and output waveforms.

4. Close the capacitor switch. Again observe the output waveform. Describe the effect that the filtering action of the capacitor has on the output-voltage waveform.

5. Measure the voltage across points 3 and 4. Open the capacitor switch. Again measure the voltage across points 3 and 4. In addition to its filtering action, what effect does the capacitor have on the value of the output voltage?

6. Observe the waveform at the output of the half-wave-rectifier circuit (points 3 and 4). Observe the waveform at the output of the two-diode full-wave-rectifier circuit (points 5 and 6). Describe the difference between these waveforms. Which of the output voltages do you think could be more easily filtered to produce a smooth output voltage? Why?

NOTE: Remove C1 from the circuit before proceeding to step 7. Otherwise, C1 would be damaged if the switch were accidentally closed.

7. Reverse the connections of diode CR1 in the half-wave-rectifier circuit (Fig. 20-19 at B). What effect does this have on the polarity of the output voltage across points 3 and 4? Why?

Test your knowledge by writing, on a separate sheet of paper, the word or words that most correctly complete the following statements:

1. The most widely used semiconductor materials are _____ and _____.
2. Impurities are added to pure semiconductor materials by a process called _____.
3. Atoms of semiconductors have _____ valence electrons.
4. The bonds between atoms in a crystal-lattice structure are called _____, or _____, bonds.
5. An n-type semiconductor contains many _____ electrons.
6. The lack of an electron in a crystal-lattice structure is called a _____.
7. A p-type semiconductor material contains many _____.
8. A hole can be considered as a _____ charged current carrier.
9. The two electrodes of a typical semiconductor diode are called the _____ and the _____.
10. A diode will conduct when it is _____-biased.
11. The highest reverse-bias voltage that can be applied to a diode is its _____ rating.

FOR REVIEW AND DISCUSSION

1. Define a semiconductor. Name two semiconductor materials.
2. What is meant by a solid-state device?
3. Describe a crystal-lattice structure.
4. Define n-type and p-type semiconductor materials.
5. Explain how n-type and p-type semiconductor materials are formed by the process of doping.
6. What is a p-n junction?
7. Explain the conditions of forward and reverse bias in a semiconductor diode.
8. Describe a basic half-wave-rectifier circuit.
9. Describe a basic full-wave-rectifier circuit.
10. What is the purpose of a filter capacitor in a rectifier circuit?
11. What is meant by the peak inverse voltage of a diode?
12. How are diodes tested?

INDIVIDUAL-STUDY ACTIVITIES

1. Demonstrate the basic rectifying action of a diode.
2. Demonstrate the difference between a half-wave- and a full-wave-rectifier circuit.
3. Demonstrate how diodes can be tested with an ohmmeter and with a diode tester.

Unit 21　Transistors

Transistors are semiconductor devices that can be used to control current or amplify an input voltage or current. In a transistor radio, for example, transistors are used to amplify the very small signal existing across the antenna coil. Thus a much larger signal, strong enough to operate the radio's loudspeaker, is created.

TRANSISTORS AND ELECTRON TUBES

In most electrical and electronic devices, transistors have almost completely replaced electron tubes. Transistors perform the same functions as electron tubes. However, they

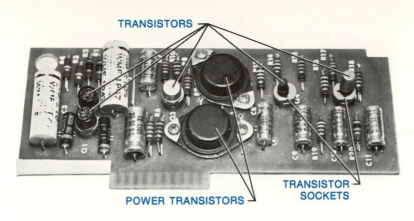

TRANSISTORS

POWER TRANSISTORS

TRANSISTOR SOCKETS

Fig. 21–1. Transistors used in an amplifier circuit (*Photograph by David C. Richards*)

also have several important advantages. They are smaller, thus making more-compact products possible (Fig. 21–1). The transistor is also more rugged than the electron tube. It usually provides better performance over a longer period of time. Most important, the transistor usually needs much less current and voltage to work. This saves energy. For example, a 12-volt automobile radio using tubes draws about 2.5 amperes. A similar transistor automobile radio draws only a fraction of an ampere. The lower power needs of transistor circuits have made possible small, lightweight, portable products that work for long periods of time on small, low-voltage batteries.

KINDS OF TRANSISTORS

The two most common kinds of transistors are the *n-p-n transistor* and the *p-n-p transistor*. They are often called *bipolar transistors*. These transistors are made by combining n-type and p-type materials. The materials are arranged as two diodes connected in a "back-to-back" fashion. This arrangement forms three regions called the *emitter*, the *base*, and the *collector*. These regions are identified by the symbols E, B, and C (Fig. 21–2). The regions of a transistor are joined to wire leads or to pins, which connect the transistor to the circuit (Fig. 21–3).

Transistors enclosed in metal cases often have a fourth lead known as the *shield lead*. This lead is internally attached to the case and connected to a common point in a circuit. The metal case shields the transistor from nearby electrostatic and magnetic fields.

Symbol Interpretation. There is a convenient way to remember whether the symbol for a junction transistor represents the n-p-n or the p-n-p type. Note in what direction the arrow that stands for the emitter is pointing. If the arrow

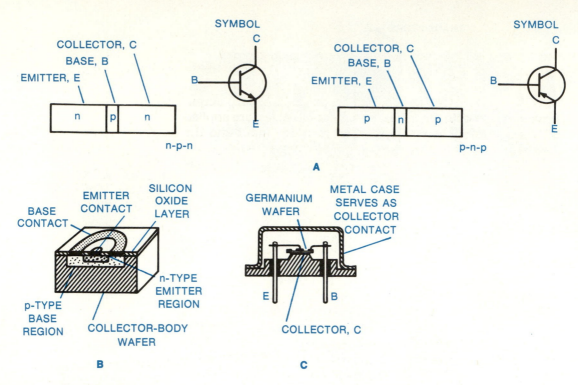

Fig. 21–2. Transistor construction: (A) basic arrangement of materials; (B) typical construction of low-power, silicon-diffused n-p-n planar transistor; (C) typical construction of germanium alloy p-n-p power transistor

points away from the base, it can be thought of as "not pointing n." The symbol thus represents an n-p-n transistor. If the arrow points toward the base, it can be thought of as "pointing n." This symbol thus represents a p-n-p transistor (see Fig. 21–2 at A).

Identification. Most transistors are identified by a number-letter code, such as 2N, followed by a series of numbers as, for example, 2N104, 2N337, and 2N2556. Other transistors are identified by a series of numbers or by a combination of numbers and a letter, such as 40050, 40404, and 4D20.

Transistor Manuals. Transistor identification codes do not indicate whether the device is of the n-p-n or the p-n-p type. Such technical data are found in *transistor manuals*. These manuals also give information about the use of transistors in different kinds of circuits.

Substitution. N-p-n and p-n-p transistors can never be directly substituted for each other. However, transistors can usually be replaced with other transistors that have different identifications but are designed to perform the same function. A listing of transistors that can substitute for other transistors is given in *transistor substitution guides* or *manuals.*

OPERATION OF THE TRANSISTOR

An *amplifier* is a device or circuit that can increase the value of a voltage or current. A transistor can act as an amplifier device. This is because a small signal in its input circuit can control a much larger signal in its output circuit. This action is called *power gain*. Voltages, called *bias voltages,* are applied to those parts of the transistor that form the input and the output circuits. When the bias voltage of the input circuit is varied, the transistor behaves as a variable resistor, the resistance of which increases or decreases.

The n-p-n Transistor. The most common transistor amplifier circuit is called a *common-emitter circuit* (Fig. 21–4 at A). It is so named because the emitter is common to, or is a part of, both the input and output circuits.

In this circuit, the base-emitter p-n junction of the input circuit is forward-biased by battery B1. Battery B2, which has a higher voltage than battery B1, is connected across the transistor from the collector to the emitter in the output circuit. This makes the collector more positive with respect to the emitter than the base is positive with respect to the emitter. The base is then negative with respect to the collector. The base-collector p-n junction is reverse-biased.

With the base-emitter p-n junction forward-biased, free electrons in the emitter move toward the junction. At the same time, holes within the base move toward the junction (Fig. 21–4 at B). Some of the holes and electrons combine in the area near the junction. This produces a current in the external base-emitter circuit. The base region is very thin. Thus, most of the electrons that move toward the base-emitter junction pass through it and into the collector. There these electrons are attracted to the positive terminal of battery B2.

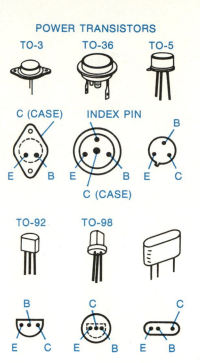

POWER TRANSISTORS

Fig. 21–3. Common n-p-n and p-n-p transistor cases and a base-view identification of leads

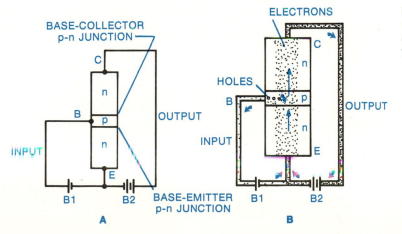

Fig. 21–4. A common-emitter n-p-n transistor amplifier circuit: (A) circuit diagram; (B) electron flow

189

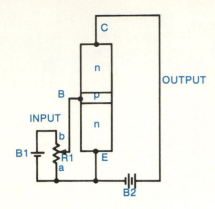

Fig. 21-5. Basic method of controlling the output current in a transistor circuit

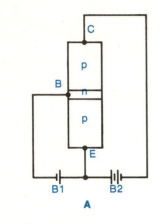

A

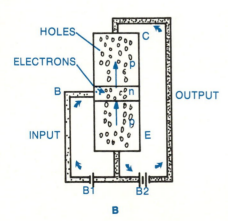

B

Fig. 21-6. A common-emitter p-n-p transistor amplifier circuit: (A) circuit diagram; (B) electron flow

At the same time, electrons from the negative terminal of battery B2 move into the emitter. The current in the external collector-emitter output circuit is much greater than the current in the input circuit. The circuit therefore acts as an amplifier.

Control of Output Current. There is another very important feature of the transistor amplifier circuit. The output current can be controlled by varying the current in the input circuit. How this can be done is shown in Fig. 21-5. Here a potentiometer R1 is connected across cell B1 in the input circuit. As the sliding arm moves toward point a, the base-emitter p-n junction becomes less and less forward-biased. As a result, the current in the input circuit decreases. This causes a greater decrease of current in the output circuit.

The p-n-p Transistor. When a p-n-p transistor is used in an amplifier circuit, the polarities of the bias voltages applied to the base-emitter and the base-collector p-n junctions are opposite to those in the n-p-n transistor circuit (Fig. 21-6 at A). As in the n-p-n transistor circuit, this causes the base-emitter junction to be forward-biased and the base-collector junction to be reverse-biased.

In this circuit, holes within the emitter move toward the base-emitter p-n junction. Electrons within the base also move toward the junction. Some of the holes and electrons combine in the area near the junction. However, most of the holes pass through the base and into the collector (Fig. 21-6 at B). Here the holes are attracted to the end of the collector connected to the negative terminal of battery B2. As this continues, electrons from the negative terminal of the battery enter the collector to combine with the holes. At the same time, an equal number of electrons within the emitter break their electron-pair bonds and enter the positive terminal of the battery. This again produces a much greater current in the external output circuit than is in the input circuit.

PRACTICAL AMPLIFIER CIRCUITS

A simple, yet practical, one-transistor amplifier circuit is shown in Fig. 21-7. Note that, in this circuit, the bias voltages applied to the p-n-p transistor are supplied by a single source of energy, battery B1. Resistor R1 acts as a current limiter to provide the proper base-bias current. As in any transistor amplifier circuit, the base-collector p-n junction is reverse-biased.

The input signal from an audio-frequency source, such as a microphone, is applied to the input circuit. It is coupled to the

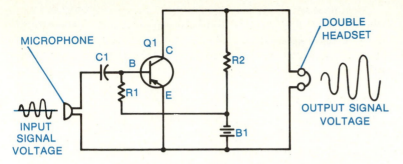

Fig. 21–7. Schematic diagram of a one-transistor common-emitter amplifier circuit. The transistor is identified by the letter Q.

base of the transistor through capacitor C1. As the input-signal voltage varies in value and reverses in polarity, it aids and opposes the forward bias applied to the base-emitter junction of the transistor by the battery. This causes the current in the input circuit to vary in direct proportion to the audio signal. It causes the larger output current to vary in the same way.

As the current in the output circuit changes in value, the voltage drop across R2 in the output circuit also changes in value. The transistor, the resistor R2, and the battery form a series circuit. Therefore, as the output current increases, the voltage drop across resistor R1 increases. The output voltage across the transistor, from collector to emitter, decreases. When the current in the output circuit decreases, the voltage drop across R1 decreases. The output voltage increases. The changes in the output-signal voltage are greater than those in the input-signal voltage. Thus, the input signal has been amplified. The amplified output voltage is then applied to the headset. There it is changed into sound waves. In more powerful amplifier circuits, two or more transistors are used. They are connected so that the output signal of each transistor is applied to the input circuit of the next transistor. Each transistor acts as an amplifier. Thus, this provides a much greater voltage amplification of the original input signal.

WORKING WITH TRANSISTORS

Although transistors are rugged devices, they can be damaged. Too much heat can permanently damage the crystal-lattice structure of the material.

Heat Sink. Transistors that must conduct large amounts of current are often mounted on *heat sinks* to keep them from overheating (Fig. 21–8). A heat sink absorbs heat from a transistor and dissipates it, or throws it off, more quickly than the transistor itself could. This lets the transistor work at a lower temperature.

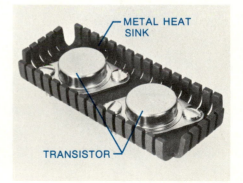

Fig. 21–8. Power transistors mounted on a heat sink (*International Electronic Research Corporation*)

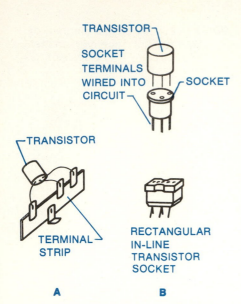

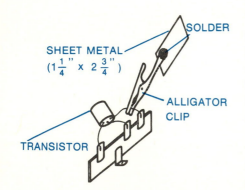

Fig. 21–9. Methods of connecting transistors: (A) leads soldered to circuit; (B) transistor plugged into socket

Fig. 21–10. Using a simple heat sink made of a small piece of sheet metal and an alligator clip

Connections. Transistors are connected to a circuit in one of two ways. Either their leads are soldered to circuit terminals, or they are plugged into transistor sockets (Fig. 21–9). Sockets make it easier to put in or take out transistors. They also eliminate the danger of overheating the transistors. Overheating can happen when the leads are soldered to the circuit.

When soldering a transistor lead, it is best to use a soldering iron that does not produce more heat than is needed to do the job. A soldering iron rated at 30 to 50 watts is usually hot enough. A heat sink should always be attached to a transistor lead that is being soldered. This can be the jaws of a longnose pliers or some other kind of heat sink (Fig. 21–10). A transistor should always be taken out of a socket when a socket terminal is being soldered or unsoldered.

Bias Voltages. Never put a transistor into a circuit until you are sure the right values of bias voltages will be applied to its terminals. Excessive voltage applied to a p-n junction of a transistor will cause the transistor to conduct more current than it can handle safely. A transistor should never be taken out of, or put into, a live circuit. Otherwise, damaging surges of current can pass through the transistor.

Correct Polarity. A transistor can be damaged if bias voltages of the wrong polarity are applied to it. The danger of such damage is much less with common-emitter circuits. Nevertheless, it is always a good idea to check carefully to see that the transistor is properly connected before applying voltage to the circuit. This is particularly true with circuits that have been built in the shop and are being used for the first time.

TESTING TRANSISTORS

Transistors often become defective because of overheating. Overheating is caused by too much current flowing through the base-to-emitter diode section, the base-to-collector diode section, both these sections, or from emitter to collector. As with diodes, overheating seriously disrupts the crystal-lattice structure. This may result in either an open or a shorted transistor.

Transistor Tester. One kind of transistor tester is shown in Fig. 21–11. Such a tester indicates transistor leakage and gain. Too much leakage occurs when a transistor conducts more than a normal amount of reverse-bias current between any two of its electrodes. The *current gain* of a transistor, referred to as *beta* (β), is the ratio of the collector current to the base current in a common-emitter circuit. Most transistor testers can be used to make either in-circuit or out-of-circuit tests.

Testing with an Ohmmeter. A transistor can be tested with an ohmmeter. If the transistor being tested is wired into a circuit, it should be isolated from the circuit by disconnecting its base lead. The ohmmeter, adjusted to a low or medium ohms range, is connected between the base and the emitter. The resistance reading is noted. The ohmmeter leads to the base and to the emitter are then reversed. If the resistance between the base and the emitter is significantly higher in one direction than in the other, the base-emitter diode section of the transistor can be assumed to be in good condition (Fig. 21–12). To test the base-collector diode section, the same procedure is followed, with the ohmmeter connected to the base and to the collector. As a final test, the ohmmeter is connected between the emitter and the collector.

If any of these tests show direct continuity, the transistor probably has a short. If any of the tests show infinite resistance, the transistor is open. If there is some doubt about the accuracy of the ohmmeter test, recheck the transistor with a transistor tester.

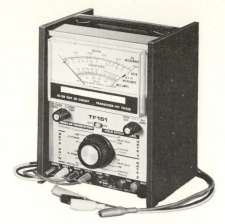

Fig. 21–11. Transistor tester (*Sencore, Inc.*)

 SAFETY
Some ohmmeters, on the R × 1 range, can supply enough voltage to damage some transistors. Also, some ohmmeters supply enough voltage on the highest range to break down transistor junctions.

LEARNING BY DOING

19. The Transistor in Action. The following procedure will let you see the effect that the base current has on the output, or collector, current of a transistor. This action makes it possible for the transistor to act as an amplifier.

Fig. 21–12. Testing n-p-n and p-n-p transistors with an ohmmeter

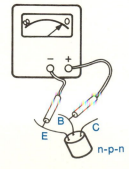
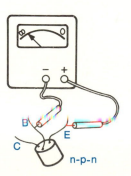
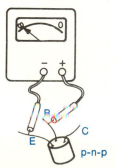
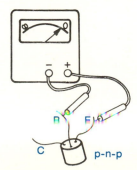

193

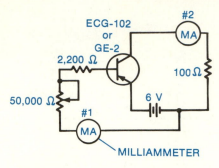

Fig. 21–13. Diagram for Learning by Doing No. 19, "The Transistor in Action"

Table 21–1. Table for Learning by Doing No. 19, "The Transistor in Action"

I_B (mA)	I_C (mA)
0.2	
0.3	
0.4	
0.5	
0.6	
0.7	
0.8	
0.9	
1.0	

circuit board, prepunched, 4 × 5 in. (102 × 127 mm)
transistor, p-n-p, type ECG-102, GE-2, or equivalent
resistor, carbon-composition or film, 100 ohms, 1 watt
resistor, carbon-composition or film, 2,200 ohms, ½ watt
potentiometer, 50,000 ohms, ½ watt
10 Fahnestock clips, 1 in. (25.4 mm), or solderless
spring terminals
6-volt battery or a low-voltage dc power supply with a
current rating of at least 0.05 ampere
2 milliammeters (or volt-ohm-milliammeters)

Procedure

1. Wire the circuit shown in Fig. 21–13.
2. Prepare a table like Table 21–1 on a separate sheet of paper.
3. Adjust the potentiometer to produce a base current I_B of 0.2 mA, as indicated by meter 1. Record the value of the resulting collector current I_C, as indicated by meter 2, in the table.
4. Repeat step 3 with the potentiometer adjusted to produce each of the base currents given in the table.
5. Explain why each adjustment of the potentiometer produces a different value of base current.
6. Prepare a graph that shows the relationships between the base currents and the collector currents.
7. Describe the effect that each small change of the base current has on the corresponding change in the value of the collector current. Explain the operation of the transistor that makes this possible.

NOTE: The ratio of a transistor's collector current to the base current is known as the *current gain*, or *beta* (β), of the transistor.

20. Transistor Amplifier Circuit. This activity lets you see how a transistor amplifier circuit works. It will also give you experience in using an oscilloscope and an audio-signal generator. Information about the oscilloscope and the signal generator is given in Unit 47.

circuit board, prepunched, 4 × 5 in. (102 × 127 mm)
transistor, p-n-p, type SK 3025 or equivalent
capacitor, electrolytic, 25 µF, 12 WVDC
resistor, carbon-composition or film, 100 ohms, 1 watt
resistor, carbon-composition or film, 22,000 ohms, ½ watt

Fahnestock clips, 1 in. (25.4 mm), or solderless spring
   terminals
6-volt battery or a low-voltage dc power supply with a
   current rating of at least 0.05 ampere
audio-signal generator
oscilloscope

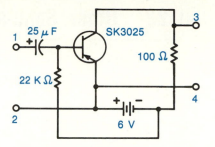

Fig. 21–14. Diagram for Learning by Doing No. 20, "Transistor Amplifier Circuit"

Procedure

1. Wire the circuit shown in Fig. 21–14.
2. Connect the signal generator to input terminals 1 and 2. Adjust it to produce a signal with a frequency of 400 Hz.
3. Connect the oscilloscope to terminals 1 and 2. Adjust the signal generator and the oscilloscope so that an undistorted sine waveform about $\frac{1}{4}$ in. (6.4 mm) high is observed.
4. Disconnect the oscilloscope from terminals 1 and 2. Connect it to output terminals 3 and 4.
5. The voltage gain, or amplification, of the amplifier circuit can be estimated by comparing the peak-to-peak height of the input waveform to the height of the output-voltage waveform. By using this method, what do you estimate the voltage gain of the amplifier circuit to be?

SELF-TEST

Test your knowledge by writing, on a separate sheet of paper, the word or words that most correctly complete the following statements:

1. N-p-n and p-n-p transistors are often called _____ transistors.
2. The three electrodes of an n-p-n or a p-n-p transistor are the _____, the _____, and the _____.
3. The arrow on the emitter of a p-n-p transistor symbol points _____ the base.
4. A transistor amplifier circuit in which the emitter is a part of both the input and output circuits is called a _____ circuit.
5. An n-p-n or a p-n-p transistor is operated with the base-collector p-n junction _____ biased.
6. The input signal to a transistor amplifier stage is applied between the _____ and the _____ of the transistor.

FOR REVIEW AND DISCUSSION

1. What is a transistor?
2. What is the purpose of the shield in a transistor?
3. What is meant by an amplifier device or circuit?
4. Explain the basic operation of an n-p-n transistor.
5. Explain the basic operation of a p-n-p transistor.
6. What is the purpose of a heat sink?
7. Describe two ways of connecting transistors into circuits.
8. What precautions must be observed when soldering transistor leads?
9. What is meant by the beta of a transistor?

INDIVIDUAL-STUDY ACTIVITY

Give a demonstration showing how transistors can be tested with an ohmmeter or with a transistor tester.

Unit 22 Other Solid-state Devices

SYMBOL

A

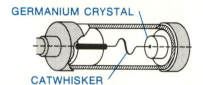

GERMANIUM CRYSTAL

CATWHISKER

B

Fig. 22–1. Germanium "crystal" detector diode: (A) typical diode; (B) interior construction (*Photo courtesy of Ohmite Manufacturing Company*)

Fig. 22–2. Diode AM detector circuit

Fig. 22–3. Silicon controlled rectifiers: (A) stud-mounted; (B) plastic-case (*International Rectifier*)

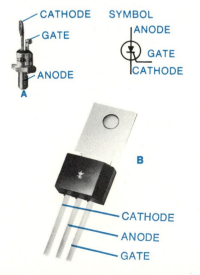

CATHODE

GATE

ANODE

SYMBOL

ANODE

GATE

CATHODE

A

B

CATHODE

ANODE

GATE

In addition to the popular semiconductor silicon diodes and junction transistors, there are other solid-state rectification and current-control devices. These include special forms of rectifiers, the field-effect transistor, and photocells. Another important solid-state product is the *integrated circuit*. An integrated circuit contains a large number of components built into tiny chips, or wafers, of certain materials.

DETECTOR DIODE

The *detector diode* is a semiconductor device used in radio and television circuits to produce a rectifying action known as *audio detection*, or *demodulation* (Fig. 22–1). This results in the separation of an audio signal (voice or music) from a high-frequency carrier signal. Carrier signals are used in radio, television, and other communications systems. In order for a carrier signal to carry information, it is *modulated*, or made to vary, according to varying voice or music signals. These signals are mixed with or placed on the carrier signal at a broadcasting station.

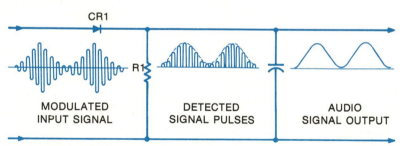

CR1

R1

MODULATED
INPUT SIGNAL

DETECTED
SIGNAL PULSES

AUDIO
SIGNAL OUTPUT

A basic amplitude-modulated (AM) detector circuit is shown in Fig. 22–2. In this circuit, the modulated carrier signal is applied to detector diode CR1. This diode acts as a rectifier. As a result, a pulsating direct current passes through resistor R1. The value of this current varies according to the variations of the modulated carrier signal. This causes a voltage that represents the audio-modulating signal to be present across the resistor and the output of the detector circuit.

SILICON CONTROLLED RECTIFIER

The *silicon controlled rectifier* (SCR) is a four-layer semiconductor device equipped with three external connections. These are the *anode,* the *cathode,* and a control electrode called the *gate* (Fig. 22–3). Like the silicon diode, the silicon controlled rectifier conducts current in only one direction, from cathode to anode. However, the exact time at which it will begin to conduct, when an alternating voltage is applied between the anode and the cathode, can be controlled by applying a positive trigger voltage to the gate.

An example of a silicon controlled rectifier circuit is shown in Fig. 22–4. In this circuit, the amount of control voltage applied to the gate can be varied by adjusting variable resistor R1. As the control voltage is varied, the silicon controlled rectifier can be made to conduct early or late during the positive alternation. Early conduction produces the largest load current. Late conduction gives less load current. Because of this current-control feature, silicon controlled rectifiers are used in lamp-dimmer circuits and in speed controls for motors.

The silicon controlled rectifier is one of a group of devices commonly called *thyristors*. These are generally used in switching or current-control circuits in which a trigger voltage is applied to their control electrode.

SELENIUM RECTIFIER

The selenium rectifier performs the same rectifying function as the silicon diode. It usually is found in older products. Modern radios, amplifiers, and television sets generally use the silicon diode.

A simple selenium rectifier cell consists of a metal baseplate (usually aluminum). One side of the baseplate is coated with a thin layer of the element selenium. A special alloy is formed over the selenium. This alloy provides for a uniform contact over the entire surface. It forms the anode for the rectifier (Fig. 22–5 at A). The cell will conduct current only

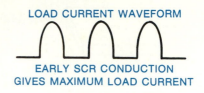

LOAD CURRENT WAVEFORM

EARLY SCR CONDUCTION
GIVES MAXIMUM LOAD CURRENT

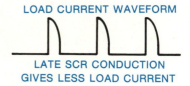

LOAD CURRENT WAVEFORM

LATE SCR CONDUCTION
GIVES LESS LOAD CURRENT

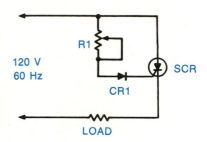

Fig. 22–4. Simple silicon controlled rectifier current-control circuit

Fig. 22–5. Selenium rectifier: (A) basic parts; (B) commercial selenium rectifier assemblies

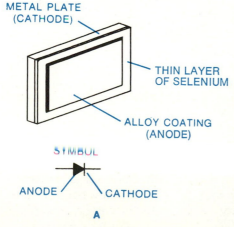

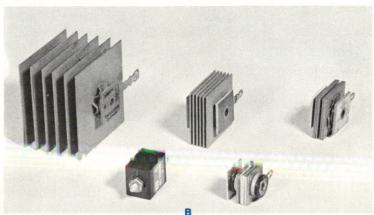

197

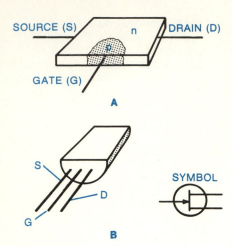

Fig. 22–6. An n-channel field-effect transistor: (A) basic construction; (B) transistor in TO–92 case

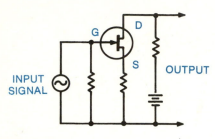

Fig. 22–7. Field-effect-transistor amplifier circuit

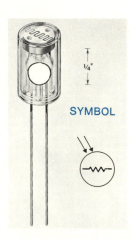

Fig. 22–8. Broad-area cadmium sulfide photoconductive cell (*Radio Corporation of America*)

from the metal baseplate to the selenium coating, or from cathode to anode. In practical selenium rectifiers, several cells are stacked and are connected in series to form the complete rectifier assembly (Fig. 22–5 at B).

FIELD-EFFECT TRANSISTOR

In one common kind of *field-effect transistor* (FET), the control electrode is the *gate*. This is a region of p-type semiconductor material that forms an island within a block of n-type semiconductor material. The two other electrodes, the *source* and the *drain*, are the ends of the n-type material (Fig. 22–6).

When a direct voltage that makes the gate negative with respect to the source is applied to the transistor, an electrostatic field is created about the gate. This field acts on free electrons within the n-type material and limits current flow through the transistor from source to drain. As the voltage between the gate and the source is decreased, making the gate less negative with respect to the source, the transistor conducts more current from source to drain.

An amplifier circuit with one field-effect transistor is shown in Fig. 22–7. In this circuit, the input signal causes variations of the voltage applied between the gate and the source. As a result, greater variations of voltage appear across the output circuit. Therefore, the transistor acts as a voltage amplifier. Field-effect transistors are used in several kinds of circuits, including those found in audio amplifiers, electronic timers, and test instruments.

PHOTOCELLS

Transducers are a group of devices that, in general, convert one form of energy into another form of energy or into a variation of some electrical quantity. *Photocells*, one of these devices, react to light energy. In some photocells, known as *photoconductive cells*, light energy causes a decrease in the resistance of the cells. Other photocells, known as *photovoltaic cells*, convert light energy into electric energy by generating a voltage. Photovoltaic cells are often called *solar cells*.

Photoconductive Cell. A typical cadmium sulfide photoconductive cell is shown in Fig. 22–8. Cadmium sulfide is a semiconductor compound consisting of cadmium and sulfur.

As light strikes the active material of the cell, some of the valence electrons in atoms of the material gain enough energy to escape from their parent atoms. They become free electrons. As the intensity of the light increases, more and more free electrons are made available. The resistance between the terminals of the cell decreases.

The use of a photoconductive cell as a light-sensing device is shown in Fig. 22–9. As light strikes the photocell, its resistance decreases. This allows more current to flow in the circuit. The reading of the ammeter is, at any instant of time, directly related to the intensity of the light.

Photoconductive cells are often used in a system that includes a light source and a relay. In such a system, when the path between the light source and the photocell is interrupted, the current in the relay circuit decreases. The relay, which acts as an on-off switch, is used to control counters, alarm systems, inspection or supervision equipment, and other devices.

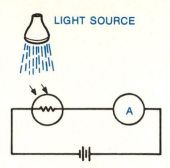

Fig. 22–9. This circuit is an example of how a photoconductive cell can be used as a light-sensing device.

THERMISTORS

Thermistors, or thermal (heat) resistors, are devices that are usually designed so that their resistance decreases with an increase in temperature. They are made of compounds called *oxides.* Oxides are combinations of oxygen and metals, such as manganese, nickel, and cobalt. Thermistors come in various shapes, some of which are shown in Fig. 22–10.

Since the resistance of a thermistor varies with temperature, it operates as a heat-controlled resistor. Because of this, a thermistor can be used as a *heat sensor.* This is a device that converts a change in temperature into a corresponding change in the value of current in a circuit.

An example of such a circuit used for temperature measurement is shown in Fig. 22–11. The thermistor is connected in series with an ordinary dry cell and an ammeter. As the temperature around the thermistor changes, the value of the current also changes. The meter scale can be calibrated, or marked off, in degrees so that a direct temperature reading can be made.

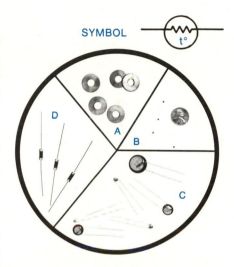

Fig. 22–10. Thermistors: (A) washer; (B) bead; (C) disk; (D) rod (*Keystone Carbon Company*)

THERMOCOUPLE

A *thermocouple* is a solid-state device used to change heat energy into voltage. It consists of two different kinds of metals joined at a junction (Fig. 22–12 at A). When the junction is heated, the electrons in one of the metals gain enough energy to become free. These electrons then move across the junction and into the other metal. This movement produces a voltage across the terminals of the thermocouple. Several combinations of metals are used to make thermocouples. These include iron and constantan, copper and constantan, and antimony and bismuth.

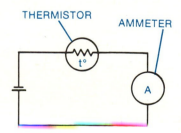

Fig. 22–11. A simple circuit using a thermistor as a heat-sensing or temperature-measurement device

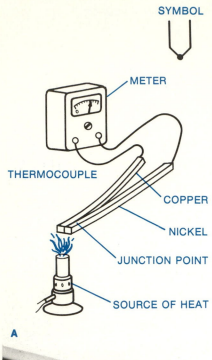

SYMBOL

METER

THERMOCOUPLE

COPPER

NICKEL

JUNCTION POINT

SOURCE OF HEAT

A

B   THERMOCOUPLE

Fig. 22–12. The thermocouple: (A) basic construction and operation; (B) typical pyrometer thermocouple (*Photo courtesy of Minneapolis-Honeywell Regulator Company*)

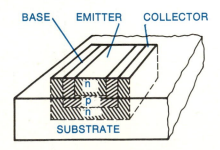

BASE   EMITTER   COLLECTOR

n
p
n

SUBSTRATE

Fig. 22–13. Basic fabrication of a monolithic n-p-n transistor

Thermocouples are used as heat sensors in thermometer-like instruments called *pyrometers* (Fig. 22–12 at B). In a pyrometer, the voltage produced by a thermocouple causes current to pass through an electric meter. The meter is calibrated to indicate temperature. A thermocouple can be put in a furnace. As the temperature of the furnace increases, the voltage produced by the thermocouple also increases. As a result, more current passes through the meter. The meter then indicates the increase of current as a higher temperature. Temperatures ranging from 2700 to above 10800°F (1500 to 6000°C) can be measured very accurately with pyrometers.

PELTIER EFFECT

A change of temperature occurs at the junction of two different conducting materials when current passes through the junction. The change of temperature (increase or decrease) depends on the direction of the current. This is known as the *Peltier effect.* It is used with various kinds of semiconductor junctions to operate some heating and cooling systems.

INTEGRATED CIRCUITS

The *integrated circuit* (IC) represents a new way of building circuits. In such a circuit, components such as diodes, transistors, resistors, and capacitors either are formed into a common block of a base material, called the *substrate,* or are formed on the surface of the substrate. This process forms a *monolithic,* or single, block of material.

In the monolithic integrated circuit, diodes and transistors are formed into the substrate by a diffusion process. This process causes impurities to spread into given areas of the substrate. This creates p-type and n-type regions within the substrate. These regions then make up the sections of the diodes and transistors (Fig. 22–13).

In the thick- or thin-film integrated circuit, components such as resistors and capacitors are formed by placing or depositing certain materials on the substrate (Fig. 22–14). Hybrid integrated circuits are made up of components of both the monolithic type and the film type.

An integrated circuit chip is shown in Fig. 22–15. Circuit assemblies of this size or smaller may have dozens of components. For this reason, integrated circuits have the very important advantages of being very small and light. This is often referred to as *microminiaturization.*

In its complete form, an integrated-circuit assembly is packaged in some kind of enclosure (Fig. 22–16). Each of

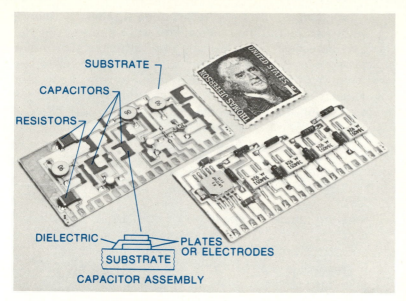

LABELS ON IMAGE 1:
SUBSTRATE
CAPACITORS
RESISTORS
DIELECTRIC
PLATES OR ELECTRODES
SUBSTRATE
CAPACITOR ASSEMBLY

Fig. 22–14. Thick-film integrated circuits (*Raytheon Company*)

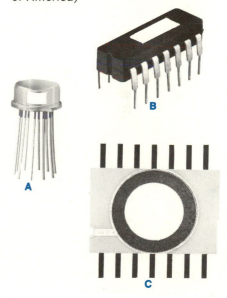

INTEGRATED CIRCUIT

Fig. 22–15. Integrated circuit on a drop of water (*Radio Corporation of America*)

these packages may contain the components of a specific circuit, such as an amplifier, or of a combination of circuits.

LEARNING BY DOING

21. A Silicon Controlled Rectifier Circuit. As you have learned, the silicon controlled rectifier is a very useful current-control device. The following procedure will let you become more familiar with the silicon controlled rectifier and see how it works.

MATERIALS NEEDED

circuit board, prepunched, 3 × 4 in. (76 × 102 mm)
1 silicon controlled rectifier, International Rectifier no. 106B1 or equivalent
1 pilot lamp, no. 40
1 miniature lamp socket, screw-base
1 dry cell, 1½ volts
6-volt ac power supply or transformer with a 6.3-volt secondary winding

Procedure

1. Connect the circuit shown in Fig. 22–17. What is the condition of the lamp at this time?
2. Touch the lead extending from the positive terminal of the dry cell to the gate (G) of the silicon controlled rectifier. What effect does this have on the lamp?
3. Explain how the circuit works.

A

B

C

Fig. 22–16. Integrated-circuit packages: (A) TO–5 case; (B) dual inline case; (C) flatpack (*Radio Corporation of America*)

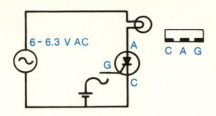

6 - 6.3 V AC

C A G

A

G

C

Fig. 22–17. Diagram for Learning by Doing No. 21, "A Silicon Controlled Rectifier Circuit"

22. Experiences with the Photoconductive Cell. By following the procedure outlined below, you will be able to see the effect an increase in light intensity has on the resistance of a typical photoconductive cell. You can then use the cell to see how it controls a circuit. The relay used in this circuit is a magnetically controlled switch. For more information about relays, see Unit 14, "Electromagnetism."

MATERIALS NEEDED

circuit board, prepunched, 4 × 6 in. (102 × 152 mm)
1 photoconductive cell, Clairex no. CL704M or equivalent
1 resistor, carbon-composition or film, 10,000 ohms, ½ watt
1 potentiometer, 100,000 ohms, ½ or 1 watt
2 transistors, p-n-p, SK-3004, GE-2, or equivalent
1 miniature sensitive relay, coil resistance 500 ohms, sensitivity about 40 milliwatts, spdt, Radio Shack no. 275 B 004 or equivalent
1 pilot lamp, no. 40
1 miniature lamp socket, screw-base
a small desk lamp with a 60-watt incandescent light bulb
9-volt transistor-radio battery or a dc power supply
6-volt battery or a dc power supply with a current rating of at least 200 mA
volt-ohm-milliammeter

Procedure

1. On a sheet of paper, prepare a table like Table 22–1.
2. Connect the photoconductive cell across the VOM. Using the instrument as an ohmmeter, cover the cell with your hand. Measure its resistance. Record this resistance in the dark-resistance column of the table.
3. Remove your hand from the cell. Measure its resistance. Record this resistance in the normal-resistance column of the table.
4. Turn the desk lamp on. Move it toward the cell. As you do this, you will note that the resistance of the cell gradually decreases to a minimum value. Record this resistance in the high-intensity column of the table.

Table 22–1. Table for Learning by Doing No. 22, "Experiences with the Photoconductive Cell"

Dark Resistance	Normal Resistance	High-intensity Resistance

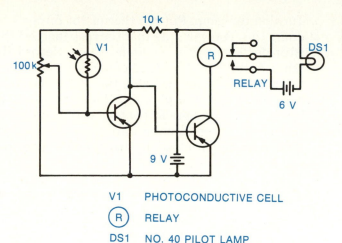

Fig. 22–18. Diagram for Learning by Doing No. 22, "Experiences with the Photoconductive Cell"

V1	PHOTOCONDUCTIVE CELL
(R)	RELAY
DS1	NO. 40 PILOT LAMP

5. Connect the circuit shown in Fig. 22–18.
6. Put the 60-watt lamp about 1 ft (300 mm) from the cell. Adjust the potentiometer until the relay contacts in the pilot-lamp circuit are closed. When the relay contacts close, the pilot lamp will light.
7. Turn off the 60-watt lamp and all other nearby lamps. This will cause the relay contacts to open. The pilot lamp will go out.
8. By moving the lighted 60-watt lamp closer to and farther away from the cell, you will note that the pilot lamp can be made to go on and off. Explain the reason for this.

23. The Thermistor as a Current-control Device. By following the procedure below, you will be able to see the effect of a temperature increase on the resistance of a typical thermistor. You can then use the thermistor to see how it controls another circuit. The relay used in this circuit is a magnetically controlled switch. For more information on relays, see Unit 14.

MATERIALS NEEDED

With the exception of the photoconductive cell, use the materials listed for *Learning by Doing* No. 22, "Experiences with the Photoconductive Cell." In place of the photoconductive cell, use 1 thermistor, Fenwal no. RB41L1 or equivalent.

Procedure

1. On a sheet of paper, prepare a table like Table 22–2.
2. Connect the volt-ohm-milliammeter, operated as an ohmmeter, to the thermistor. Measure the resistance of the thermistor at ordinary room temperature. Record this resistance in the normal-resistance column of the table.

Table 22–2. Table for Learning by Doing No. 23, "The Thermistor as a Current-Control Device"

Normal Resistance	Hot Resistance

3. Turn on the lamp. Put the thermistor on the bulb. After doing this, you will note that the resistance of the thermistor decreases to a minimum value. Record this resistance in the hot-resistance column of the table.

4. Connect the circuit shown in Fig. 22–18. Use the thermistor in place of the photoconductive cell.

5. Bring the thermistor in contact with the 60-watt lamp. Adjust the potentiometer until the relay contacts in the pilot-lamp circuit close. In this condition, the pilot lamp will light.

6. Turn the 60-watt lamp off. Let the thermistor cool. After a short period of time, the relay contacts in the pilot-lamp circuit will open. The lamp will go out. Explain why the relay contacts open.

24. Making Thermocouples. This activity shows how a voltage can be generated by heating the junction of two different kinds of metals.

MATERIALS NEEDED

1 piece each of sheet copper, tinplate, galvanized steel, brass, and zinc, 20 to 26 gage, 1 × 4 in.
(25 × 102 mm)
1 wax candle or electric heating element
galvanometer with center-zero scale

Procedures

1. Lightly tin both surfaces to be joined. Then solder the copper and tinplate together as shown in Fig. 22–19.

2. Connect the unsoldered ends of the thermocouple to the galvanometer.

3. Heat the junction of the metals with the candle or electric heating element. Observe the reading of the galvanometer. Do not apply heat any longer than is needed for the pointer of the galvanometer to reach a steady, maximum position. Otherwise, the solder may melt.

4. Prepare on a piece of paper a table like the one in Fig. 22–19.

5. Record the maximum reading of the galvanometer on the table. Also indicate on the table the voltage polarity of the thermocouple, as, for example, copper ($+$) and tinplate ($-$).

6. Repeat procedures 1 to 5 with the other combinations of the metals given in the table.

7. Refer to the data given in the table. Which of the metal combinations produced the most current, as indicated by the reading of the galvanometer?

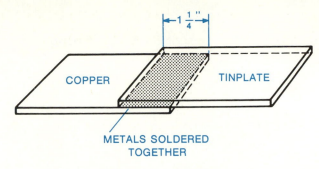

Fig. 22-19. Thermocouple assembly and table for Learning by Doing No. 24, "Making Thermocouples"

COPPER

TINPLATE

1 1/4"

METALS SOLDERED TOGETHER

Metals	Maximum Reading of Galvanometer	Polarity	
Copper-tinplate		copper ( )	tinplate ( )
Copper-galv. steel		copper ( )	galv. steel ( )
Copper-brass		copper ( )	brass ( )
Copper-zinc		copper ( )	zinc ( )
Tinplate-galv. steel		tinplate ( )	galv. steel ( )
Tinplate-brass		tinplate ( )	brass ( )
Tinplate-zinc		tinplate ( )	zinc ( )
Galv. steel-brass		galv. steel ( )	brass ( )
Galv. steel-zinc		galv. steel ( )	zinc ( )
Brass-zinc		brass ( )	zinc ( )

SELF-TEST

Test your knowledge by writing, on a separate sheet of paper, the word or words that most correctly complete the following statements:

1. Separating an audio signal from a radio-frequency carrier signal is called _____.
2. A detector diode acts as a _____.
3. The electrodes of a silicon controlled rectifier are the _____, the _____, and the _____.
4. Silicon controlled rectifiers are also commonly called _____.
5. The electrodes of a field-effect transistor are the _____, the _____, and the _____.
6. In a field-effect-transistor amplifier stage, the input signal is applied between the _____ and the _____.
7. Photocells belong to a group of devices often called _____.
8. The energy used to activate a photocell is _____.
9. The word *thermal* means _____.
10. Compounds of oxygen and metals are called _____.
11. A thermistor operates as a _____-controlled resistor.
12. A thermocouple is a solid-state device used to change _____ energy into _____.

1. Explain how a detector diode performs the function of detection.
2. Describe a silicon controlled rectifier. State the purpose for which this device is commonly used.
3. Draw the schematic diagram of a basic silicon controlled rectifier circuit. Explain how this circuit works.
4. Describe a field-effect transistor.
5. Draw the schematic diagram of a basic field-effect-transistor amplifier circuit. Explain how this circuit works.
6. What characteristic of a photoconductive cell enables it to act as a light sensor?
7. Describe the operation of a light-sensing device containing a photoconductive cell.
8. Explain the operation of a thermistor.
9. Describe the construction of a thermocouple. Explain how this device works.
10. For what purpose are thermocouples commonly used?
11. What is an integrated circuit?

INDIVIDUAL-STUDY ACTIVITIES

1. Give a demonstration showing the basic working action of a silicon controlled rectifier, a photoelectric cell, a thermistor, or a thermocouple.
2. Visit a bakery, a bottling plant, or a packaging company. Learn what electronic control devices are used for counting, inspecting, heat control, and so on. Report to your class on these devices.

Unit 23 Printed Circuits

A *printed circuit* has conductors that are thin strips of metal, usually copper, bonded to a baseboard (Fig. 23–1 at A). The base is made of laminations, or layers, of special paper or glass mat joined by phenolic or epoxy resins. The electronic components are mounted on the other side of the base. They are connected to the conductors by soldering (Fig. 23–1 at B). Such a circuit assembly is more compact than one that is hand wired. It can also be mass-produced efficiently. Because of these advantages, printed circuits are being used in all kinds of electrical and electronic products.

In this unit, you will learn about the basic processes used in the industrial manufacture of printed circuits. You will also learn how to make your own printed circuits.

MANUFACTURE OF PRINTED CIRCUITS

In its original form, at least one entire surface of a printed-circuit board is covered with copper foil that is bonded to it. This forms a copper-clad laminate (Fig. 23–2). For some printed circuits, both sides of the baseboard are copper clad.

Layout Diagram. Before a printed-circuit board is made, a layout diagram must be prepared. This diagram is a pattern

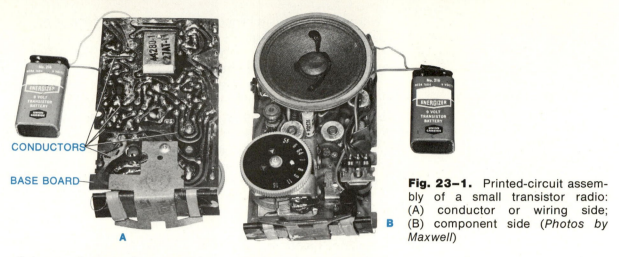

CONDUCTORS

BASE BOARD

A

B

Fig. 23–1. Printed-circuit assembly of a small transistor radio: (A) conductor or wiring side; (B) component side (*Photos by Maxwell*)

of the conductors that are to be on the board (Fig. 23–3). It shows which surfaces of the copper are to be protected from an etching solution. The solution is used to remove the unwanted copper foil from the board.

The layout diagram shown in Fig. 23–3 was made with adhesive printed-circuit tape. This tape comes in several different sizes and shapes. The tape is applied to a surface in a pattern representing the layout diagram. A layout diagram can also be drawn with ink.

Transfer of the Pattern. The simplest way to transfer a layout diagram to the copper-clad laminate is to lay out the diagram directly on the surface of the laminate with tape. In industrial work, the pattern is usually transferred to the surface of the laminate by screen printing or by a photographic process.

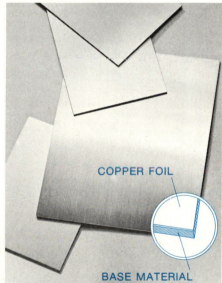

COPPER FOIL

BASE MATERIAL

Fig. 23–2. Small sheets of copper-clad laminate with copper on one side (*Graymark Enterprises, Inc.*)

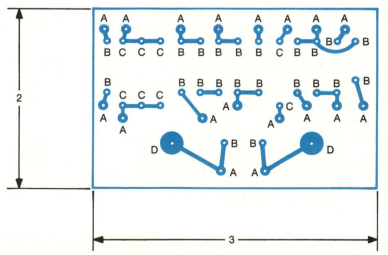

Fig. 23–3. Printed-circuit layout diagram made with tape

ORIGINAL ART

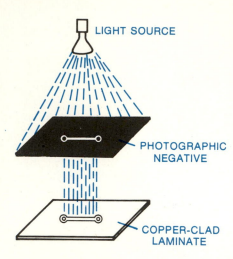

LIGHT SOURCE

PHOTOGRAPHIC NEGATIVE

COPPER-CLAD LAMINATE

Fig. 23–4. Basic photoengraving method of transferring a printed-circuit layout diagram to photosensitive copper-clad laminate

Screen printing is a process by which a *resist* is put on the copper surface through a stencil of the layout diagram. The unwanted copper, the part not protected by the resist, is removed from the laminate with an etching solution.

A layout diagram can also be transferred to a printed circuit board by using copper-clad material coated with a photosensitive resist compound. First, a picture of the layout diagram is taken. Then the negative of this picture is placed on the copper-clad laminate and exposed to strong light (Fig. 23–4). Light passes through the light-colored areas of the negative representing the diagram. The light strikes the copper surface in a pattern identical to the layout diagram. These surfaces then resist the effects of the etching solution.

Etching, Cleaning, and Drilling. After the layout diagram has been transferred to the copper surface, the copper-clad laminate is put in an acid or alkaline etching solution. This solution eats through the unprotected copper, leaving only the copper under the layout (Fig. 23–5). Next, the tape or resist is simply removed from the base. The remaining conductors are then cleaned. As needed, holes are drilled or punched through the conductors and the base laminate for component wires. The baseboard is now ready for the components to be mounted.

Soldering. When all components have been mounted on the printed-circuit board, the assembly is ready for soldering. This is commonly done in a one-step process known as *wave soldering.* Waves of melted solder are washed into each of the joints that connect component leads to the conductors (Fig. 23–6). This kind of soldering takes only a short time. It is, therefore, well suited for use in mass production. After soldering, the assembly is ready for final inspection and testing.

MODULES

A single unit or board of a printed circuit is often called a *module* or circuit pack. A module may contain the components of a specific circuit, such as a rectifier or an amplifier, or of a combination of circuits. In many electronic devices, modules are designed to be plugged into a circuit. Modules of this kind often have plated terminals that are inserted into clip sockets (Fig. 23–7).

Modules can be removed conveniently for inspection and testing. If a module proves to be defective, it can then be repaired or replaced without disturbing any other parts of the circuit system.

MAKING YOUR OWN PRINTED CIRCUIT

If the printed-circuit board you are making is simple, you can draw the conductor layout directly on the copper foil with a soft lead pencil. However, if the conductor pattern is complicated, it is best to prepare a layout diagram first. You can then transfer it to the copper-clad laminate. Preparing the diagram is often the most difficult part of making a printed-circuit board. In any case, the final form of the printed-circuit conductor pattern will be determined by the size and number of components, by the overall circuit layout, and by design decisions.

Preparing the Layout Diagram. Though there are no rigid rules for making a layout diagram, the following guidelines should prove helpful:

1. Conductors must not cross each other.
2. The conductors should be about $\frac{1}{16}$ to $\frac{1}{8}$ in. (1.5 to 3.0 mm) wide.
3. There should be a space of at least $\frac{1}{8}$ in. (3.0 mm) between conductors.
4. The conductors, which may be straight or curved, should not be made longer than needed.
5. Conductor terminal points, through which holes will be drilled, should be at least $\frac{3}{16}$ in. (4.5 mm) wide.
6. The circuit layout should allow enough working space on the board without wasting laminate.

A good first step in preparing a layout diagram is to arrange the components of the circuit on a sheet of graph paper according to the schematic diagram. If this arrangement seems suitable for the printed-circuit board, the layout is then roughed in on paper with a pencil or pen. If this proves acceptable, the diagram is put into finished form.

A schematic diagram of a circuit and the layout diagram for the printed-circuit board of this project are shown in Fig. 23–8. Studying and comparing these diagrams will help you design your own layout diagrams.

Cutting to Size. When the layout diagram has been prepared, the graph paper is cut to the final size. A border of at least $\frac{1}{4}$ in. (6.5 mm) should be left between any edge of the paper and conductor.

If a sheet of copper-clad laminate of the same size as the diagram paper is not available, it can be cut from a larger sheet. This is best done by using sheet-metal squaring shears or a fine-tooth hacksaw. Copper-clad laminate can be cut with tin snips. However, this will often cause the base lami-

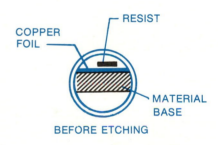

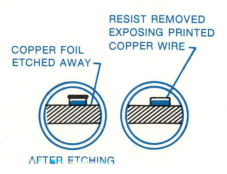

Fig. 23–5. The effect of the etching process

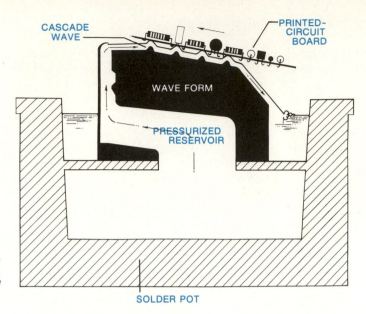

Fig. 23–6. The process of wave soldering being used with a printed-circuit board (*ITT Industrial and Automation Systems*)

CASCADE WAVE

PRINTED-CIRCUIT BOARD

WAVE FORM

PRESSURIZED RESERVOIR

SOLDER POT

Fig. 23–7. Shown are both sides of a plug-in printed-circuit module used in a computer. (*AT&T*)

PLATED TERMINALS

PRINTED WIRING PATHWAYS (CONDUCTORS)

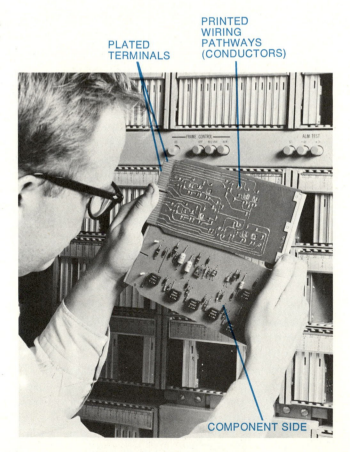

COMPONENT SIDE

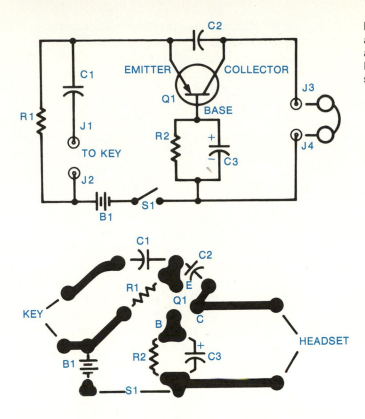

Fig. 23-8. Schematic diagram of a code-practice oscillator circuit and a sample of the printed-circuit layout diagram prepared for its construction

nate material to chip away at the edges. After you have cut the laminate to size, you should clean the copper surface thoroughly by rubbing it lightly with fine-grade steel wool.

Transfer of the Pattern. You can transfer the layout diagram to the copper-clad laminate simply by redrawing the pattern on the copper surface, using a resist material. If done carefully, this will be accurate enough for most printed-circuit projects.

If tape resist is used, the correct lengths and shapes of tape are applied to the copper surface. Follow the layout diagram as closely as possible. The best way to apply an asphalt-base (tar) resist is to use a ball-point resist pen (Fig. 23-9). Asphalt-base resist is also available in liquid form. This resist can best be applied to the copper surface with a small artist's brush. The resist should be applied to the copper surface rather heavily. Let it dry for several hours before etching.

Etching. Etching compounds are available either as ready-to-use solutions or as powders that must be dissolved in water. For safe and good etching, never use an etching solution until you know the procedure. Safety glasses should be worn while working with an etching solution.

Fig. 23-9. Resist ink pen (*Injectorall Electronics Corporation*)

For most etching, a glass pan about $1\frac{1}{2} \times 6 \times 10$ in. ($38 \times 152 \times 254$ mm) will serve as a good container for the etching solution. Put the copper-clad laminate into the pan with the copper side up. Then slowly pour the etching solution into the pan to a depth of about 1 in. (25 mm). For best results, the solution should be heated to a temperature of about 100°F (38°C).

After the extra copper has been completely removed from the base laminate, take the printed-circuit board out of the solution. Wash it with water. Let it dry thoroughly. Next, remove the tape, or if asphalt-base resist has been used, remove this resist from the conductors with a suitable solvent. The surfaces of the conductors should then be cleaned by light rubbing with fine steel wool.

Before being etched, a copper-clad laminate is often a little warped. This is usually corrected by the etching process.

Tinning and Drilling. After the conductors of the printed-circuit board have been cleaned, the terminal points of the conductors are lightly tinned. The centers of the terminal points through which holes will be drilled should be center-punched to keep the drill from slipping over the copper surface. When doing this, use light pressure on the punch so that you will not damage the laminate. The holes are then drilled with a high-speed drill no larger than $\frac{1}{16}$ in. (1.59 mm) in diameter.

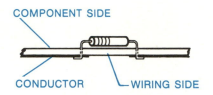

Fig. 23–10. Mounting components on a printed-circuit board

Mounting of Components. When mounting small components such as fixed resistors and capacitors on the printed-circuit board, first insert their full-length leads through the proper mounting holes. Be sure that the components are mounted so that any information written on them can be easily read. This will help you determine the specifications of any component if you should have to replace it.

The leads are then cut so that the ends extend about $\frac{1}{8}$ in. (3 mm) beyond the wiring side of the board. The lead ends are then bent at right angles. This is so that a larger surface of each end will be in contact with the terminal-point surface to which it is to be soldered (Fig. 23–10).

When mounting semiconductor devices such as diodes and transistors, leave their leads long enough to extend at least $\frac{3}{8}$ in. (10 mm) above the component side of the board. This will let you put a heat sink on each lead while soldering it.

Components such as transformers, potentiometers, variable capacitors, and multisection electrolytic capacitors that are designed for use with printed circuits have pin terminals (Fig. 23–11). These terminals are simply put into holes drilled in the board for them and soldered.

Fig. 23–11. The transformer used on printed-circuit boards has pin terminals.

Soldering. Be careful not to use too much heat or solder when soldering component leads to printed-circuit conductors. Enough heat should be applied to the joint to let the solder flow onto all surfaces being joined. Too much heat may cause the conductor to separate from the baseboard (Fig. 23–12 at A). For most printed-circuit soldering, a light-duty soldering iron (25 or 40 watts) will provide enough heat.

Using too much solder may cause it to flow to another conductor or terminal (Fig. 23–12 at B). Printed-circuit solders are in wire form. Fine work usually requires small-diameter solder wire. A 60/40 solder (60 percent tin, 40 percent lead) is normally acceptable for printed-circuit work.

REPAIR OF PRINTED CIRCUITS

It often takes a trained technician to troubleshoot and repair printed circuits. However, as a part of your electricity and electronics course, you may have an opportunity to do some printed-circuit repair work. This can give you interesting and valuable experience.

Repair jobs commonly done on printed circuits are replacing defective components, resoldering loose solder joints, and repairing broken conductors. Loose joints and conductor breaks are usually caused by dropping or bending a printed-circuit board.

The most important rule for doing any printed-circuit repair is not to use too much heat while soldering or desoldering. Too much heat will very often seriously damage conductors. Again, a light-duty soldering iron of 25 or 40 watts is best.

Replacing Components. Usually, the hardest part of replacing components is removing the defective components from the printed-circuit board. This must be done very carefully to keep from damaging conductors.

One way to remove a component such as a fixed resistor or capacitor is simply to cut its leads as near to the "body" of the component as possible. The leads of the replacement component are then soldered to the wire stubs (Fig. 23–13). This method is particularly good when the wiring side of the board cannot be conveniently reached.

When the wiring side of a printed-circuit board can be reached, a component can be replaced by desoldering its leads and slowly pulling them from the mounting holes. The leads of the replacement component are then put into the mounting holes and soldered into place (see Fig. 23–10). It may be necessary to melt the solder around the mounting

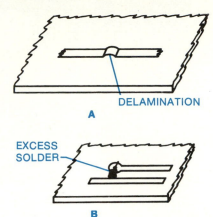

Fig. 23–12. Printed-circuit defects caused by excessive heat or solder: (A) delamination, or separation, of the conductor from the surface of a board; (B) excess solder can short adjacent conductors.

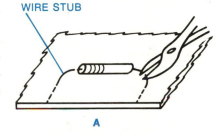

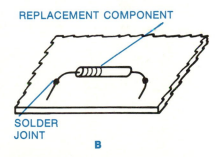

Fig. 23–13. Replacing a defective printed-circuit component: (A) cut the leads of the defective component; (B) solder the leads of the replacement component to the wire stubs.

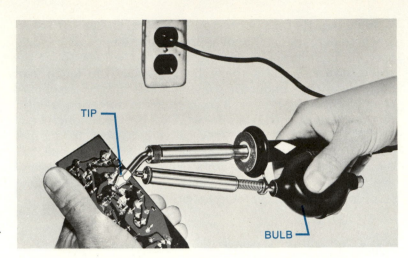

TIP

BULB

Fig. 23–14. Using a desoldering/resoldering tool (*Enterprise Development Corporation*)

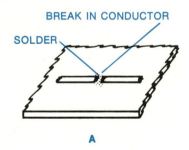

BREAK IN CONDUCTOR

SOLDER

A

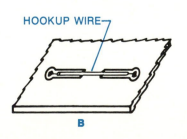

HOOKUP WIRE

B

Fig. 23–15. Repairing broken conductors on a printed-circuit board

hole before putting in the leads. This method makes for a neater job.

Removing components such as potentiometers and multi-section electrolytic capacitors that have fixed terminals is a more difficult job. In this case, the molten solder must be completely removed from all the terminal joints before the component itself can be taken out. Otherwise, a desoldered terminal joint will become resoldered while another joint is being heated. This problem can be solved by using a *desoldering tool.* It consists of a soldering iron and a suction device. First, the solder on a joint is melted with the hollow-tip soldering iron. The molten solder is then removed by suction into the rubber syringe bulb (Fig. 23–14). After all the terminals of a component have been desoldered in this way, the component can be easily taken out of the printed-circuit board.

Loose Solder Joints. Loose solder joints on a printed-circuit board can often be repaired by reheating the joint without using more solder. If more solder must be put on a joint, this should be done carefully to keep any from flowing to a nearby terminal or conductor.

Broken Conductors. If the break in a conductor is a thin crack, it can usually be filled in with solder (Fig. 23–15 at A). On some printed-circuit boards, the conductors are coated with a kind of varnish that must be taken off before the crack can be soldered. This can be done by light rubbing with fine sandpaper or by light scraping with a knife blade.

A broken printed-circuit conductor can also be repaired by "bridging" over the break with hookup wire connected to the conductor terminal points (Fig. 23–15 at B). This method is best when a piece of a conductor has been destroyed or badly damaged.

---

25. Making a Printed-circuit Radio. Making this one-transistor radio receiver will give you interesting experience in designing and building a printed circuit with a variety of components. Best results will be obtained if the radio is operated with an outside antenna at least 50 ft (15.25 m) long and with a ground connection. Any permanent outside antenna should include a lightning arrester.

MATERIALS NEEDED

copper-clad laminate, one-side copper
C1, capacitor, 270 pF, 100 WVDC
C2, capacitor, 0.047 μF, 100 WVDC
C3, capacitor, 0.001 μF, 100 WVDC
CR1, germanium crystal diode, type 1N34A or equivalent
L1, ferrite antenna coil, Vari-Loopstick or equivalent with mounting bracket and knob
Q1, transistor, p-n-p, 2N109 or equivalent
R1, carbon-composition or film resistor, 220,000 ohms, $\frac{1}{2}$ watt
S1, switch, spst, slide or toggle type
headset, 2,000 ohms
9-volt transistor-radio battery
battery connector
4 Fahnestock clips (Insulated tip jacks can be substituted for the clips.)
4 solder lugs, no. 8
printed-circuit resist or tape, and etching compound

Procedure

1. The printed-circuit board should be designed to include only those conductors used for connecting the components shown within the dotted lines of Fig. 23–16. The board should have enough space for mounting the antenna coil, battery, switch, and clips needed for the outside antenna, ground, and headset connections.

NOTE: You may prefer to design a separate panel for mounting the antenna coil, battery, and switch.

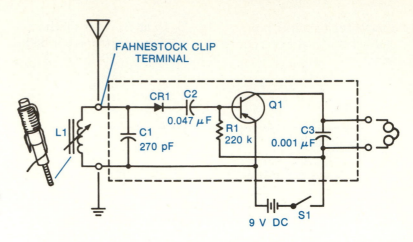

Fig. 23-16. Schematic diagram for Learning by Doing No. 25, "Making a Printed-circuit Radio"

2. Prepare the layout diagram. Transfer this pattern to the copper-clad laminate.
3. Etch the printed-circuit board. Drill the needed mounting holes.
4. Mount components and wire the circuit. Use hookup wire for connections to components not connected by the printed-circuit conductors.

SELF-TEST

Test your knowledge by writing, on a separate sheet of paper, the word or words that most correctly complete the following statements:

1. The conductors of a printed circuit are usually thin _____ of copper that are _____ to a baseboard.
2. Printed-circuit assemblies can be _____ efficiently.
3. The original form of the base on which a printed circuit is to be built is called _____.
4. The layout diagram is a pattern of the _____ that are to be used on a printed-circuit board.
5. In industrial work, the layout diagram is transferred to the surface of copper-clad laminate by _____ or by a _____ process.
6. The process for removing extra copper from a printed-circuit board is called _____.

7. A one-step manufacturing process for soldering components to a printed-circuit board is called _____ soldering.
8. A single printed-circuit unit is often called a _____.
9. A _____ is often used to coat conductor surfaces prior to etching.
10. Too much heat can cause a conductor on a printed-circuit board to _____ from the base.

FOR REVIEW AND DISCUSSION

1. Describe a printed-circuit assembly.
2. State two advantages of printed-circuit assemblies, as compared to those constructed by conventional hand wiring.
3. What is the purpose of a printed-circuit layout diagram?
4. State three ways in which a layout diagram can be transferred to copper-clad laminate.
5. What is a resist? How is it used in making a printed-circuit board?

Electrical Wiring— Materials, Tools, and Processes

Unit 24   Project Planning and Construction

Building a shop project that you choose can be one of the more interesting and useful activities in an electricity and electronics course (Fig. 24–1). Such a project will let you use some of the theoretical and practical ideas you have learned in this course. It will also give you an opportunity to do planning—the very important steps involved in "thinking through" a job. A project will allow you to work with devices and processes used in different kinds of electrical and electronics activities. Finally, successfully finishing a project can be very satisfying to you. It shows that you were both able and willing to do a job in the responsible way of an efficient worker.

One kind of project covered in this unit is what might be called a construction project. It often involves wiring from a schematic diagram and making certain components. You probably will build a chassis or some other form of base. On that, you will assemble or mount a circuit. Other kinds of projects use materials from do-it-yourself kits. These often involve doing experiments. Still other project activities may center around repairing existing circuits. Each of these will give you valuable experiences.

CHOOSING THE PROJECT

The choice of a construction project should be based on several things: (1) the function of the finished project, (2) your interest and ability, (3) the availability of plans and materials, (4) the availability of tools and facilities, and (5) the cost.

Fig. 24–1. Constructing a project can be a valuable experience since it makes use of students' skills, initiative, and knowledge of correct work procedures.

217

Function. The project should be *functional.* That means that it should satisfy the need for a particular product. This will increase your desire to build the project and the satisfaction that you will get from it.

Interest and Ability. You should choose a project that you will find interesting. In addition, it should be something that you will be able to build. These two things are related. Being very interested in a job will often encourage you to gain a new skill. However, some projects may turn out to be too difficult. This can result in a waste of time and materials. It may discourage you from taking part in other shop activities. If you are not sure that you can make a certain project, choose a simpler one. You can then build on this experience and go on to do more challenging projects.

Plans and Materials. Unless a project is of your own design, its plans should include a schematic or pictorial diagram. It should also include a list of materials needed and any special hints for doing the job. Many project suggestions or plans do not include a chassis layout diagram. If the project is a printed circuit, a printed-circuit layout diagram should be prepared. Designing a chassis or a printed-circuit board can be an interesting and valuable experience.

Before making a final choice of a project, you should make sure that all the components and materials needed for it are available. Some of these can probably be found in your school shop. Others can be bought from local stores or ordered from catalogs.

Tools and Facilities. The choice of a project should depend on the availability of tools, equipment, space, and processes. Making the project should not require special tools, hazardous operations or processes, or any equipment not likely to be found in the school shop.

Cost. The total cost of a project should be included in any plan. This is important since, after starting a project, you may discover that it will cost more than you are either willing or able to pay.

PLANNING AND DESIGNING A PROJECT

Planning and designing an original project gives you a good opportunity to use theoretical information, knowledge of devices and materials, and skills related to problem solving. Even the simplest original project can test your ability to transform an idea into a practical product (Fig. 24-2). Engineers must have this ability. However, such ability is important for all who do technical work. Have a questioning attitude about the project.

What is it to be?
How can it be done?
How can it be done better?
Can it be done in a simpler way?

These are questions that challenge the mind. In design work, as in many other kinds of technical jobs, imagination is valuable.

The basic steps in planning and designing a small, low-cost radio are shown in a project-idea sketch (Fig. 24-3). Here you see the way in which an idea is expanded and refined until an overall final plan is developed. This process is really one of choosing. You test several different ideas on paper until you find what seems to be the most practical solution to the problem.

An example of a project plan sheet that can be used for making a project is shown in Fig. 24-4. When this plan sheet is filled out, it will have all the information needed for building the project according to the ideas developed on the project-idea sketch sheet (Fig. 24-3). These planning sheets can, in one form or another, be used with any project.

Fig. 24-2. Planning a project means thinking through the problem to find a practical solution.

PROJECT KITS

In recent years, many project materials have become available in kits. A kit usually has all the materials needed for a project. The kit generally has instructions that show and tell how the project is to be assembled and wired (Fig. 24-5). The instructions often include both pictorial and schematic diagrams of the circuit.

EXPERIMENTAL ACTIVITIES

The main purpose of an experimental activity is to show how a circuit or device works. This kind of activity can be very useful for demonstrating, in a practical way, how the theory and practice are related.

A number of ideas for experiments are given in the *Learning by Doing* activities included in many units of this textbook. You will find that building these projects and showing them to your class will be interesting and informative.

More experiments are given in the *Laboratory Activities Manual to Accompany Understanding Electricity and Elec-*

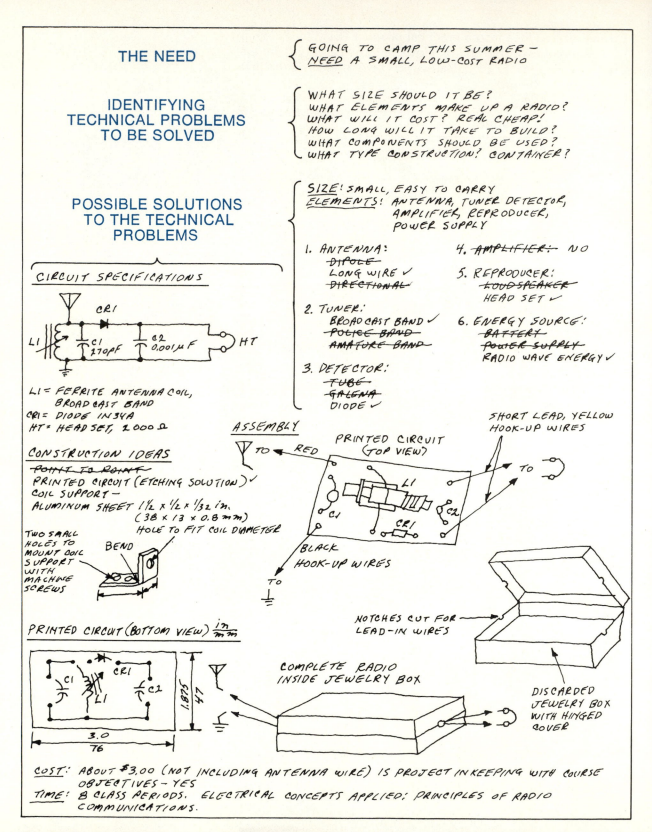

THE NEED { GOING TO CAMP THIS SUMMER — <u>NEED</u> A SMALL, LOW-COST RADIO

IDENTIFYING TECHNICAL PROBLEMS TO BE SOLVED { WHAT SIZE SHOULD IT BE?
WHAT ELEMENTS MAKE UP A RADIO?
WHAT WILL IT COST? REAL CHEAP!
HOW LONG WILL IT TAKE TO BUILD?
WHAT COMPONENTS SHOULD BE USED?
WHAT TYPE CONSTRUCTION? CONTAINER?

POSSIBLE SOLUTIONS TO THE TECHNICAL PROBLEMS { <u>SIZE</u>: SMALL, EASY TO CARRY
<u>ELEMENTS</u>: ANTENNA, TUNER DETECTOR, AMPLIFIER, REPRODUCER, POWER SUPPLY

1. ANTENNA:
   ~~DIPOLE~~
   LONG WIRE ✓
   ~~DIRECTIONAL~~

2. TUNER:
   BROADCAST BAND ✓
   ~~POLICE BAND~~
   ~~AMATURE BAND~~

3. DETECTOR:
   ~~TUBE~~
   ~~GALENA~~
   DIODE ✓

4. ~~AMPLIFIER:~~ NO

5. REPRODUCER:
   ~~LOUDSPEAKER~~
   HEAD SET ✓

6. ENERGY SOURCE:
   ~~BATTERY~~
   ~~POWER SUPPLY~~
   RADIO WAVE ENERGY ✓

<u>CIRCUIT SPECIFICATIONS</u>

CR1
L1
C1 270pF
C2 0.001μF
HT

L1 = FERRITE ANTENNA COIL, BROADCAST BAND
CR1 = DIODE IN34A
HT = HEAD SET, 2,000 Ω

<u>CONSTRUCTION IDEAS</u>
~~POINT TO POINT~~
PRINTED CIRCUIT (ETCHING SOLUTION) ✓
COIL SUPPORT —
ALUMINUM SHEET 1½ × ½ × 1/32 in.
(38 × 13 × 0.8 mm)
HOLE TO FIT COIL DIAMETER

TWO SMALL HOLES TO MOUNT COIL SUPPORT WITH MACHINE SCREWS
BEND

ASSEMBLY
TO ◄ RED
PRINTED CIRCUIT (TOP VIEW)
L1
C1
CR1
C2
TO
BLACK HOOK-UP WIRES
TO ⏚

SHORT LEAD, YELLOW HOOK-UP WIRES
TO

<u>PRINTED CIRCUIT (BOTTOM VIEW)</u> in/mm

C1
CR1
L1
C2
1.875
47
3.0
76

COMPLETE RADIO INSIDE JEWELRY BOX

NOTCHES CUT FOR LEAD-IN WIRES

DISCARDED JEWELRY BOX WITH HINGED COVER

<u>COST</u>: ABOUT $3.00 (NOT INCLUDING ANTENNA WIRE) IS PROJECT IN KEEPING WITH COURSE OBJECTIVES — YES
<u>TIME</u>: 8 CLASS PERIODS. ELECTRICAL CONCEPTS APPLIED: PRINCIPLES OF RADIO COMMUNICATIONS.

Fig. 24–3. An example of a project-idea sketch sheet

Name ______________________ Date ____________

Class ____________________ Project ____________

Name of School
City, State

PROJECT PLAN SHEET

Quantity	Material Specifications	Unit Cost	Total Cost
		20% Cost of Finish	
		Total	

Instructor's O.K. ______

Approximate Number of Hours to Construct ____

Grade ________

- -

Steps to follow in making this project (attach sketch or drawing to this sheet showing all dimensions and indicating special construction features):

Specification change: (Must be o.k.'d by instructor before change is made.)

1. ______________________________________

2. ______________________________________

3. ______________________________________

Electrical Theory Involved: (Clip other sheets to plan if needed.)

Fig. 24–4. A project plan sheet

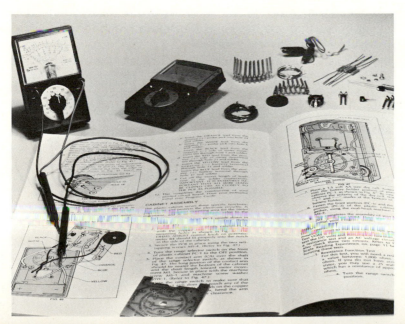

Fig. 24–5. Components of a volt-ohm-milliammeter kit and pages of the associated instruction manual (*Graymark Enterprises, Inc.*)

```
+------------------------------------------------------------------+
|                      REPAIR-JOB REPORT                           |
|                                                                  |
|                                                                  |
|   Name _____  Date _____       |
|   Product or Circuit Worked On _____       |
|   _____       |
|                                                                  |
|   For Whom This Job Was Done _____       |
|   _____       |
|                                                                  |
|   Description of Defect or Defects _____       |
|   _____       |
|   _____       |
|   _____       |
|                                                                  |
|   General Procedures Used in Doing the Job _____       |
|   _____       |
|   _____       |
|   _____       |
|   _____       |
|   _____       |
|   _____       |
|   _____       |
|   _____       |
|   _____       |
|   _____       |
|   _____       |
|                                                                  |
|   Instruments Used  _____    _____           |
|                     _____    _____           |
|   Time Spent in Doing the Job _____      |
|   Approved     Yes _____  No _____                 |
+------------------------------------------------------------------+
```

Fig. 24–6. Repair-job report

tronics. The manual first gives directions for making a circuit board, a power supply, and coils that may be used in doing many experiments. These experiments will give you valuable experiences that will increase your understanding of electricity and electronics. Working with circuits, devices, and common test instruments will help you see how they can be used in many ways. All these experiences will help make your electricity and electronics course both interesting and useful.

REPAIR PROJECTS

Repairing an existing circuit or device can give you much satisfaction and useful experience. The satisfaction comes from applying your knowledge to a problem and solving it. In doing repair jobs, you will learn about troubleshooting, or finding problems in components or circuits. This is similar to what professionals do. Thus, your experience can be put to use in a future career.

Most repair jobs require a knowledge of theory, components and materials, tools, test instruments, and testing methods. Procedures for using the basic test instrument, the volt-ohm-milliammeter, are given in various units. These procedures include the testing of devices and components such as resistors, capacitors, diodes, and transistors. Test instruments and troubleshooting methods are discussed throughout the book.

Repair projects should include a report (Fig. 24–6). In addition to being a record of your work, such a report gives you an opportunity to describe a job in writing. This can be a very useful experience. Written reports must be prepared as part of many technical jobs.

SAFETY

Safety must be considered when designing and building any project. A project should always be made so that it is free of shock and fire hazards. If instructions are given, they should be clear, so that the project can be used easily.

FOR REVIEW AND DISCUSSION

1. Explain how construction of a project can provide a valuable learning experience.
2. Name five things that should be considered when choosing a construction project.
3. What are some of the factors involved in planning and designing a construction project?
4. Describe an experiment project.
5. What information is given on a repair-job report?

INDIVIDUAL-STUDY ACTIVITIES

1. Prepare a written or an oral report describing the general procedures for planning, designing, and constructing a project.
2. Prepare a list of books, magazines, and other publications that include information about the construction of electrical and electronic projects. Report to your class about the project materials found in each.
3. Show your class a project you constructed. Demonstrate its operation. Describe any special construction features.
4. Ask your teacher about any project fairs or exhibits that are to be held in your area. If possible, plan to build a project for the occasion.

Unit 25   Wires and Cables

This unit is about wires and cables that are used as conductors in electrical and electronics circuits. It will help you choose the kind of wire or cable best suited for your job.

Wires used in electrical and electronics work are usually made of soft, annealed copper and are round. They may be solid or stranded, and bare or insulated (Fig. 25–1 at A). *Stranded wire* is made up of many smaller wires twisted together. It is more flexible than solid wire. Therefore, it is less likely to break if frequently bent or twisted. Several insulated wires contained within a single covering form a *cable* (Fig. 25–1 at B). Some kinds of flexible cables are called *cords*.

WIRE SIZE

An American wire gage (AWG) number tells the size of bare, round wire. This is based on its diameter. The larger the AWG number, the smaller the wire (Fig. 25–2). The American wire gage is also known as the Brown and Sharpe (B & S) wire gage.

Wire Gage. The AWG number and the diameter of wires can be conveniently found with a wire gage (Fig. 25–3). To use this gage, put the wire to be measured into the slot that gives the snuggest fit without binding (Fig. 25–4). You then read the AWG number of the wire at the bottom of the slot on one side of the gage. The diameter of the wire is given on the other side.

Circular-mil Area. The cross-sectional area of round wires is often stated in terms of circular mils. One *mil* is one-thousandth (0.001) of an inch (0.0254 mm). One *circular mil* is equal to the area of a circle with a diameter of 0.001 in. (0.0254 mm). To find the circular-mil area of a wire, square its diameter, given in mils. Thus, a no. 20 AWG wire (closest standard size—0.80 mm), which has a diameter of 0.03196 in. (0.81 mm), has a circular-mil area of 31.96^2, or about 1,022 circular mils.

SAFE CURRENT-CARRYING CAPACITY

The safe current-carrying capacity of a wire is that amount of current in amperes that the wire will conduct without overheating or damaging its insulation. The larger the cross-sec-

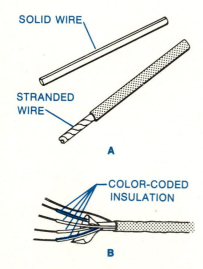

Fig. 25–1. Wires and cable

SOLID WIRE

STRANDED WIRE

A

COLOR-CODED INSULATION

B

Fig. 25-2. Actual American wire gage sizes

tional area of a wire, the greater its current-carrying capacity. The safe current-carrying capacity of a wire also depends on its insulating material. For example, a wire insulated with heat-resistant plastic has a greater current-carrying capacity than a wire of the same size insulated with ordinary rubber. The safe current-carrying capacity of a wire is sometimes called the *ampacity* of the wire.

WIRE TABLE

Information about some of the physical and electrical characteristics of wires can be found in a wire table (Table 25-1). Other wire tables give the safe current-carrying capacities of wires insulated with different materials.

COMMON WIRES

Many different kinds of wires are used for electrical and electronics wiring jobs. Each of these is best suited for a particular purpose because of its size, insulation, or composition. Some wires are *tinned,* or covered with a thin coat of solder. This prevents corrosion and makes it easier for one wire to be soldered to another.

Hookup Wire. Hookup wire is used mostly for wiring the components in a circuit (Fig. 25-5). It is either solid or stranded copper wire with a plastic insulation. The most common sizes range from no. 18 to no. 24 AWG (1.00 to 0.50 mm).

Magnet Wire. Magnet wire is solid wire. It is usually round and is insulated with a thin, tough coating of a varnishlike, plastic compound. This kind of wire is often referred to as *plain enamel* (PE) *magnet wire.* The most common sizes range from no. 14 to no. 40 AWG (1.60 to 0.08 mm). It is called magnet wire because it is used for winding the coils in electromagnets, solenoids, transformers, motors, and generators (Fig. 25-6).

Bus-bar Wire. Bus-bar wire is solid, tinned, and uninsulated copper wire. Usual sizes range from no. 12 to no. 30 AWG (2.00 to 0.25 mm). It is used as a common terminal for several wires (Fig. 25-7).

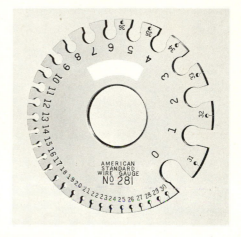

Fig. 25-3. American standard wire gage (*The L. S. Starrett Company*)

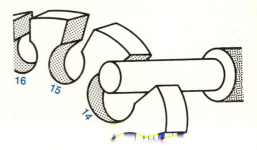

Fig. 25-4. Using the wire gage

225

Table 25-1. Wire Table for Standard Annealed Copper

AWG (B & S) Gage	Standard Metric Size (mm)	Diameter in mils	Cross-sectional Area		Ohms per 1,000 ft at 20°C (68°F)	Lbs per 1,000 ft	Ft per lb
			circular mils	square inches			
0000	11.8	460.0	211,600	0.1662	0.04901	640.5	1.561
000	10.0	409.6	167,800	0.1318	0.06180	507.9	1.968
00	9.0	364.8	133,100	0.1045	0.07793	402.8	2.482
0	8.0	324.9	105,500	0.08289	0.09827	319.5	3.130
1	7.1	289.3	83,690	0.06573	0.1239	253.3	3.947
2	6.3	257.6	66,370	0.05213	0.1563	200.9	4.977
3	5.6	229.4	52,640	0.04134	0.1970	159.3	6.276
4	5.0	204.3	41,740	0.03278	0.2485	126.4	7.914
5	4.5	181.9	33,100	0.02600	0.3133	100.2	9.980
6	4.0	162.0	26,250	0.02062	0.3951	79.46	12.58
7	3.55	144.3	20,820	0.01635	0.4982	63.02	15.87
8	3.15	128.5	16,510	0.01297	0.6282	49.98	20.01
9	2.80	114.4	13,090	0.01028	0.7921	39.63	25.23
10	2.50	101.9	10,380	0.008155	0.9989	31.43	31.82
11	2.24	90.74	8,234	0.006467	1.260	24.92	40.12
12	2.00	80.81	6,530	0.005129	1.588	19.77	50.59
13	1.80	71.96	5,178	0.004067	2.003	15.68	63.80
14	1.60	64.08	4,107	0.003225	2.525	12.43	80.44
15	1.40	57.07	3,257	0.002558	3.184	9.858	101.4
16	1.25	50.82	2,583	0.002028	4.016	7.818	127.9
17	1.12	45.26	2,048	0.001609	5.064	6.200	161.3
18	1.00	40.30	1,624	0.001276	6.385	4.917	203.4
19	0.90	35.89	1,288	0.001012	8.051	3.899	256.5
20	0.80	31.96	1,022	0.0008023	10.15	3.092	323.4
21	0.71	28.46	810.1	0.0006363	12.80	2.452	407.8
22	0.63	25.35	642.4	0.0005046	16.14	1.945	514.2
23	0.56	22.57	509.5	0.0004002	20.36	1.542	648.4
24	0.50	20.10	404.0	0.0003173	25.67	1.223	817.7
25	0.45	17.90	320.4	0.0002517	32.37	0.9699	1,031.0
26	0.40	15.94	254.1	0.0001996	40.81	0.7692	1,300
27	0.355	14.20	201.5	0.0001583	51.47	0.6100	1,639
28	0.315	12.64	159.8	0.0001255	64.90	0.4837	2,067
29	0.280	11.26	126.7	0.00009953	81.83	0.3836	2,607
30	0.250	10.03	100.5	0.00007894	103.2	0.3042	3,287
31	0.224	8.928	79.70	0.00006260	130.1	0.2413	4,145
32	0.200	7.950	63.21	0.00004964	164.1	0.1913	5,227
33	0.180	7.080	50.13	0.00003937	206.9	0.1517	6,591
34	0.160	6.305	39.75	0.00003122	260.9	0.1203	8,310
35	0.140	5.615	31.52	0.00002476	329.0	0.09542	10,480
36	0.125	5.000	25.00	0.00001964	414.8	0.07568	13,210
37	0.112	4.453	19.83	0.00001557	523.1	0.06001	16,660
38	0.100	3.965	15.72	0.00001235	659.6	0.04759	21,010
39	0.090	3.531	12.47	0.000009793	831.8	0.03774	26,500
40	0.080	3.145	9.888	0.000007766	1049.0	0.02993	33,410

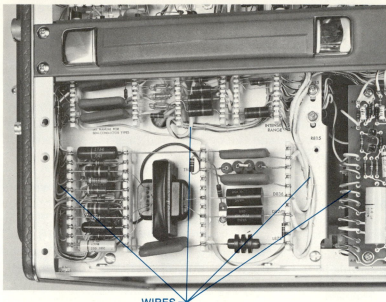

WIRES

Fig. 25–5. Hookup wire used in an oscilloscope circuit (*Tektronix, Inc.*)

Test-prod Wire. Test-prod wire is a very flexible, stranded wire. It is insulated with rubber or with a combination of rubber and other materials. The most common sizes are no. 18 and no. 20 AWG (1.00 and 0.80 mm). It is used for making the test leads of instruments such as the volt-ohm-milliammeter.

Antenna Wire. Radio-antenna wire is a stranded wire that may be either bare or insulated. Each strand is usually made of Copperweld, which consists of a steel core coated with copper. This gives the wire a high tensile strength, which keeps it from sagging when a length of it is suspended between supports.

Television Lead-in Wire. This is a twin-lead, or two-conductor, wire used for connecting a television receiver to its antenna. The wires are most often made of Copperweld insulated with polyethylene plastic. They are usually contained within an outer jacket, also made of polyethylene (Fig. 25–8). This allows for precise spacing between the wires. Because of that, they can conduct the maximum amount of signal energy from the antenna to the receiver. Twin-lead wire is also used to connect a frequency-modulation (FM) radio receiver with its antenna.

Grounding Wire. Grounding wire is usually a solid, bare wire made of copper or of aluminum. It is used for connecting metal cabinets and certain parts of circuits, such as those in a house wiring system, to water pipes or to ground rods.

MAGNET WIRES

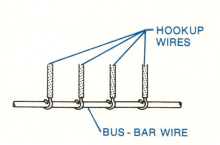

Fig. 25–6. Magnet wire used in a fan motor (*Delco Products Division of General Motors Corporation*)

HOOKUP WIRES

BUS - BAR WIRE

Fig. 25–7. Bus bar wire used as a common connecting point in a circuit

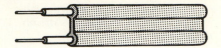

Fig. 25–8. Television twin-lead lead-in wire (*Alpha Wire Division of Loral Corporation*)

Building Wire and Cable. Building wire and cable are used to connect the devices and equipment in building wiring systems. These cables and wires are discussed in Unit 31.

ELECTRICAL CORDS

Cords are used to connect electrical and electronic equipment and appliances to outlets. Cords are flexible. Thus, they are able to withstand the bending and twisting that occur as portable products are used.

Lamp Cord. The common lamp cord is made up of two parallel, stranded wires. Each wire is covered with rubber or plastic insulation that is joined into a single unit (Fig. 25–9 at A). The wires in a lamp cord are usually no. 16 or no. 18 AWG (1.25 or 1.00 mm). Lamp cord is used with products such as portable lamp fixtures, light-duty appliances, radios, and television sets.

Service Cords. Service cords have two or three stranded, bare-copper wires. These wires are insulated with rubber or plastic and put in a round outer jacket, also made of rubber or plastic (Fig. 25–9 at B). The wires usually range in size from no. 10 to no. 18 AWG (2.50 to 1.00 mm). The most common service cords are listed by the Underwriters' Laboratories as types S, SV, and SJ. These cords are most often used with portable power tools and with medium- and heavy-duty appliances. Service cords often have a separate grounding conductor.

Heater Cord. The wires in a heater cord are insulated with a material that can withstand high temperatures. Such cords are used with heat-producing appliances and with soldering irons and guns. Heater-cord wires are stranded and are usually no. 16 or no. 18 AWG (1.25 or 1.00 mm).

One common kind of heater cord looks much like ordinary rubber-insulated lamp cord. However, its insulation is made of a thermosetting plastic compound that can withstand high temperatures.

Fig. 25–9. Electrical cords: (A) rubber-insulated lamp cord; (B) service cord (*General Electric Company*)

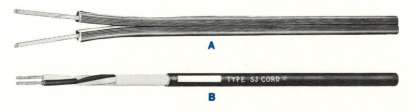

COMMUNICATION CABLES

Cables are most often used for connecting components or products that are some distance apart. Cables may be unshielded or shielded. A shielded cable is made up of one or more insulated wires. These are often contained within a sheath of copper wires woven together to form a braid (Fig. 25–10 at A). This protects the inner wire or wires from undesirable external electric or magnetic fields. It is placed directly beneath the outer covering jacket of a shielded cable. It may or may not serve as one conductor of the cable.

Audio Cable. Audio cable is commonly used in telephone circuits, in intercommunications circuits, and in circuits connecting the loudspeakers of public address systems to amplifiers. This cable is made up of two or more solid or stranded wires. These are color-coded and placed in an outer jacket of plastic insulation (Fig. 25–10 at B). Wire sizes range from no. 20 to no. 24 AWG (0.80 to 0.50 mm).

Coaxial Cable. Most kinds of coaxial cable are made up of one stranded copper wire insulated with polyethylene and placed in the center of a copper-braid shield. The outer jacket of the cable is vinyl-plastic insulation (Fig. 25–10 at C). The braid is one conductor of the cable. Coaxial cable is used, for example, to connect a microphone to an audio amplifier. It is so named because the conductors in the cable have a common axis, or center.

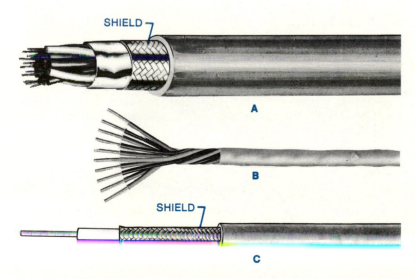

SHIELD

A

B

SHIELD

C

Fig. 25–10. Communication cables: (A) shielded cable; (B) audio cable; (C) coaxial shielded cable (*Alpha Wire Division of Loral Corporation*)

26. The Safe Current-carrying Capacity of Wires. The amount of current that any wire can conduct safely depends mainly on the wire's cross-sectional area. If too much current flows for that size of wire, the wire becomes overheated. This can damage the insulation and perhaps cause a fire. A simple experiment can show the fire hazard created when undersized wires are used to carry the load current of a circuit.

MATERIALS NEEDED

3½ in. (89 mm) PE magnet wire, no. 18 (1.00 mm)
4 in. (102 mm) PE magnet wire, no. 21 (0.71 mm)
4 in. (102 mm) PE magnet wire, no. 25 (0.45 mm)
4 in. (102 mm) PE magnet wire, no. 30 (0.25 mm)
2 in. (51 mm) PE magnet wire, no. 36 (0.125 mm)
1 no. 6 dry cell, 1½ volts

Procedure

1. Remove ½ in. (1 mm) of insulation from each end of the wires. Join them as shown in Fig. 25–11 at A.
2. Connect the joined wires to the dry cell as shown in Fig. 25–11 at B. This connection short-circuits the cell. Therefore, do not leave the wires connected any longer than needed to finish this procedure. Otherwise, the cell will be quickly ruined.

 CAUTION: Touch the no. 36 (0.125-mm) wire quickly and carefully. It will be hot enough to cause painful burns.

Fig. 25–11. Wiring diagram for Learning by Doing No. 26, "The Safe Current-carrying Capacity of Wires"

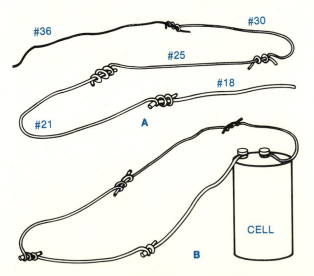

3. After a few seconds, touch the no. 18 (1.00-mm), the no. 21 (0.71-mm), the no. 25 (0.45-mm), the no. 30 (0.25-mm) and the no. 36 (0.125-mm) wires. Which of these wires is the hottest? Why?
4. Using the power formula $P = I^2R$, explain why each of the wires is heated to a different temperature although the current is the same.

SELF-TEST

Test your knowledge by writing, on a separate sheet of paper, the word or words that most correctly complete the following statements:

1. Stranded wire is more _____ than solid wire.
2. A combination of insulated wires enclosed in a single covering is called a _____.
3. Some kinds of flexible cables are called _____.
4. The abbreviation for the _____ is AWG.
5. As the AWG size numbers of round wires increase, the diameters of the wires _____.
6. The AWG numbers and the diameters of round wires are conveniently found by using a _____.
7. The circular-mil area of a round wire is found by _____ the diameter of the wire expressed in _____.
8. The safe current-carrying capacity of a wire is that amount of _____ that the wire will conduct without becoming _____ or _____.
9. Wires are often tinned to prevent _____.
10. Copperweld wire consists of a _____ core that is coated with _____.
11. Television-antenna lead-in wire is a _____, or two-conductor, wire.
12. Grounding wires are solid wires made of _____ or of _____.
13. Heater cords are often insulated with a material that can withstand _____.

14. Some cables have shields to protect the enclosed conductors from undesirable external _____ or _____ fields.
15. In a coaxial cable, all the conductors have a common _____.

FOR REVIEW AND DISCUSSION

1. State the difference between a solid wire and a stranded wire.
2. What is meant by a cable?
3. Define and describe the American wire gage.
4. How is a wire gage used?
5. How is the circular-mil area of a round wire determined?
6. What is meant by the safe current-carrying capacity of a wire?
7. Why are some wires tinned?
8. How is hookup wire used?
9. Why is it important for electrical cords to be flexible?
10. Name three kinds of electrical cords and state the purpose for which each is commonly used.
11. Describe a coaxial cable.

INDIVIDUAL-STUDY ACTIVITY

Collect examples of as many different kinds of wires and cables as you can. Mount these on a panel and identify each. Exhibit this to your class. Describe the uses of each wire and cable. Small samples of wires and cables can often be obtained from local dealers or from manufacturers.

Unit 26   Wiring Tools and Devices

Hand wiring involves using wires to connect the terminals of sockets and other devices from point to point. Integrated circuits and printed circuits have greatly lessened the need for wiring by hand. However, hand wiring is still used in manufacturing and in the repair and servicing of many products. To do hand wiring correctly, you must know about wires. You must also be familiar with the tools and the wiring devices used with circuits. Such knowledge will help you do new wiring or a repair job in a neat, efficient way.

PLIERS

Pliers are an essential wiring tool. There are many different kinds of pliers. The three kinds discussed below are generally used when working with wires.

Diagonal-cutting Pliers. Diagonal-cutting pliers are only for cutting wires (Fig. 26–1 at A). They should never be used as a gripping tool or to notch sheet metal. This can damage the cutting jaws. The jaws can also be damaged if they are used to cut hardened iron or steel wires. As with other kinds of pliers, diagonal-cutting pliers are sized according to their length. The common sizes are 4, 5, and 6 in. (102, 127, and 152 mm).

Longnose Pliers. Longnose pliers are used most for gripping and bending small wires. They sometimes have wire-cutting jaws (Fig. 26–1 at B). The relatively weak jaws of these pliers may be bent out of shape if too much pressure is used to grip objects. Common sizes are 5 and 6 in. (127 and 152 mm).

Because of the shape of the *nose,* or gripping jaw, longnose pliers are very useful for bending wires into a loop. This is often done in fastening the end of a wire to a screw terminal. Pliers having longer and thinner gripping jaws than longnose pliers are often called *needlenose pliers.*

Side-cutting Pliers. Side-cutting pliers are also known as *electrician's pliers.* This tool has heavy-duty gripping and wire-cutting jaws. These are very useful for bending, twisting, and cutting large wires (Fig. 26–1 at C). Common sizes are 6, 7, and 8 in. (152, 178, and 203 mm).

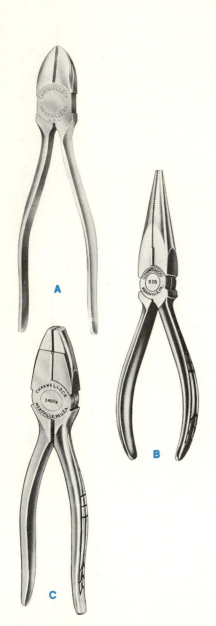

Fig. 26–1. Pliers: (A) diagonal-cutting; (B) longnose; (C) side-cutting (*Channellock, Inc.*)

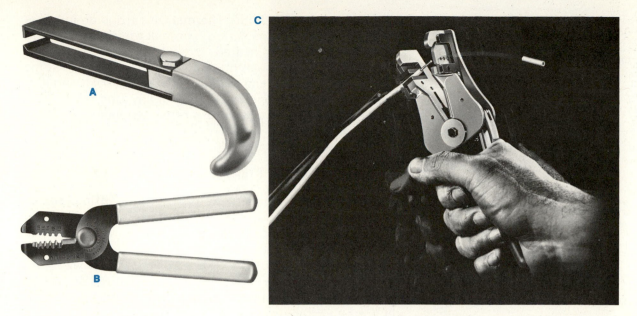

Fig. 26–2. Wire strippers: (A) basic wire stripper; (B) notch-type wire stripper; (C) using the automatic wire stripper (*Ideal Industries, Inc.*)

WIRE STRIPPERS

Wire strippers are tools used to remove insulation from wires that are to be joined or connected.

Basic Wire Stripper. A simple wire stripper is shown in Fig. 26–2 at A. To use this tool, press the blades together until they cut through the insulation of the wire. You then pull the tool away from the wire, thereby removing the insulation. When using this kind of wire stripper, be careful not to press the blades together more than is needed to cut through the insulation. Otherwise, the blades may cut into the wire and weaken it.

Another common kind of wire stripper is shown in Fig. 26–2 at B. In removing insulation with this tool, it is important to use the right size stripping notch to keep from damaging the wires.

Automatic Wire Stripper. To use an automatic wire stripper, put the wire into the right size notch on the cutting jaws. Then squeeze the handles. This will cause the jaws to separate. The cutting jaws will cut through the insulation. At the same time, the clamping jaws will pull the insulation from the wire (Fig. 26–2 at C). Automatic wire strippers are very useful for removing tough plastic or fabric insulation from wires. Since they do not require any pulling, they are also good to use with wires that have one end connected to a terminal.

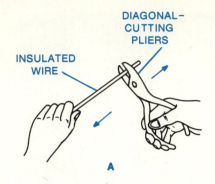

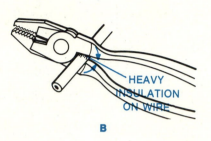

Fig. 26–3. Using pliers as wire strippers

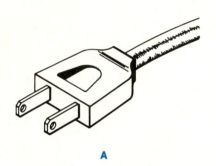

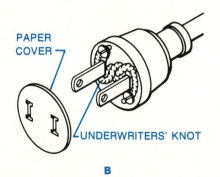

Fig. 26–4. Attachment plugs, or plug caps: (A) molded plug; (B) plug with screw terminals

Other Stripping Methods. Thermal wire stripping involves using heat to remove insulation. Thermal wire strippers have adjustable, electrically heated "jaws." These melt through the insulation. The insulation of some wires, including some magnet wires, can be removed by the heat of a soldering iron without further stripping or cleaning of any kind. Wires with enamel or varnish insulation are often stripped chemically with a paint solvent.

USING PLIERS AS WIRE STRIPPERS

With practice, you can learn to use the wire-cutting jaws of pliers as a wire stripper, particularly on wires with a thin plastic insulation. To do this, first put a wire between the jaws. Then apply enough pressure to the handles to allow the jaws to cut through the insulation. Then pull the pliers away from the wire to remove the insulation (Fig. 26–3 at A). Be sure to use just the right amount of pressure so that the wire itself will not be nicked or cut.

When pliers are used to strip a wire covered with a heavy plastic or fabric insulation, it is often necessary to first crush the insulation. With most pliers, this can be done by inserting the wire between the handles just behind the pivot point and by applying enough pressure (Fig. 26–3 at B). The crushed insulation can then be either picked off the wire or removed with the cutting jaws.

PLUGS AND JACKS

Electrical plugs provide a convenient way to connect devices to a circuit. This is done by inserting a plug into a jack or a receptacle wired to the circuit.

Attachment Plugs. Attachment plugs are used with lamp and service cords for making connections to outlets. On many products, molded attachment plugs are used (Fig. 26–4 at A). On others, attachment plugs are fastened to a cord by pressure contacts or screw terminals. Attachment plugs with screw terminals are wired as shown in Fig. 26–4 at B). To do this, first remove about $\frac{1}{2}$ in. (1 mm) of the insulation from each wire of the cord. Then twist each bare end of the stranded wire tightly so that the strands form a compact bundle. Apply solder to the tips to keep them from separating. Otherwise, a loose strand might cause a short circuit. Wrap each wire around a terminal screw in a clockwise direction. This tends to keep the wire under the screwhead as the screw is tightened.

FIRST STEP   SECOND STEP   THIRD STEP   FOURTH STEP

Fig. 26–5. Tying an Underwriters' knot

When space permits and when an attachment plug does not have a cord clamp, tie the two wires in an *Underwriters' knot.* This will keep the wires from pulling loose from the screw terminals if the cord is pulled (Fig. 26–5).

Banana Plugs and Jacks. Banana plugs and jacks are commonly used for connecting test leads to instruments and for other connections (Fig. 26–6). Wire leads are held in the plugs by set screws, pressure, or soldering.

Tip Plugs and Jacks. Tip plugs and jacks are used with test instruments (Fig. 26–7 at A). Because they are small, these devices are often used in spaces where other kinds of connecting devices could not fit.

Phono Plugs and Jacks. Phono plugs and jacks are often used for connecting the phonograph cartridge output to the input of the amplifier and for making other audio-line connections (Fig. 26–7 at B). These devices are usually connected to shielded cables by soldering.

Phone Plugs and Jacks. Standard phone plugs and jacks are used for connecting headsets to radio receivers and audio amplifiers (Fig. 26–7 at C). The standard plug prong and the jack socket of these devices are $\frac{1}{4}$ in. (6.35 mm) in diameter.

Miniature phone plugs and jacks are used for the same purposes. They are also similar in construction. However, the prong and socket are $\frac{1}{8}$ in. (3.18 mm) in diameter.

Open-circuit and Closed-circuit Phone Jacks. Phone jacks may be either open-circuit or closed-circuit (Fig. 26–8 at A). The closed-circuit jack allows a circuit to be completed through the jack contacts when a plug is not inserted into it. If, for example, a headset is plugged into a closed-circuit jack of a radio receiver, the circuit is completed through the headset. If, on the other hand, the headset is not plugged into the jack, the radio circuit is completed to the loudspeaker (Fig. 26–8 at B).

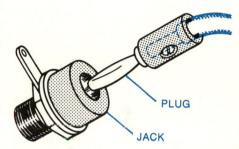

Fig. 26–6. Banana plug and jack

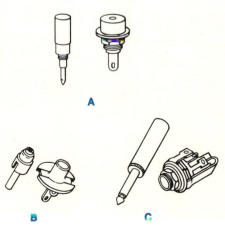

A

B   C

Fig. 26–7. Plugs and jacks: (A) tip; (B) phono; (C) phone

235

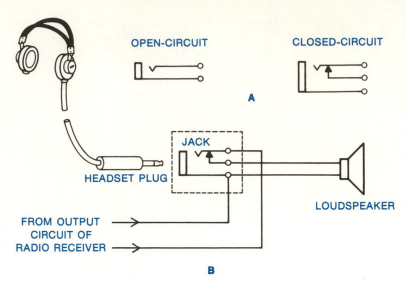

OPEN-CIRCUIT   CLOSED-CIRCUIT

A

JACK

HEADSET PLUG

LOUDSPEAKER

FROM OUTPUT
CIRCUIT OF
RADIO RECEIVER

B

Fig. 26–8. Phone jacks: (A) symbols for open- and closed-circuit phone jacks; (B) operation of a closed-circuit phone jack

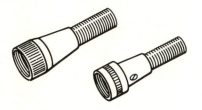

Fig. 26–9. Microphone connectors

Fig. 26–10. Wire nuts

MICROPHONE CONNECTORS

Microphone connectors are used for connecting the coaxial cables of microphone cords to amplifiers (Fig. 26–9). They are also commonly used for connecting leads to various kinds of test instruments.

WIRE NUTS

Wire nuts are devices used for making solderless connections in house wiring systems, motors, and various appliances (Fig. 26–10 at A).

To use a wire nut, strip about ⅜ in. (10 mm) of insulation from the ends of the wires to be connected. Put the wires into the wire nut. Then firmly screw the nut over the wires (Fig. 26–10 at B). Wire nuts come in different sizes. The size

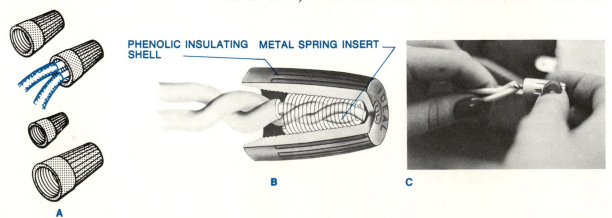

PHENOLIC INSULATING   METAL SPRING INSERT
SHELL

A   B   C

needed depends on the number and sizes of wire to be connected. Figure 26–10 at C shows two wires being inserted into a wire nut.

CLIPS

Various kinds of clips are used for making temporary wire connections to circuits. *Alligator clips* have spring-loaded jaws. These allow a firm wire connection to be made to a terminal point (Fig. 26–11 at A). They are used with instrument test leads and other temporary wiring systems.

Fahnestock clips have a spring lever. This is pushed down while a wire is inserted into the clip hasp (Fig. 26–11 at B). For a good connection, a wire must be thoroughly cleaned before it is inserted into such a clip. Fahnestock clips are often used for making temporary connections to experimental models.

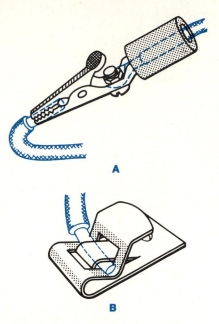

Fig. 26–11. Clips: (A) alligator; (B) Fahnestock

TERMINAL STRIPS AND LUGS

Lug *terminal strips* consist of solder lugs and a mounting lug or lugs, usually fastened to a thin phenolic strip (Fig. 26–12 at A). They provide convenient solder terminals for wires and component leads (Fig. 26–12 at B). Whenever possible, soldered connections should be made on terminal strips.

Terminal, or *wire*, *lugs* are used for connecting wires to screw terminals. These lugs come in solder and solderless versions (Fig. 26–13 at A and B). Solderless lugs are fastened to the ends of wires with a terminal, or crimping, tool, as shown in Fig. 26–13 at C.

CLAMPING AND STRAPPING DEVICES

Clamping and strapping devices are used to hold wires and cables in place. They make the job look neater and lessen the physical strain on individual wires.

Cable clamps are made of either metal or plastic (Fig. 26–14 at A). They are used to secure wires and cables to the surfaces of metal, wood, or plastic.

Insulated staples are a form of wire or cable clamp with an insulating saddle. They are used to secure wires and cables to wood surfaces (Fig. 26–14 at B).

Plastic tie straps are used to secure a bundle of wires and cables (Fig. 26–14 at C). Wires and cables are also commonly tied or laced together with heavy lacing string, as shown in Fig. 26–14 at D.

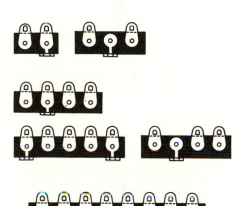

A

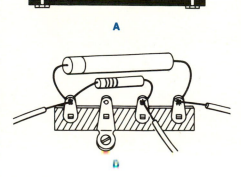

D

Fig. 26–12. Terminal strips

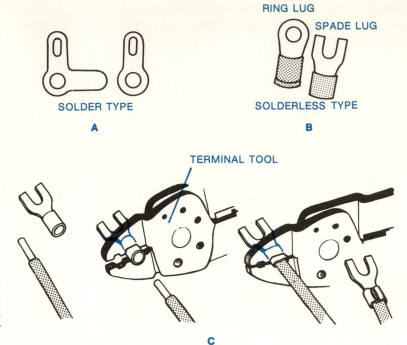

SOLDER TYPE

A

RING LUG

SPADE LUG

SOLDERLESS TYPE

B

TERMINAL TOOL

C

Fig. 26–13. Terminal or wire lugs: (A) solder lugs; (B) solderless lugs; (C) fastening a solderless lug to the end of a wire with a terminal, or crimping, tool

Fig. 26–14. Clamping and strapping devices: (A) cable clamps; (B) insulated staple; (C) plastic tie straps; (D) laced cables

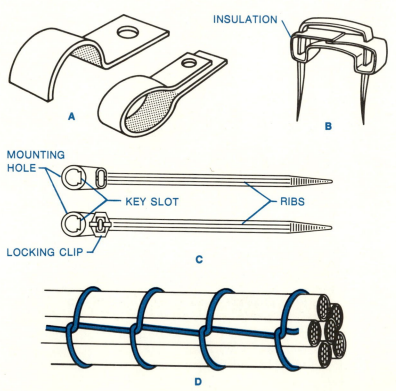

INSULATION

A

B

MOUNTING HOLE

KEY SLOT

RIBS

LOCKING CLIP

C

D

BREADBOARDS

In the early days of radio, a flat wooden board was used as a base for mounting components. Fahnestock clips, held fast to the board with wood screws, provided a convenient way to make temporary connections. This made it easy to experiment with different circuits. The flat wooden board was called a *breadboard* because of its appearance. Modern "breadboards" have many forms.

Solderless Connections. Figure 26–15 shows a solderless spring terminal. Mounted on a breadboard, this kind of terminal makes it possible to connect several leads from different components to a single terminal. Several terminals usually are mounted on a single breadboard.

The breadboard shown in Fig. 26–16 uses ordinary solid hookup wires. The wire size can vary between no. 22 and no. 30 AWG. Wire jumpers are measured from point to point. Short leads should be used. About ⅜ in. (10 mm) of insulation must be stripped from the wires. The bare wire lead is put into a hole in the board that contains a flat spring clip. The flat spring clip grips the wire tightly, making a solderless connection. Five of these spring clips are placed in a row of five holes. Then any lead pushed into a hole is connected automatically with all other holes in the row. Therefore, five connections, called a *tie point*, can be made in any row of five holes. Note that both discrete and integrated circuits are connected to the solderless breadboard. *Discrete circuits* are

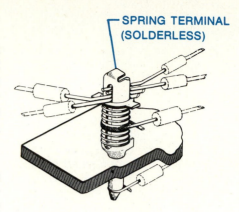

SPRING TERMINAL (SOLDERLESS)

Fig. 26–15. Solderless spring terminal (*Vector Electronic Company, Inc.*)

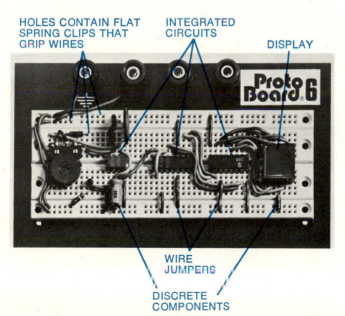

HOLES CONTAIN FLAT SPRING CLIPS THAT GRIP WIRES

INTEGRATED CIRCUITS

DISPLAY

WIRE JUMPERS

DISCRETE COMPONENTS

Fig. 26–16. Decade counter circuit with display arranged on a solderless breadboard (*Continental Specialties Corporation*)

CIRCUIT BOARD

Fig. 26–17. Components mounted on a prepunched circuit board (*Vero Electronics, Inc.*)

made up of separate components connected by leads. *Integrated circuits* contain many interconnected components in a single block of material.

Soldered Connections. Figure 26–17 shows a prepunched circuit board. This board is made of a phenolic material and comes in several sizes. Prepunched circuit boards make a convenient base for mounting and wiring circuit components. These circuit boards are often used as a breadboard. Sometimes a special pin is pushed into a hole, as in Fig. 26–18. These pins form convenient contact points for soldering the leads of components.

Manufacturers' surplus printed-circuit boards make another very useful and inexpensive breadboard. These can be found in many radio and electronic supply stores. Many of these boards do not have the components soldered in place. If they do, the components can be desoldered and removed. Remove the components carefully so they can be used again.

Although the circuit on the board is not the one you want for your experimental circuit, you can modify it. Figure 26–19 shows a breadboard made from a surplus printed-circuit board. This breadboard circuit is an experimental radio-frequency (rf) oscillator. The components are soldered to any convenient printed circuit. Even the predrilled mounting holes can be used to mount the circuit. Note that a tube socket is used in place of a regular crystal socket. Note also that jumper wires are used to modify the previously printed circuit to meet specific needs. Using surplus printed-circuit boards and old components saves time and money. Also, it avoids any mess in etching the circuits.

Advantages of Breadboard Circuits. Breadboards allow technicians and electronic engineers to design, build, and test circuits quickly. Breadboard circuits are used to make industrial *prototypes*, or first models, of circuits to see if they will work as planned. After being checked out completely, a pro-

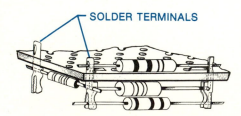

SOLDER TERMINALS

Fig. 26–18. Using push-in terminals that need to be soldered (*Vector Electronic Company, Inc.*)

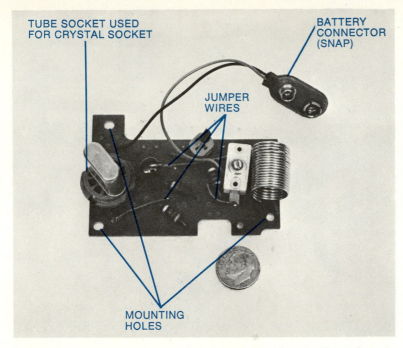

TUBE SOCKET USED
FOR CRYSTAL SOCKET

BATTERY
CONNECTOR
(SNAP)

JUMPER
WIRES

MOUNTING
HOLES

Fig. 26–19. Breadboard circuit of an rf oscillator mounted on a surplus printed-circuit board

totype is rearranged as needed before the finished product is made.

Breadboards can help you build circuits more quickly, more easily, and less expensively. They also can help you test and improve your circuit.

INSULATING TAPES, DEVICES, AND MATERIALS

Insulation is used to prevent undesirable electrical contacts. *Electrical tapes* are made of plastic, rubber, or cloth, usually ¾ in. (19 mm) wide (Fig. 26–20). Plastic tape is thinner than rubber or cloth tape. It is also more adhesive and has a higher dielectric strength. Electrical tape made of black, coated cloth is generally called *friction tape*.

Tubing. Plastic insulating tubing, or *spaghetti*, provides a very convenient way to insulate wires (Fig. 26–21). The size of this tubing is usually specified by its inside diameter and the largest wire size that can fit into it.

Grommets. Grommets made of rubber or plastic are used to protect wires and cords that must pass through holes in metal (Fig. 26–22). The size of a grommet is the diameter of the mounting hole into which it will fit snugly.

Fiber Washers. Fiber shoulder washers are used to insulate devices such as jacks and mounting screws from metal sur-

Fig. 26–20. Electrical insulating tapes: (A) plastic tape; (B) friction tape; (C) combination rubber-friction tape (*B. F. Goodrich Company*)

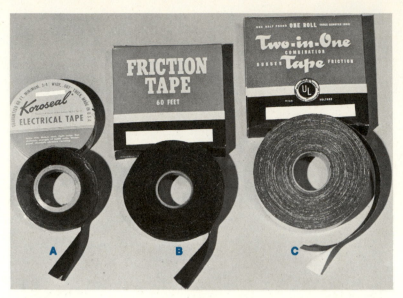

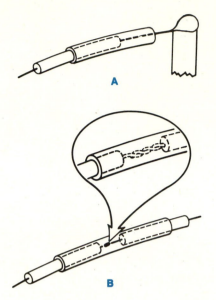

Fig. 26–21. (A) Using insulating tubing to cover exposed wiring at a terminal point; (B) tubing used to cover a splice

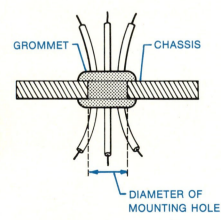

Fig. 26–22. A grommet being used to protect wires passing through a hole in a chassis

Fig. 26–23. Fiber washers: (A) shoulder and flat washers; (B) use of a shoulder and flat washer

faces (Fig. 26–23 at A). The shoulder of the washer insulates the device from the inside surface of the mounting hole. The use of a shoulder washer and a flat fiber washer to insulate a banana plug from a metal chassis is shown in Fig. 26–23 at B. The size of a shoulder washer is usually given by the inside and the outside diameters of its shoulder.

Insulating Papers, Cloths, and Varnishes. Rag and kraft papers of various thicknesses and shapes are often used as insulation in machines such as motors and generators (Fig. 26–24). Insulating papers and insulating cloths made of glass fiber or varnished cotton, silk, or linen are also used to insu-

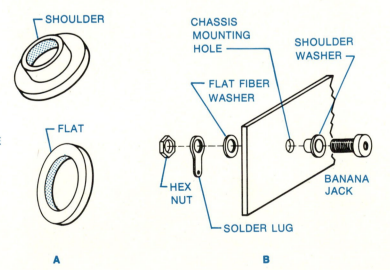

late the windings of transformer coils from the metal cores around which they are wound. Varnished linen is known as *varnished cambric.*

In addition to being applied to papers and cloths, insulating varnishes are used to strengthen the insulation of magnet wires and to repair defective insulation on these and other wires. These varnishes are applied by dipping, spraying, or brushing. Insulating varnishes are air-dried or are dried by being baked at a temperature of 220 to 270°F (104 to 132°C) for several hours.

WIRE MARKERS

Wire markers are used to identify wires, cables, and terminals in various wiring systems. They make it possible to quickly find and trace wires in a circuit. The marking of wires and terminals is also very useful in testing certain circuits and in replacing wires in them.

Many wires, individual or in cables, are color-coded. For more marking, wire markers with a letter or number are used (Fig. 26–25 at A). Other ways of marking wires are shown in Fig. 26–25 at B.

WIRE WRAPPING AND UNWRAPPING

Making an electrical connection by tightly coiling a wire around a metal terminal is called *wire wrapping.* This process is used because soldering is slow and has some disadvantages. For example, soldering uses heat. This can be dangerous to a worker and to the components being soldered. A soldered connection takes up a lot of space. Modern elec-

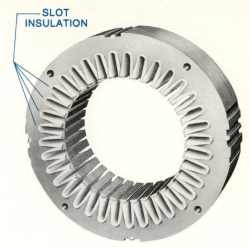

Fig. 26–24. Insulating paper used to line the winding slots in the laminated stator of an electric motor (*Brook Motor Company*)

Fig. 26–25. Marking wire: (A) applying a clip-sleeve wire marker (*W. H. Brady Company*); (B) other practical methods of marking wires

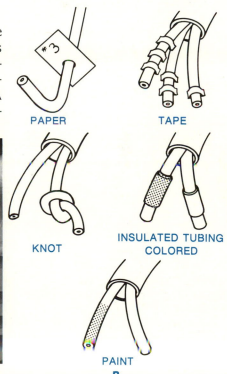

PAPER   TAPE

KNOT   INSULATED TUBING COLORED

PAINT

B

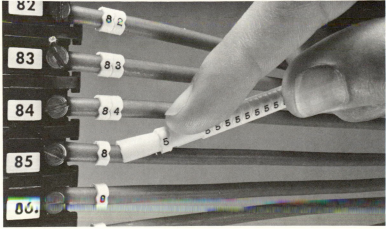

A

243

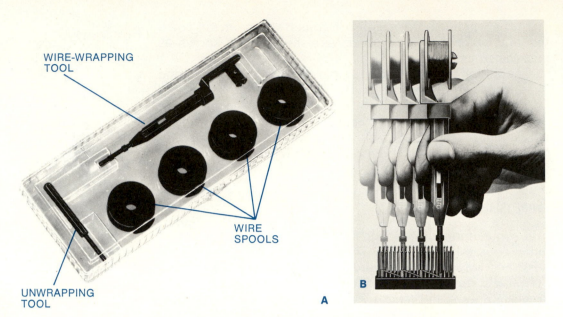

Fig. 26–26. (A) Wire-wrapping and unwrapping tools; (B) using the wire-wrapping tool (*OK Machine and Tool Corporation*)

tronic circuits are very compact, with little space available. Also, unsoldering a connection is difficult. Wire wrapping and unwrapping meet the needs of many modern electronic circuits.

Terminal Post. To make a wire-wrapping connection, you will need a terminal post that has at least two sharp edges. Most terminal posts have four sharp edges. Terminal posts used in wire wrapping are generally square with each side 0.025×0.025 in. (0.64×0.64 mm). Insulated solid copper wire from no. 26 to no. 32 AWG (0.40 to 0.20 mm) is commonly used for wire wrapping.

Wire-wrapping and Unwrapping Tools. Figure 26–26 at A shows a package containing a wire-wrapping tool, an unwrapping tool, and four spools of wire. Fig. 26–26 at B illustrates four wire-wrapped connections made to four terminal posts. Each wire wrap is made separately. Then the tool is moved to the next terminal post without cutting or stripping the wire.

The wire-wrapping tool shown in Fig. 26–26 uses a solid copper wire with a special insulation that can be pierced by the sharp corners of terminal posts. Thus, the bare copper wire from no. 26 to no. 32 AWG (0.40 to 0.20 mm) is commonly used for wire wrapping.

Using the Wire-wrapping Tool. When using a hand wire-wrapping tool, be sure the tool is designed to wrap around the terminal posts being used. In using the wire-wrapping tool, you must move the slide to the "wrap" position. Then feed the wire through the center hole in the body of the tool, through the eye, and through the wrapping bit. Allow about $\frac{1}{4}$ in. (6 mm) of wire to stick out of the wrapping bit (Fig. 26–27).

Put the tool over the terminal post. Rotate the tool clockwise about 8 to 10 turns. Keep a firm, but gentle, pressure on the tool while turning it. During the turning, the wire will be pulled down from the spool and wrapped tightly around the terminal. To make a point-to-point connection, lift the tool off the first terminal and put it over the next terminal. Repeat the clockwise turning, thereby making another connection. To cut the wire, move the slide to the "cut" position. Keep turning the tool in a clockwise direction. This cuts the wire close to the terminal.

Unwrapping. To unwrap a connection, put the *unwrapping tool* over the wire-wrapped terminal post so that it engages the post. Turn the unwrapping tool in a counterclockwise direction. This causes the wire to free itself from the terminal post. Thus, the connection is easily removed. Because wire is nicked by wrapping, it should be thrown away and only new wire used for wrapping.

Wire wrapping and unwrapping make it easy to change connections in a circuit. Therefore, these methods are often used in developing prototype circuits on breadboards.

Types of Wire-wrapping Tools. A variety of wire-wrapping tools and unwrapping tools are available. Figure 26–28

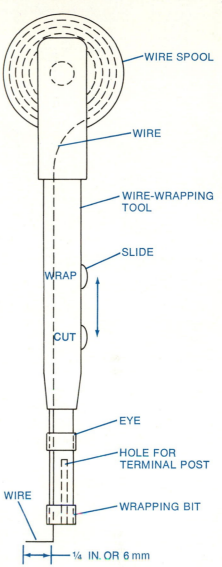

Fig. 26–27. Wire inserted in the wire-wrapping tool

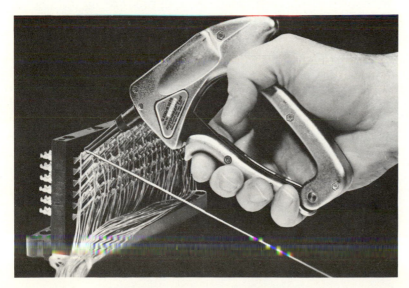

Fig. 26–28. Wire wrapping using a fast-action tool (*OK Machine and Tool Corporation*)

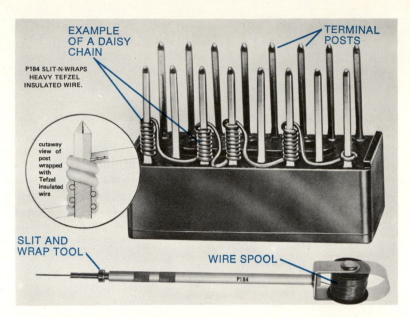

Fig. 26–29. Wire wrapping with a tool that slits the insulation before wrapping (*Vector Electronic Company, Inc.*)

shows a fast-action tool. Figure 26–29 shows another kind of wire-wrapping tool that slits the insulation on the wire just before wrapping it. With some wire-wrapping tools, you must strip the ends of the wire before wrapping. Usually, about 1 in. (25 mm) of insulation is stripped away.

Advantages of Wire Wrapping. These are some of the advantages of wire wrapping: (1) a wire-wrapped connection can be removed as easily as it is made; (2) the tools are simple to use and require little training; (3) wire wrapping can be done in tight spaces; (4) wire-wrapped connections are dependable; (5) wire-wrapped connections are as durable as most connections; (6) wire wrapping is economical.

IC TEST CLIP

An integrated circuit (IC) can be conveniently checked with an *IC test clip.* Figure 26–30 shows a test clip with a prewired cable. The test clip is clipped like a wide clothespin to the terminals of the integrated circuit. This test clip will make contact with 24 terminals—12 on each side of an integrated circuit. The ribbonlike cable brings the leads away from the work area. This makes it convenient for testing the integrated circuit or connecting it to another device.

Fig. 26–30. Integrated-circuit (IC) test clip with prewired cable (*Continental Specialties Corporation*)

SELF-TEST

Test your knowledge by writing, on a separate sheet of paper, the word or words that most correctly complete the following statements:

1. Diagonal-cutting pliers should be used only for _____ wires.
2. Side-cutting pliers are also known as _____ pliers.
3. The size of pliers is given by their _____.
4. The _____ is a tool used to remove insulation from wires.
5. Attachment plugs are fastened to a cord by _____ or _____.
6. The _____ knot is used to keep wires from being pulled loose in an attachment plug.
7. The _____ phone jack allows a circuit to be completed through the jack contacts when a plug is not inserted into it.
8. A _____ nut is a device used for making solderless connections.
9. A terminal, or crimping, tool is used to fasten _____ to the ends of wires.
10. Plastic tie strips are used to secure a _____ of wires and cables.
11. _____ circuits are made up of separate components connected by leads.
12. A _____ is used by technicians and electronic engineers to design, build, and test circuits quickly.
13. Electrical tape made of black, coated cloth is called _____.
14. Plastic insulating tubing is also known as _____.
15. Wires and cords that pass through holes in metal are protected by _____.
16. Insulating varnishes are _____ or are dried by being _____.
17. Integrated circuits can be checked conveniently with an _____.

FOR REVIEW AND DISCUSSION

1. Name three common kinds of pliers. Describe their uses.
2. How is the size of pliers given?
3. What do you have to be careful of when using a wire stripper?
4. Describe how to remove insulation from wires with the wire-cutting jaws of ordinary pliers.
5. How are attachment plugs used? Describe how to attach wires to the screw terminals of a plug.
6. Why is an Underwriters' knot used in attachment plugs?
7. Name three common kinds of plugs and jacks. Tell how they are used.
8. Explain the difference between an open-circuit and a closed-circuit phone jack.
9. What is the purpose of a microphone connector?
10. What is a wire nut? How is it used?
11. How are terminal strips and lugs used?
12. Name three common clamping and strapping devices.
13. List the advantages of developing prototype circuits on breadboards.
14. How are wires and cables laced?
15. How are prepunched circuit boards used?
16. Describe where and how insulating tubing (spaghetti) is used.
17. What is a grommet? Why is it used?
18. Describe the different kinds of wire-wrapping tools and unwrapping tools available today.

INDIVIDUAL-STUDY ACTIVITIES

1. Give a demonstration showing various kinds of wiring tools and devices. Tell how each is used.
2. Write a report on how wire wrapping and unwrapping is used in the electronic industries today.

Unit 27   Soldering

Soldering is the process of joining metals with another metal that has a low melting point. This metal is called *solder*. In electricity and electronics, the metals being joined are usually copper wires, lugs, terminal points, and the like. Electrical soldering is called *soft soldering*. Its main purpose is to make a good electrical contact between the soldered surfaces.

SOLDER AND SOLDER FLUX

The solder used for electrical and electronic soldering is usually in the form of a wire. This wire-shaped solder has one or more cores of rosin *flux* (Fig. 27–1). When the solder is melted, the flux flows over the surfaces to be soldered. It acts as a cleaner to remove any oxides coating the surface. These oxides are produced when oxygen in the air combines chemically with the metal. They must be removed because solder will not stick to surfaces covered with oxides.

An acid flux should never be used when soldering copper wires. Acid causes copper to corrode. This produces a weak solder joint and a high resistance between the connected wires.

Composition and Melting Point of Solder. Soft solder is an alloy of tin and lead. It is usually 40 percent tin and 60 percent lead, 50 percent tin and 50 percent lead, or 60 percent tin and 40 percent lead. Of these three, the 60 percent-tin–40 percent-lead solder has the lowest melting point. It ranges from about 360 to 370°F (182 to 188°C). A solder with a low melting point makes it possible to solder with less heat. This lessens the danger of damaging components or insulation. This is very important in connecting semiconductor components.

Eutectic Solder. Eutectic solder is 63 percent tin and 37 percent lead. It has the lowest melting point of any tin-lead combination, 360°F (182°C).

Size. In addition to composition, solders are specified by diameter or by an AWG number. The most popular solders have diameters of 1/16 in. or 3/32 in. (1.59 or 2.38 mm).

SOLDERING IRONS

A *soldering iron* is a tool used to melt solder. In an electric soldering iron, heat is produced as current passes through a

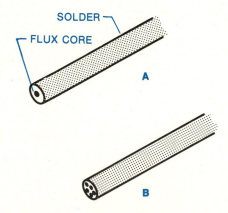

Fig. 27–1. Core solder: (A) single core; (B) multicore

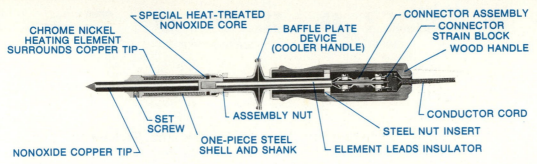

CHROME NICKEL HEATING ELEMENT SURROUNDS COPPER TIP — SPECIAL HEAT-TREATED NONOXIDE CORE — BAFFLE PLATE DEVICE (COOLER HANDLE) — CONNECTOR ASSEMBLY — CONNECTOR STRAIN BLOCK — WOOD HANDLE — SET SCREW — ASSEMBLY NUT — CONDUCTOR CORD — STEEL NUT INSERT — NONOXIDE COPPER TIP — ONE-PIECE STEEL SHELL AND SHANK — ELEMENT LEADS INSULATOR

heating element. The heating element is usually a coil wound with high-resistance wire (Fig. 27–2). The amount of heat produced is proportional to the wattage rating of the heating element. For electrical and electronic work, soldering-iron sizes range from 25 to 100 watts. Heavy-duty soldering irons used for soldering very large wires or for soldering wires to large objects may have a wattage rating of 250 watts or more. A general rule is to choose a soldering iron that does not produce more heat than is needed for the job.

Soldering-iron Holders. You should always keep a heated soldering iron in a holder or support while not using it (Fig. 27–3). This will keep work surfaces from being charred or burned by a carelessly placed iron. Holders enclosed with a perforated metal guard give the best protection against burning flesh or clothing. A heated soldering iron can cause a fire if not handled properly. Thus, you should be very careful when using any soldering iron.

Tips. Soldering-iron tips come in several different shapes (Fig. 27–4). They are usually made of copper. However, some are nickel plated or iron clad to reduce corrosion. Most soldering irons have replaceable tips. These are either screwed into the barrel socket or held in place by a set screw.

After a period of time, a soldering-iron tip will become coated with a thick layer of oxide. This reduces the amount of heat that is transferred from the iron to the tip. The oxide will

Fig. 27–2. Construction features of a typical electric soldering iron (*American Electrical Heater Company*)

Fig. 27–3. Soldering-iron holders (*Hexacon Electric Company*)

DIAMOND   CHISEL   SCREWDRIVER   CONE

TURNED DOWN

Fig. 27–4. Soldering-iron tips (*American Electrical Heater Company*)

ROSIN-CORE SOLDER

A

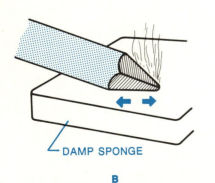

DAMP SPONGE

B

Fig. 27–5. Preparing the soldering-iron tip for use: (A) tinning; (B) removing excess solder

also cause the tip to stick in the iron. To keep these things from happening, a tip should be removed from a soldering iron after it has been used for some time. It should be cleaned with sandpaper before being put back into the iron. When putting a tip into a soldering iron, make sure that it is fully inserted into the socket of the iron.

Tinning and Cleaning. Before being used, a soldering-iron tip should be *tinned*. This involves melting a thin coat of solder over its surfaces (Fig. 27–5 at A). This allows the most heat to be conducted from the tip to the surfaces being soldered. After tinning a tip, you should remove any extra solder from it by carefully rubbing it over a slightly damp sponge (Fig. 27–5 at B).

After being used for some time, the faces, or flat surfaces, of the tip of a soldering iron become coated with the burned residue of solder flux. This coating lessens the amount of heat delivered to the surfaces being soldered. When using a soldering iron, always keep the tip clean and shiny by rubbing it over a damp sponge. This is necessary if a good soldering job is to be done. A plated tip should never be cleaned with a file or with a wire brush. This can damage its protective coating.

SOLDERING GUNS

A *soldering gun* is a solder-melting tool that has a step-down transformer. This applies a low voltage across the tip. The current passing through the tip produces enough heat to melt solder (Fig. 27–6). The gun is controlled by an on-off trigger switch. The switch often has two trigger positions. This allows the soldering gun to be operated at two heat levels. The wattage ratings for the two heats could be, for example, 100 and 140. After the switch is depressed to the first trigger position, the tip becomes heated to soldering temperature in a few seconds. To solder, touch the tip to the metal. Feed a little solder to the tip to release the flux. Then apply solder to the work until the solder flows freely. Withdraw the gun immediately and release the trigger. The second trigger position produces more heat for heavier jobs. Most soldering guns have one or more lamps that light up when the trigger is depressed. The lamps illuminate the surfaces being soldered and show that the iron is on.

Soldering guns have replaceable tips that have a variety of shapes. These are tinned and cleaned in the same way as those used with soldering irons.

A soldering gun with an open tip as shown in Fig. 27–6 should never be used for soldering semiconductor devices such as transistors and integrated circuits. This is because a strong magnetic field surrounds the tip. This field can seriously disturb the atomic arrangement within the devices, damaging or ruining them.

PREPARING WIRES FOR SOLDERING

For most soldering jobs, about $\frac{1}{2}$ in. (13 mm) of insulation should be removed from the wire. Wires from which insulation has just been removed are usually clean enough for soldering. However, bare wires such as capacitor and resistor leads may be coated with oil, dirt, or an oxide coating. These wires should be cleaned with an ink eraser, with fine sandpaper, or with the blade of a knife (Fig. 27–7). When cleaning any wire, you should be careful not to nick the wire or to remove too much copper.

After the insulation has been removed from a stranded wire, the strands should be twisted together to form a firm bundle. This will keep the loose strands from making contact with other wires and causing a short circuit.

Some magnet wires are coated with an insulation that will burn off as the wires are soldered. The insulation on all other magnet wires can be removed with fine sandpaper, a knife blade, or a chemical solvent.

Tinning. Bare copper wires should always be tinned before soldering. This will allow quicker and more effective soldering with less heat. Pretinned wires can often be soldered more quickly if they are retinned just before soldering.

A wire can be tinned quickly by using a soldering-iron holder, as shown in Fig. 27–8. To do this, first melt a drop of solder on the soldering-iron tip. Put the wire in the solder. Melt a small amount of additional solder on the tip, turning the wire until all its surfaces are thinly coated with solder.

Most terminals and lugs to which wires are soldered are pretinned. They do not need more tinning. Any bare surface to which a solder connection is to be made should be tinned beforehand.

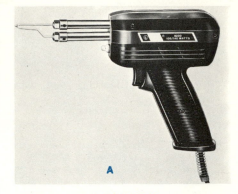

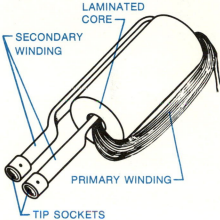

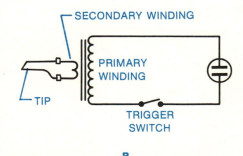

Fig. 27–6. (A) A typical soldering gun; (B) construction details and schematic diagram of a common soldering gun (*Photo courtesy of Weller Electric Corporation*)

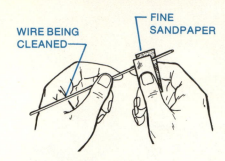

Fig. 27-7. Cleaning wire with sandpaper

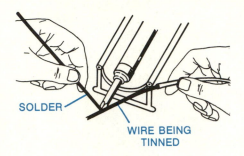

Fig. 27-8. Using a soldering-iron holder when tinning a wire

Fig. 27-9. Preparing shielded coaxial cable for soldering

Fig. 27-10. Soldering a wire to a lug

Coaxial Cable. Figure 27-9 shows how to prepare shielded coaxial cable for soldering when the braid shield must be connected into a circuit. When working with coaxial cable, be careful not to damage the insulation between the shield and the inner conductor. This is particularly important when soldering a wire lead to the shield. If too much heat is applied to such a joint, the insulation may be melted enough to cause a short circuit between it and the inner conductor.

SOLDERING WIRES TO LUGS

A wire that is to be soldered to a lug should first be mechanically fastened to the lug to hold it in place (Fig. 27-10 at A). If the wire is the lead of a semiconductor device such as a diode or a transistor, attach a heat sink to it (Fig. 27-10 at B).

Next melt a drop of solder on the tip of the soldering iron or gun. Press the tip to one side of the joint (Fig. 27-10 at C). The melted solder will allow more heat to be conducted from the tip to the joint. Briefly heat the joint. Then press one end of a length of solder wire to the other side of the lug until a small amount of the solder has melted (Fig. 27-10 at D). Let

1. REMOVE OUTER JACKET FOR A DISTANCE OF APPROXIMATELY 1½" INCHES.

2. UNBRAID THE SHIELD WITH A SCRIBER OR OTHER SHARP POINTED TOOL.

3. TWIST UNBRAIDED PORTION OF SHIELD TOGETHER AND REMOVE A SHORT LENGTH OF THE INSULATION WHICH COVERS THE CONDUCTOR.

TO SHIELD CONNECTION

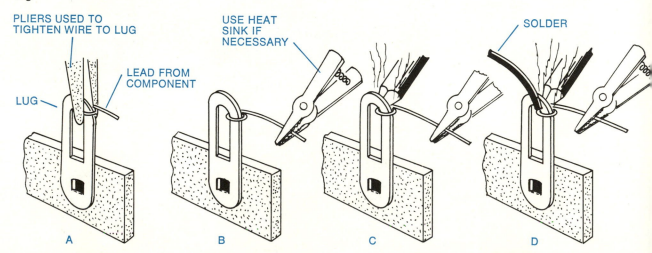

this solder flow thoroughly into the joint. Hold the tip in place a while longer. Then remove it from the joint. After the joint has cooled, remove the heat sink.

Use only enough solder to cover all surfaces of the joint without hiding the shape of the joint (Fig. 27–11). This will enable you to inspect the joint. It will also keep extra solder from causing a short circuit with nearby wires or terminals.

Do not move or otherwise handle a newly soldered wire until the solder has cooled enough. If a warm connection is moved, the solder will suddenly become less shiny. A dull-colored solder joint indicates that not enough heat was used. Such a joint, commonly called a *cold solder joint,* does not provide a good electrical contact.

Always handle a soldering iron or gun with care to keep from burning the insulation or damaging the components. Burned or charred insulation does not look good. In addition, it may cause a short circuit or ground.

SOLDERING WIRES TO HOLLOW TERMINALS

Soldering a wire to a hollow terminal, such as a phono plug, is shown in Fig. 27–12. To solder, first melt a small amount of solder on the tip of the soldering iron or gun. Hold the tip of the plug in place until the solder has been drawn up into it. Then add more solder if needed. Any extra solder or wire extending beyond the tip of the plug should be cut or filed away. This will keep the plug from binding when inserted into a jack.

SOLDERING WIRES TOGETHER

Wires are usually soldered at terminal points, such as lugs. However, it is sometimes necessary or desirable to solder

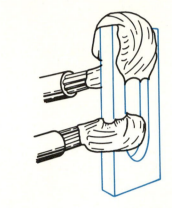

Fig. 27–11. Solder joints showing the correct amount of solder to be used

HOLD VERTICALLY

SOLDER

APPLY SOLDER TO TIP OF HEATED PIN. SOLDER WILL FLOW UP INTO PIN BY CAPILLARY ACTION.

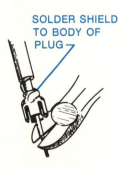

SOLDER SHIELD TO BODY OF PLUG

REMOVE EXCESS WIRE AND SOLDER.

Fig. 27–12. Soldering a wire to a hollow-terminal device such as a phono plug

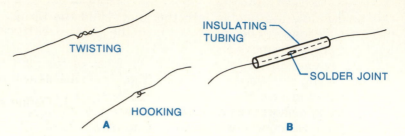

Fig. 27–13. Soldering two wires together: (A) wires are twisted or hooked together tightly to provide mechanical strength before soldering; (B) solder joint covered with insulating tubing.

two wires together without a lug. The wires are first twisted or hooked together tightly (Fig. 27–13 at A). They are then soldered.

The solder joint should be as short as possible. It should be covered with insulating tubing or other insulation (Fig. 27–13 at B). The tubing insulates the joint and makes the job look neater.

LEARNING BY DOING

27. Experiences in Soldering. Soldering is done to ensure good electrical connections. The following soldering jobs will give you experience in this important wiring activity.

MATERIALS NEEDED

1 three- or four-lug terminal strip
1 phono plug
6 in. (150 mm) untinned insulated wire, any size, no. 18 through no. 22 AWG (1.00 through 0.63 mm)
18 in. (460 mm) tinned or untinned insulated wire, any size, no. 18 through no. 22 AWG (1.00 through 0.63 mm)
6 in. (150 mm) microphone cable
soldering iron, 40 or 60 watts
solder wire, rosin core

Procedure

Do one or more of the following soldering jobs:

1. tinning a wire (Fig. 27–8)
2. soldering two wires together (Fig. 27–13)
3. soldering wires to the lugs of a terminal strip (Fig. 27–10)
4. soldering a microphone cable to a phono plug (Fig. 27–12)

After you have finished these, show them to your instructor for approval. The terminal strip can be conveniently mounted on a baseboard.

Test your knowledge by writing, on a separate sheet of paper, the word or words that most correctly complete the following statements:

1. Solder flux is used as a cleaner to remove any _____ coating from the surfaces to be soldered.
2. Never use _____ flux when soldering copper wires.
3. Soft solder is generally an alloy of _____ and _____ .
4. The melting point of 60/40 solder ranges from about _____ to _____ °F.
5. The most popular solder wires have diameters of _____ or _____ in.
6. The electrical size of a soldering iron or gun is given by its _____ rating.
7. In doing a soldering job, it is always best to use a soldering iron or gun that does not produce more _____ than is needed.
8. A soldering-iron or -gun tip is tinned by coating its surfaces with _____ .
9. Bare copper wires should always be _____ before soldering.
10. A wire that is to be soldered to a lug is first _____ fastened to the lug to hold it in place.
11. A solder joint to which enough heat has not been applied is called a _____ solder joint.
12. Always handle a soldering iron or gun with care to keep from burning the _____ or _____ the components.

FOR REVIEW AND DISCUSSION

1. Define soldering.
2. Describe the solder wire used for common electrical and electronics soldering jobs.
3. What is the purpose of soldering flux?
4. Why is an acid flux never used when soldering copper wires?
5. How is the electrical size of a soldering iron given?
6. What is the purpose of a soldering-iron holder?
7. Why is it important to keep the tip of a soldering iron well tinned?
8. How is a soldering-iron tip cleaned?
9. Why is a wire tinned before soldering?
10. Describe the process of soldering a wire to a lug.
11. What is meant by a cold solder joint?

INDIVIDUAL-STUDY ACTIVITIES

1. Prepare a written or an oral report on the different kinds of soldering materials and processes used in wiring.
2. Give a demonstration showing the right way to solder several common kinds of jobs.

Unit 28   Making a Sheet-metal Chassis

Sheet metal is used to build a chassis for mounting and enclosing components (Fig. 28–1). This unit will be very helpful if you choose to build a project that requires mounting components on a sheet-metal chassis. You will find designing and building such a chassis an interesting and valuable experience. You will also learn how to use several tools and devices.

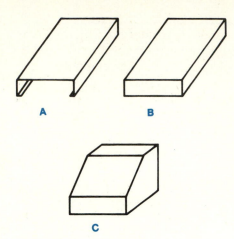

Fig. 28-1. Types of chassis: (A) open-end; (B) box; (C) sloping-panel

The most common sheet metals used for building chassis are aluminum and galvanized steel. Galvanized sheet metal is mild steel with a coating of zinc.

The thickness of ferrous sheet metals such as iron and steel is given by a United States Standard (USS) gage number. A *ferrous metal* is one made of iron or one that contains iron. The thickness of nonferrous sheet metals such as copper and aluminum is most often given by an American wire gage (AWG) number. In both cases, as the gage number increases, the thickness of the metal decreases. For example, a no. 10 (0.135-in., or 3.43-mm) sheet metal is thicker than no. 12 (0.105-in., or 2.69-mm) sheet metal. The sheet metals used for making chassis are usually from no. 14 to no. 20 gage thick (0.080 in., or 2.03 mm, to 0.035 in., or 0.89 mm).

CHASSIS LAYOUT DIAGRAM

A chassis layout diagram shows the size of the sheet metal to be used, the location and shape of the holes to be drilled or punched, and the places where the metal is to be bent (Fig. 28–2). The diagram is usually drawn to scale. That way its main features can be seen in proper relation to each other.

The diagram should be prepared carefully. You should consider all the circuit components to be used with the chassis. A good first step in making the layout diagram is to arrange the components in a suitable pattern on a sheet of unruled paper. This sheet stands for the chassis (Fig. 28–3). Designing a chassis will give you good experience in solving design problems that can have more than one solution.

LOCATION OF COMPONENTS

There are no standard rules for placing the components of a circuit on a chassis. However, you should consider the following when determining the locations.

1. Transformers, transistors, and can-shaped electrolytic capacitors are usually mounted on the top surface of the chassis, with the capacitors upright.
2. Switches, input and output jacks, and receptacles are often mounted on the front or back of the chassis.
3. The shafts of potentiometers and variable capacitors are brought out at the front of a chassis.
4. Related components should be put in the same general area. (For example, a rectifier transformer should be mounted near the on-off switch and the electrolytic filter capacitor of the rectifier circuit.)
5. All components should be placed so that they can be

connected with the shortest lengths and as little cross-over of wires as possible.

6. The chassis should provide enough space for the mounting of wiring devices such as terminal strips and grommets.
7. There should be enough space to drill holes for wires connecting components on top of and underneath the chassis.
8. The chassis should be deep enough that components mounted beneath it will not extend below the sides.

CUTTING OUT THE CHASSIS BLANK

After preparing the chassis layout diagram, you have to choose the sheet metal for the chassis. The metal should be heavy enough to support the components without bending or buckling. You will probably cut the blank for the chassis from a larger sheet of metal after drawing the layout diagram on it. A soft lead pencil or a scriber and a square can be used for making the layout (Fig. 28–4).

Sheet metal is cut with tin snips or with squaring shears (Fig. 28–5). With tin snips, a small piece of sheet metal can be cut from a larger sheet without disturbing other surfaces of the sheet. Squaring shears make a square, cleaner cut. However, when you use them, you have to cut completely across a larger sheet. This may result in too much waste of sheet metal.

in.	mm.
1/4	6
5/16	8
3/8	10
1/2	13
3/4	19
7/8	22
1 1/8	28
1 1/2	38
1 3/4	44
2 3/4	70

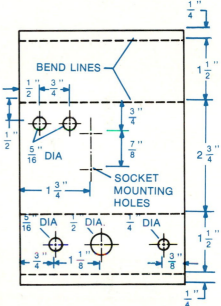

Fig. 28–2. Chassis layout diagram. Metric sizes are above.

Fig. 28–3. Student arranging components on a sheet of paper before making a layout diagram for a sheet-metal chassis

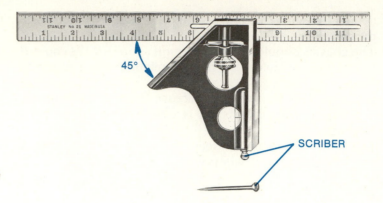

Fig. 28-4. Combination square with scriber

Fig. 28-5. Cutting sheet metal: (A) tin snips; (B) squaring shears (*Photo of squaring shears, courtesy of The Peck, Stow, and Wilcox Company*)

LAYING OUT THE CHASSIS BLANK

When you have cut out the chassis blank, transfer information from the layout diagram to it. This information generally includes the location of bend lines and notch-cutting lines, the outline of square or rectangular holes, and the centers of round holes. The transfer can be done with a soft lead pencil or with a scriber. The centers of holes can then be marked with a prick punch. The prick-punch marks should then be enlarged with a center punch (Fig. 28-6). When using a center punch, hit the punch just hard enough to make a sharp indentation on the surface of the metal without causing it to bulge out on the other side.

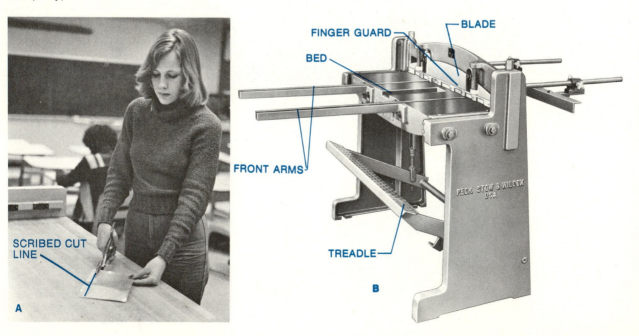

DRILLING HOLES

Round holes up to ½ in. (12.7 mm) in diameter are best cut through sheet metal with twist drills (Fig. 28–7). These are used with a portable electric drill or with a drill press (Fig. 28–8). The machines should be operated at medium speeds and with just enough pressure to cause the twist drill to cut into the metal. The diameter of a twist drill is given by a fraction, whole number, or letter stamped on its shank.

 CAUTION: Always wear safety glasses while drilling holes in sheet metal. Never try to drill a hole unless the metal is securely clamped.

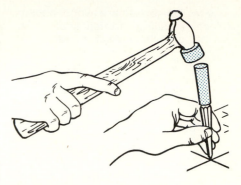

Fig. 28–6. Marking the center of a hole to be drilled in sheet metal

A safe way to drill a hole with a portable electric drill is shown in Fig. 28–9. The wood backing helps hold the metal in place. It also keeps the metal from bending as drill pressure is applied to it. Put a piece of wood or thick paper between the metal and the front vise jaw to keep the metal from being marked.

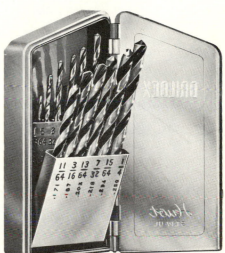

Fig. 28–7. Set of fractional-sized twist drills

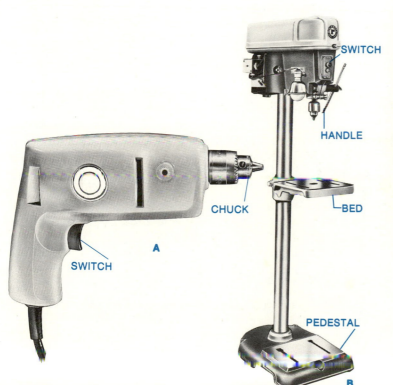

Fig. 28–8. (A) Portable electric drill; (B) drill press (*Rockwell Manufacturing Company*)

259

Fig. 28–9. Use a wood backing when drilling through sheet metal with a portable electric drill.

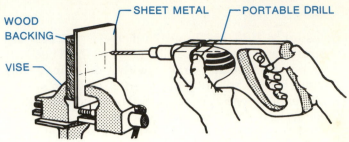

WOOD BACKING — SHEET METAL — PORTABLE DRILL — VISE

DRIVE SCREW
PUNCH
DIE
1"
DRIVE SCREW HEAD (SQUARE)

Fig. 28–10. A chassis punch for making a 1-inch (25-mm) diameter hole

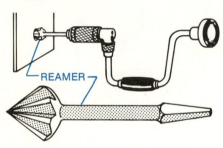

REAMER

Fig. 28–11. Countersink used as a reamer

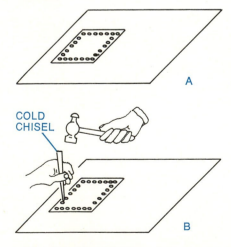

A

COLD CHISEL

B

Fig. 28–12. Making a large opening in sheet metal

CHASSIS PUNCH

A *chassis punch* is a tool that cuts clean, accurate holes in sheet metal (Fig. 28–10). Round chassis punches come in sizes from $\frac{1}{2}$ to $3\frac{1}{8}$ in. (12.7 to 79.4 mm) in diameter. They are also available in square and other shapes.

To use a chassis punch, first drill a hole, a little larger in diameter than the drive screw of the punch, through the metal. Then assemble the tool with the sheet metal sandwiched between the die and punch. The drive screw passes through the die and is threaded into the punch. Turn the drive-screw nut with a wrench. This forces the cutting edges of the punch through the sheet metal and into the die. A hole of the same size and shape as the punch is cut.

FILES AND REAMERS

Rough edges of sheet metal can be smoothed with a flat file. A 10- or 12-in. (254- or 305-mm) second-cut file is best.

Round metal files, often called *rattail files,* are very useful for removing burrs from the edges of drilled holes and for enlarging holes in sheet metal.

Burrs can also be removed with a *countersink bit,* used with a bit brace (Fig. 28–11). Another kind of reamer, the *taper reamer,* is very useful for enlarging holes in sheet metal. When using a taper reamer, use just enough pressure to let its cutting edges remove thin shavings of metal from the edges of the hole.

MAKING LARGE OPENINGS

A large opening in a sheet-metal blank can be made by first drilling small holes close to each other along the outline of the opening (Fig. 28–12 at A). The piece being removed is then completely cut out with a cold chisel (Fig. 28–12 at B). When doing this, use just enough hammer pressure on the chisel to cut through the metal without causing it to bulge on

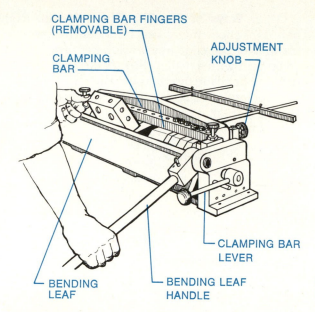

CLAMPING BAR FINGERS (REMOVABLE)

CLAMPING BAR

ADJUSTMENT KNOB

CLAMPING BAR LEVER

BENDING LEAF

BENDING LEAF HANDLE

Fig. 28–13. Main parts of the box-and-pan brake

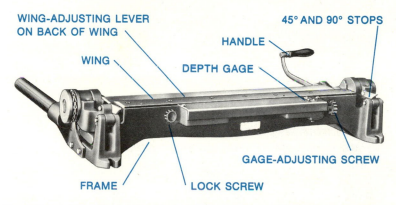

WING-ADJUSTING LEVER ON BACK OF WING

WING

45° AND 90° STOPS

HANDLE

DEPTH GAGE

GAGE-ADJUSTING SCREW

FRAME

LOCK SCREW

Fig. 28–14. Main parts of the bar folder (*The Peck, Stow, and Wilcox Company*)

the other side. Use a hardwood or soft-metal plate under the sheet metal during the cutting. Then smooth the edges of the hole with a file.

FORMING THE CHASSIS

After all cutting and drilling has been done on the chassis blank, the flat sheet metal is ready to be formed or bent into shape. Two machines are used for this, the *box-and-pan brake* and the *bar folder* (Figs. 28–13 and 28–14). With practice, you can use these machines to make almost any shape you desire.

Hand Seamer. The *hand seamer* is a convenient tool for making simple bends or flanges in sheet metal (Fig. 28–15). It can be adjusted to bend flanges of up to 1 in. (25 mm).

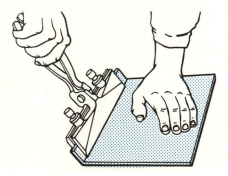

Fig. 28–15. Using the hand seamer to bend a flange

Other Ways of Bending Sheet Metal. Right-angle (90°) bends in sheet metal can be made by clamping the chassis blank over the edge of a work bench and hammering the part to be bent with a mallet (Fig. 28–16 at A). A bend of up to 90° can also be made by clamping the sheet metal between pieces of wood and then shaping it to the desired angle with a mallet (Fig. 28–16 at B).

FINISHING THE CHASSIS

An attractive and inexpensive finish can be given to a sheet-metal chassis. This can be done by simply scratching the outside surface with coarse steel wool, medium-grit sandpaper, or a wire brush. If a painted surface is desired, many kinds of paints can be applied by either brushing or spraying. Among these is one that, when dry, produces a crinkled surface that is very attractive. Always consider chassis grounds (electrical connections to the chassis) before applying any finish. Most finishing materials act as insulators.

MACHINE SCREWS

Machine screws are used with nuts and lock washers for mounting components on the sheet-metal chassis (Fig. 28–17). A special kind of machine-screw nut, called a *locknut*, has a plastic insert. Under normal conditions, this prevents the nut from becoming loose and eliminates the need for a separate lock washer. Machine screws are usually made of metal. However, nylon screws are often used for electric and electronic circuit assemblies to insulate exposed hardware. The metal screws may be plain steel or brass. Screws may be plated with nickel or cadmium to prevent corrosion.

The U.S. customary size of a machine screw is given by the diameter of its threaded part, the number of threads per inch, and the length of the threaded portion. For sizes less than $\frac{1}{4}$

Fig. 28–16. Bending small pieces of sheet metal

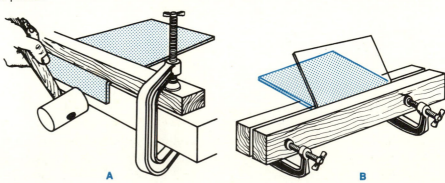

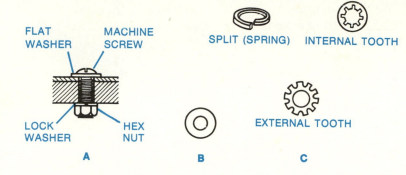

FLAT WASHER MACHINE SCREW

SPLIT (SPRING) INTERNAL TOOTH

EXTERNAL TOOTH

LOCK WASHER HEX NUT

A B C

Fig. 28–17. Machine screws and metal washers: (A) complete assembly; (B) flat washer; (C) types of lock washers

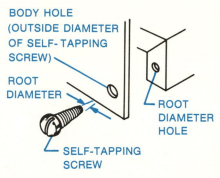

BODY HOLE (OUTSIDE DIAMETER OF SELF-TAPPING SCREW)

ROOT DIAMETER

ROOT DIAMETER HOLE

SELF-TAPPING SCREW

Fig. 28–18. Using a self-tapping screw

inch, the diameter is given by a gage number, such as 4, 6, 8, 10, or 12. The larger the number, the greater the diameter. An example of a complete specification for a machine screw is no. 6-32 RH-1 inch. Here, the 6 indicates the diameter, the 32 indicates threads per inch, RH means roundhead, and 1 inch gives the length of the threaded part. Other common sizes are 4-40, 8-32, and 10-32. The metric sizes of a machine screw are given by its nominal size (basic major diameter) and pitch. Both sizes are given in millimeters. Include the letter M in front of the nominal size to show it is metric. Use $\times$ between the nominal size and the pitch. An example would be M6 $\times$ 1. This size is close to a customary $\frac{1}{4}$-20. For a more complete explanation of metric threads, see a metric handbook.

SELF-TAPPING SCREWS

Self-tapping screws are a kind of sheet-metal screw. They are used for mounting components on a chassis and for fastening the panels and other parts of a chassis. These screws cut the threads in the hole into which they are screwed (Fig. 28–18). The size of self-tapping screws is given by a number that indicates the diameter and by the length of the threaded part.

SCREWDRIVERS

Standard screwdrivers are used with slotted-head screws (Fig. 28–19 at A). The size is usually given by the diameter and the length of the blade.

Phillips-head screwdrivers are used with cross-slot screw heads. They have a stronger driving action than standard screwdrivers (Fig. 28–19 at B). The size is most often given by a number, such as 1, 2, 3, or 4. The larger the number, the larger the point of the screwdriver.

Allen screwdrivers fit the hexagonal holes found on some set screws (Fig. 28–20 at A). The size is given by the length of

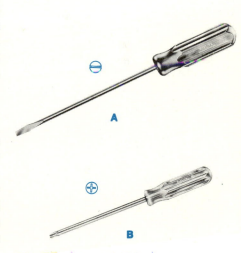

A

B

Fig. 28–19. (A) Standard screwdriver; (B) Phillips-head screwdriver (*Xcelite, Incorporated*)

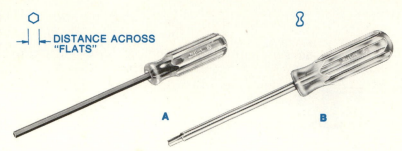

Fig. 28–20. Special screwdrivers: (A) Allen screwdriver; (B) clutch-head screwdriver (*Xcelite, Incorporated*)

DISTANCE ACROSS "FLATS"

A

B

A

Handle Color	Size	
	inches	millimeters
black	$\frac{3}{16}$	4.76
brown	$\frac{7}{32}$	5.55
red	$\frac{1}{4}$	6.35
orange	$\frac{9}{32}$	7.14
amber	$\frac{5}{16}$	7.94
green	$\frac{11}{32}$	8.73
blue	$\frac{3}{8}$	9.53

B

Fig. 28–21. Hollow-shaft nutdrivers: (A) typical set of nutdrivers; (B) color code used with nutdrivers (*Photo courtesy of Xcelite, Incorporated*)

the blade and the distance across the *flats,* or parallel sides, of the blade.

Clutch-head screwdrivers are used with clutch-head screws to get a stronger driving action (Fig. 28–20 at B). They are sized according to the tip and the diameter of the blade.

NUTDRIVERS

Nutdrivers have a shaft and handle like a screwdriver and a hexagonal socket tip for use with hexagonal-head screws and nuts (Fig. 28–21 at A). The size of a nutdriver is given by the distance across the *flats,* or any two parallel sides, of its socket. Some tool manufacturers color-code the handles for easy size identification (Fig. 28–21 at B).

SELF-TEST

Test your knowledge by writing, on a separate sheet of paper, the word or words that most correctly complete the following statements:

1. Galvanized sheet metal consists of a base of mild _____ coated with _____.
2. The thickness of sheet metal is given by a _____ gage number or by an _____ gage number. As the gage number increases, the thickness of the metal _____.
3. A chassis layout diagram shows the _____ of the sheet metal to be used, the _____ and _____ of the holes to be drilled or punched, and the places where the metal is to be _____.
4. Sheet metal is most often cut with _____ or with _____.
5. The centers of holes to be drilled through sheet metal should be marked with a _____.
6. The diameter of a twist drill is given by a _____, by a whole _____, or by a _____.
7. _____ should always be worn while drilling holes.
8. Never try to drill a hole in sheet metal unless the metal is securely _____.
9. Round holes from $\frac{1}{2}$ to $3\frac{1}{8}$ in. (13 to 79 mm) in diameter are often cut through sheet metal with a _____ punch.
10. Two common tools used for reaming in sheet-metal work are the _____ and the _____ reamer.
11. All cutting and drilling operations should be done on a chassis blank before it is _____ or _____ into its final shape.
12. Two machines used for forming and bending sheet metal are the _____ and the _____.
13. The hand seamer is used for making simple _____ or _____ in sheet metal.
14. The customary size of a machine screw is given by the _____ of its threaded part, the _____ of threads per inch, and the _____ of the threaded portion.
15. Self-tapping screws are a kind of _____ screw.
16. The size of a nutdriver is given by the distance across the _____ of its socket.

FOR REVIEW AND DISCUSSION

1. How is the thickness of sheet metal given?
2. Describe a chassis layout diagram.
3. Describe the first step in making a chassis layout diagram.
4. State five guidelines that are useful in determining the location of components on a chassis.
5. With what should you enlarge prick-punch marks?
6. How is the size of a twist drill given?
7. Describe the safe method of drilling a hole through sheet metal with a portable electric drill.
8. How is a chassis punch used?
9. Name two kinds of reaming tools. Tell the purpose for which each is used.
10. Describe the process of cutting a large opening through sheet metal.
11. Name two machines that are used for forming and bending sheet metal.
12. For what purpose is a hand seamer used?
13. How is the customary size of a machine screw given?
14. What are self-tapping screws?

INDIVIDUAL-STUDY ACTIVITY

Demonstrate the use of the tools used in working with sheet metal. Include in the demonstration common kinds of screws and screwdrivers.

6 Electricity at Work

Unit 29   Lamps and Lighting

Electric energy can be changed into light energy in a number of different ways. The first practical device for producing light by electricity was the carbon-filament lamp invented by Thomas Edison in 1879. Since then, many kinds of lamps and other lighting devices have been developed.

INCANDESCENT LAMP

The word *incandescent* means "glowing from intense heat." In an *incandescent lamp*, current passes through a filament of the metal tungsten, causing the filament to become incandescent, or white hot (Fig. 29-1).

When an incandescent lamp is made, most of the air is removed from the glass bulb before it is sealed. This is done to keep oxygen from coming into contact with the filament. That would cause it to burn out quickly. A mixture of the gases nitrogen and argon is placed in the bulb to lessen the evaporation of the filament caused by its high temperature. This lengthens the life of the lamp.

The life of general-service incandescent lamps for home use ranges from 750 to 1,500 hours. The average life of a lamp is printed on the carton.

Watts and Lumens. Incandescent lamps come in a variety of sizes and shapes (Fig. 29-2). The electrical size of a general-service incandescent lamp is given in watts and in lumens. The wattage rating indicates the amount of energy the lamp uses. The *lumen* rating indicates the amount of light it produces. A 100-watt light bulb gives off about 1,000 lumens.

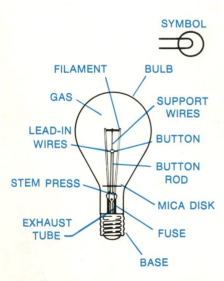

SYMBOL

FILAMENT   BULB

GAS   SUPPORT WIRES

LEAD-IN WIRES   BUTTON

BUTTON ROD

STEM PRESS

MICA DISK

EXHAUST TUBE   FUSE

BASE

Fig. 29-1. Construction of a typical general-purpose incandescent lamp

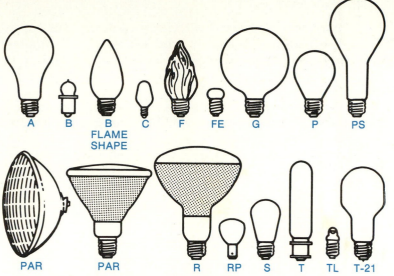

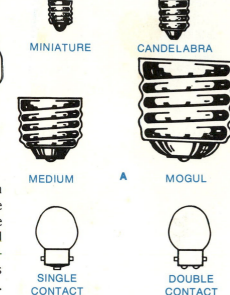

Fig. 29–2. Sizes and shapes of common incandescent lamps. The numbers and letters beneath the lamps refer to manufacturers' standard types.

MINIATURE CANDELABRA

MEDIUM **A** MOGUL

SINGLE CONTACT DOUBLE CONTACT

B

Fig. 29–3. Incandescent-lamp bases: (A) screw; (B) bayonet

Lamp Bases. Screw-base incandescent lamps are made in several sizes (Fig. 29–3 at A). Small lamps, such as those used in flashlights and as pilot lights, usually have miniature bases. The candelabra base is found on decorative lamps and large pilot lights. The medium base is used with general-purpose lamps rated at 300 watts or less. The mogul base is used with lamps having a wattage of more than 300 watts. Lamps rated at more than 1,500 watts have special bases.

Miniature- and candelabra-base lamps often have a bayonet base (Fig. 29–3 at B). To put such a lamp in a socket, line up the pins on the base with the grooves or slots in the socket. Then press the lamp in and turn it clockwise until it locks into place.

THREE-LITE LAMP

A *three-lite lamp* can produce three different amounts of light. This kind of lamp is used in floor and table fixtures. It has two separate filaments. In the 50-100-150–watt lamp, for example, one of these is a 50-watt filament. The other is a 100-watt filament (Fig. 29–4). The common wire of the filament is connected to the *shell,* the threaded part of the base. The second filament lead is connected to a ring on the bottom of the base. The third filament lead is connected to the center contact of the base. A special selector switch in the socket connects either or both of the filaments to the circuit. Thus, the lamp can be operated at 50, 100, or 150 (50 + 100) watts.

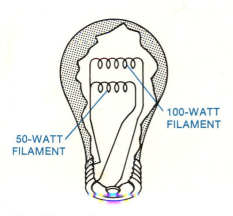

Fig. 29–4. The three-lite lamp

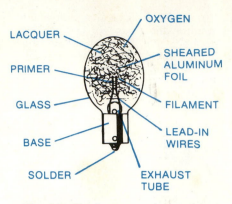

LACQUER
OXYGEN
PRIMER
SHEARED ALUMINUM FOIL
GLASS
FILAMENT
BASE
LEAD-IN WIRES
SOLDER
EXHAUST TUBE

Fig. 29–5. A common large photo-flash lamp (*General Electric Company*)

PHOTOFLASH LAMP

The *photoflash lamp,* used with flash cameras, is a special kind of incandescent lamp. It produces a large amount of light for a very short time. The lamp bulb contains pure oxygen and thin strips of aluminum or magnesium foil (Fig. 29–5).

The tungsten filament of the lamp is made to burn out with a spark when current passes through it. The spark ignites the aluminum or the magnesium, which burns very rapidly in the oxygen. This produces a very bright light that lasts for only a fraction of a second.

FLUORESCENT LAMP

The *fluorescent lamp* belongs to a group of light sources known as *electric-discharge lamps.* Such lamps produce light by passing current through a gas. The typical fluorescent lamp is a glass tube with a tungsten filament sealed into each of its ends. The inner surface of the tube is coated with a chemical called a *phosphor* (Fig. 29–6). When the lamp is made, most of the air is removed from the tube. A small amount of argon gas and mercury is sealed inside it.

Lamp Circuit. To operate what is called a *preheat* fluorescent lamp, a ballast and a starter must be used (Fig. 29–7 at A). The *ballast* is a coil of insulated magnet wire wound around an iron core. The *starter* acts as a switch that is automatically turned off when its moving element, a bimetal strip,

Fig. 29–6. View of one end of a typical preheat fluorescent lamp (*Westinghouse Electric Corporation*)

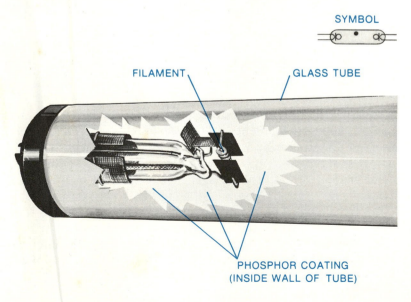

SYMBOL

FILAMENT
GLASS TUBE

PHOSPHOR COATING
(INSIDE WALL OF TUBE)

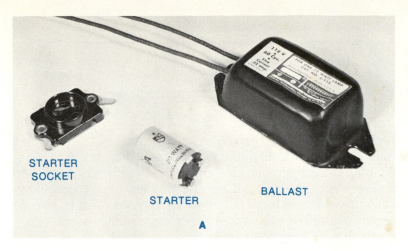

STARTER SOCKET

STARTER

BALLAST

A

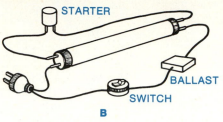

STARTER

BALLAST

SWITCH

B

Fig. 29–7. The preheat fluorescent-lamp circuit: (A) auxiliary components; (B) complete lamp circuit

reaches a certain temperature. Bimetal-strip switches (thermostats) are discusssed in Unit 33.

In the lamp circuit, the filaments, the ballast, and the starter are connected in series (Fig. 29–7 at B). When the main lamp switch is turned on, current passes through the circuit and heats the filaments. As a result, the temperature in the lamp increases. This causes the mercury to turn to gas. The current is also heating the bimetal strip in the starter. After a brief preheat period, the starter opens the circuit.

The sudden stoppage of current produces several thousand volts across the terminals of the ballast by self-inductance. This high voltage also appears across the ends of the fluorescent tube. The high voltage ionizes the gas within the tube. This causes the gas to become a good conductor. Because of its high inductance, the ballast also limits the current through the lamp circuit to a safe value.

In an *instant-start* fluorescent lamp, a step-up transformer produces the high voltage needed to ionize the gas in the lamp. Most instant-start lamps have single-pin bases.

How Light Is Produced. Current flowing through the ionized gas mixture in the lamp excites the mercury atoms (Fig. 29–8). Their energy is released in the form of *ultraviolet light*, which people cannot see. When this light strikes the inside surface of the lamp tube, the phosphor *fluoresces*, or glows. This glow is the visible light produced by the fluorescent lamp.

Efficiency. Fluorescent lamps have been available since 1938. They have become a very common source of light in homes, stores, offices, factories, and schools. One of the

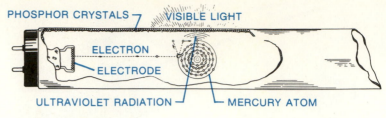

PHOSPHOR CRYSTALS — VISIBLE LIGHT

ELECTRON
ELECTRODE

ULTRAVIOLET RADIATION — — MERCURY ATOM

Fig. 29–8. How light is produced by a fluorescent lamp (*Westinghouse Electric Corporation*)

main reasons for their popularity is their efficiency. A fluorescent lamp wastes much less energy in the form of heat than does an incandescent lamp. The modern fluorescent-lamp system (which includes a ballast or transformer) is able to produce approximately three times more light than an incandescent lamp with the same wattage rating.

MERCURY LAMP

The *mercury,* or *mercury vapor, lamp* is widely used for producing large amounts of light (Fig. 29–9). Mercury lamps are used for street and bridge lighting, parking-area lighting, and lighting for other places where the lamp must be mounted high above the ground. The *arc tube,* or inner bulb, of this lamp is made of quartz, or rock crystal. The arc tube contains argon gas and a small amount of mercury. The outer bulb regulates and maintains the temperatue of the inner bulb while the lamp is working.

To start the lamp, a high voltage from a transformer, called a ballast, is applied to the lower main electrode and to the starting electrode next to it. This produces a glow discharge between these electrodes that heats the mercury, causing it to turn to gas. The gas mercury presents a low-resistance path for current through the arc tube from one main electrode to the other. The current, which is seen as a brilliant arc, is the source of light.

GLOW LAMPS

In a *glow lamp,* two electrodes are placed very close to one another in a sealed glass bulb containing argon or neon gas (Fig. 29–10 at A). When the right dc voltage is applied to the electrodes, some valence electrons within atoms of the gas gain enough energy to escape from their parent atoms. As a result of this ionization, positive gas ions are produced. As these ions move toward the negative electrode, some of them collide with and regain electrons. This causes energy in the form of visible light to be released near the negative electrode. When a glow lamp is operated with alternating voltage,

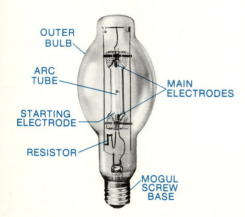

OUTER
BULB

ARC
TUBE

MAIN
ELECTRODES

STARTING
ELECTRODE

RESISTOR

MOGUL
SCREW
BASE

Fig. 29–9. The mercury-vapor lamp (*Westinghouse Electric Corporation*)

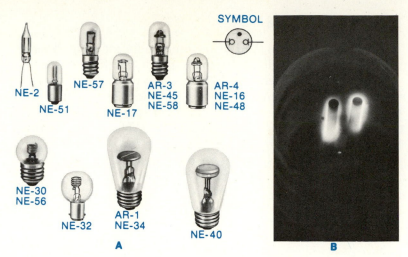

SYMBOL

NE-2
NE-51
NE-57
NE-17
AR-3
NE-45
NE-58
AR-4
NE-16
NE-48

NE-30
NE-56
NE-32
AR-1
NE-34
NE-40

A

B

Fig. 29–10. Neon and argon glow lamps (A). Both electrodes (at B) seem to glow when the lamp is operated on alternating voltage. The numbers and letters beneath the lamps refer to manufacturers' standard types. (*General Electric Company*)

the electrodes become alternately negative and positive. This happens so quickly that both electrodes seem to be glowing at the same time (Fig. 29–10 at B).

The voltage needed to start a glow lamp is called the *starting,* the *striking,* or the *ionization* voltage. The typical glow lamp has a starting voltage of about 60 volts.

Neon glow lamps of from $\frac{1}{25}$ to 3 watts in size are used as test and pilot lights. One advantage of these lamps is that they start instantly when the right voltage is applied to them. For this reason, glow lamps are very often used as high-speed indicators. When operated at 120 volts, small glow lamps such as the neon NE-2 require a current-limiting resistor. This resistor has a resistance of about 100,000 ohms. It is connected in series with one of the lamp leads or terminals. The resistor may be connected externally. It may also be contained in an indicator-light assembly. Larger glow lamps often have high-resistance, current-limiting coils in their bases.

NEON SIGNS

In a *neon sign,* an electrode is sealed into each end of a glass tube filled with neon gas. The longer the tube, the higher the voltage that must be applied to the electrodes to sufficiently ionize the gas. In common neon signs, a voltage of 10,000 volts or more may be needed. This high voltage is supplied by a step-up transformer (Fig. 29–11).

By using argon, helium, or a mixture of these and other gases, light of different colors can be obtained. Different colors are also obtained by using tinted glass tubes or coatings on the inside surface of the tubes.

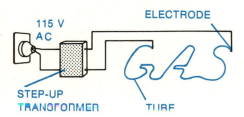

Fig. 29–11. A step-up transformer is used with a neon sign.

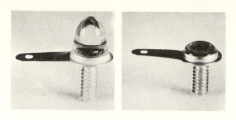

Fig. 29–12. These red-light-emitting diodes made of diffused gallium arsenide phosphide have an extremely long service life. They operate at 1.65 volts and 50 milliamperes. (*Monsanto Commercial Products Company*)

LIGHT-EMITTING DIODES

Light-emitting diodes (LED) are semiconductor devices used as pilot and display lights in many products (Fig. 29–12).

One kind of light-emitting diode is a p-n junction arrangement. Its basic element is gallium arsenide. When such a junction is forward-biased, electrons from the n section move across the junction and into the p section. There they recombine with holes. As a result, energy in the form of visible light or of infrared light is radiated from the junction area.

LASER

The word *laser* comes from the phrase *l*ight *a*mplification by *s*timulated *e*mission of *r*adiation. Thus, a laser system is a means of producing and amplifying light.

Light is a form of electromagnetic energy radiated from a source of energy. Except that it has a much higher range of frequencies, its properties are like those of the electromagnetic waves radiated from the antenna of a radio or television transmitter.

The color of light depends on the frequency or frequencies of the radiation that reaches our eyes. The frequency range of visible light extends from about 400 million MHz (red) to about 700 million MHz (violet) (Fig. 29–13).

Incoherent Light. The light produced by a common incandescent lamp consists of electromagnetic radiations with a wide range of frequencies. The visible light from the lamp is then actually made up of a combination of colors. Our eyes see this as white light. Such light is called *incoherent,* or *polychromatic,* light. An important characteristic of this light is that it spreads out very quickly after leaving the source (Fig. 29–14 at A). As a result, this light loses much of its energy when it travels over a long distance.

Coherent Light. The light produced by a laser consists of radiations that are very nearly of the same frequency. They are also in *phase*, or in step, with each other (Fig. 29–14 at B).

Fig. 29–13. The frequency range of visible-light radiations. This is sometimes called the electromagnetic spectrum of light.

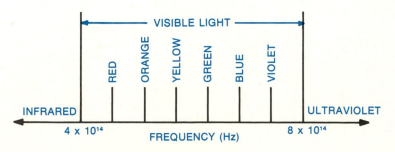

This is called *coherent light.* A very important characteristic of coherent light is that it does not spread out very much even when it is far from its source. Because of this, the light stays highly concentrated. Thus, it has much energy as it strikes a target area.

Laser Beam. The light output from a laser is in the form of a narrow beam of light (Fig. 29–15). The color of the beam depends on the kind of laser used. A laser beam can be modulated by many different electrical signals over long distances. Thus, it can be used in communications systems. (Modulation is explained in Unit 40.) The laser beam is also used for locating objects and measuring their distances from fixed points. More powerful laser beams are used for welding, metal cutting, and surgical cutting.

Operation. The working principle of a laser is that electrons in atoms can be raised from one energy level to another

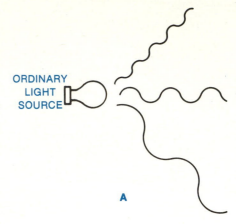

ORDINARY LIGHT SOURCE

A

LASER BEAM

B

Fig. 29–14. Patterns of light radiation: (A) incoherent light; (B) coherent-light radiation of a laser beam

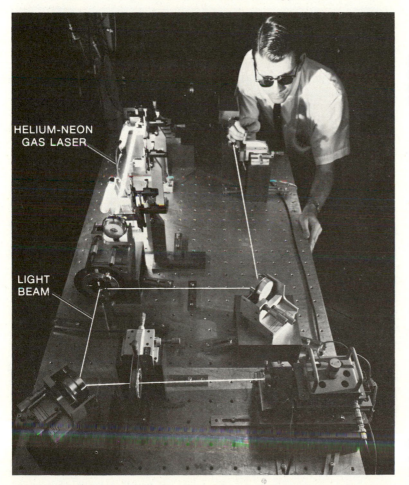

HELIUM-NEON GAS LASER

LIGHT BEAM

Fig. 29–15. Laser beam used for an optical communications experiment (*Bell Telephone Laboratories*)

by applying external energy. Electrons that are raised from one energy level to the next higher energy level are said to be *excited*. Excited electrons tend to become deexcited and return to their original energy levels when external energy is no longer applied to their atoms. When this occurs in certain materials, energy in the form of light is released.

Laser materials that react in this way include ruby rods, combinations of gases, and special kinds of semiconductor p-n junctions. In several laser systems, the material is assembled into an *optical cavity*, which has mirrorlike ends.

The energy used to excite electrons in the laser is called *pumping energy*. In the ruby laser, the pumping energy is visible light supplied by a lamp. In the gas laser, it is a high-frequency electrostatic field. In the semiconductor laser, it is usually a high dc voltage.

SELF-TEST

Test your knowledge by writing, on a separate sheet of paper, the word or words that most correctly complete the following statements:

1. The word *incandescent* means "_____ from intense heat."
2. The filament of a common incandescent lamp is made of _____ .
3. Nitrogen gas and argon gas are put inside the bulb of an incandescent lamp to lessen the _____ of the _____ .
4. The life of general-service incandescent lamps usually ranges from _____ to _____ hours.
5. The lumen rating indicates the _____ of light a lamp produces.
6. The four main kinds of lamp bases are _____, _____, _____, and _____ .
7. The fluorescent lamp belongs to a group of light sources known as _____ .
8. An instant-start fluorescent lamp is started by a _____ .
9. The light-emitting diode produces light when its p-n junction is _____-biased.
10. The word *laser* comes from the phrase _____ .
11. The laser beam is an example of _____ light.

FOR REVIEW AND DISCUSSION

1. Define the word *incandescent*. Explain how an incandescent lamp works.
2. Name the two units used to denote the size of general-service incandescent lamps.
3. Describe the construction of a common fluorescent-lamp tube.
4. Name the parts of a preheat-fluorescent-lamp circuit and explain how they work.
5. For what purposes are mercury lamps used?
6. What is meant by the starting voltage of a glow lamp?
7. Why must a step-up transformer be used with a neon sign?
8. Explain the origin of the term *laser*.
9. What factor determines the color of visible light?
10. Explain the difference between coherent and incoherent light.
11. What is meant by the pumping energy of a laser system?

INDIVIDUAL-STUDY ACTIVITIES

1. Prepare a written or an oral report about the nature of visible light.
2. Prepare a written or an oral report on the development and applications of lasers.

274

Unit 30   Producing Chemical Reactions

A chemical reaction usually results in changes to the materials taking part in the reaction. For example, when oxygen reacts with iron, a new substance called *iron oxide* (rust) is formed. When paper is burned, it changes into ashes and smoke. In each case, the composition of the molecules in the substance is changed. New substances having different properties are produced.

The chemical reactions that take place when electric energy is applied to a substance result from the movement of charges through the substance. A liquid that ionizes to form positive and negative ions and that can conduct these charges is called an *electrolyte.*

In this unit, you will learn how electrolytes cause chemical changes by a process known as *electrolysis.* This process is widely used in industry for producing certain gases and chemicals, for electroplating, for refining copper, and for extracting metals such as aluminum and magnesium from their ores.

ELECTROLYSIS OF WATER

Pure water is a very poor conductor of electricity because it does not ionize easily. However, when certain chemical compounds are added to pure water, the solution becomes a good conductor. Applying a voltage to such a solution produces a process called *electrolysis.* Electrolysis is the breaking down of water into its two chemical elements, the gases hydrogen and oxygen.

In one industrial process for making hydrogen and oxygen gases, sulfuric acid is added to pure water. Each molecule of the acid ionizes into two positive hydrogen ions (H^+) and one negative sulfate ion having two negative charges (SO_4^{--}). The positive hydrogen ions are attracted to the cathode (negative electrode). At the same time, the negative sulfate ions are attracted to the anode (positive electrode) (Fig. 30–1). At the cathode, each hydrogen ion obtains an electron. It thereby becomes a neutral atom of hydrogen. Hydrogen, being a gas, rises from the solution. It collects at the cathode. At the anode, the negative sulfate ions deposit their excess electrons and become neutral. However, the sulfate reacts at once with the hydrogen of the water, forming sulfuric acid again. The oxygen in the water, which remains after the hydrogen has been removed, rises through the solution. It collects at the anode. This process continues as the sulfuric acid is recycled through the ionization and sulfate stages.

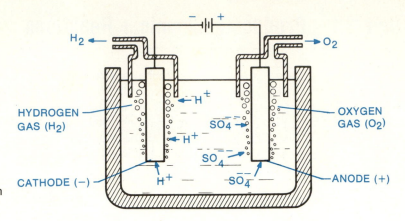

Fig. 30–1. Producing hydrogen and oxygen gases by electrolysis

ELECTROPLATING

Electroplating is a form of electrolysis used to coat a base material (usually a metal) with a thin layer of another metal. The base metals can be iron, steel, brass, or similar metals. They can be electroplated with one or more metals, which include cadmium, chromium, copper, gold, silver, and tin.

Cadmium plating will protect the base metal from corrosion (Fig. 30–2). Chromium is used to improve the appearance of base metals and to provide a much harder surface. Copper plating is generally used to provide an undercoating over which other metals are plated. Gold plating is used for decorative purposes and for improving the conductivity of

Fig. 30–2. Electroplating unit in an industrial plant (*Pioneer-Central Division of Bendix Aviation Corporation*)

contact points in some electrical equipment. Tin plating is used on surfaces likely to come in contact with food and to allow easy soldering of surfaces.

The Plating Solution. Electroplating requires the use of a plating solution, or electrolyte, that contains a compound of the plating metal. In copper plating, for example, the electrolyte consists mainly of a solution of water and copper sulfate ($CuSO_4$).

Basic Operation. The construction and operation of a basic copper-plating assembly is shown in Fig. 30–3. Here the object to be plated (the cathode) and a piece of pure copper (the anode) are suspended in the electrolyte and connected to a battery.

As the copper sulfate in the solution dissolves, it ionizes into positive copper ions (Cu^{++}) and negative sulfate ions (SO_4^{--}). The copper ions are attracted to the cathode. There they obtain the needed electrons and become metallic copper. This is deposited on the object being plated. The sulfate ions are attracted to the anode. There they combine with copper atoms to form copper sulfate. This goes into solution and ionizes. Thus, the copper sulfate replaces the copper removed from, or plated out of, the electrolyte. During this

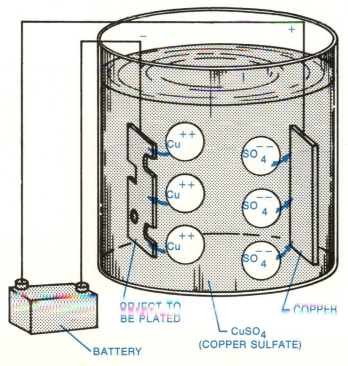

Fig. 30–3. The copper-plating circuit

time, the current in the external circuit consists of electrons that move into the cathode from the negative terminal of the battery and electrons that move into the positive terminal of the battery from the anode.

As the electroplating process continues, the pure copper that forms the anode is gradually used up and must be replaced. At the same time, the coating of copper deposited on the object being plated gets thicker.

The time needed to electroplate a base metal to a given thickness depends on the kind of plating metal used and the density of the ions in the electrolyte. Ion density is indicated by the value of the current in the external circuit. For this reason, a practical electroplating assembly has meters that show the current in the external circuit and the voltage at which the circuit is operating. The circuit also has controls for adjusting the plating current as needed for electroplating different metals.

EXTRACTING METALS FROM ORES

The production of aluminum is the most common process in which electrolysis is used to extract metals from their ores. The first step in producing aluminum is changing the ore (bauxite) into a white powder called *alumina* (aluminum oxide). Then, in a method known as the *Hall process,* alumina is dissolved in molten cryolite (sodium aluminum fluoride) to form the electrolyte. The heat needed to keep the electrolyte liquid is produced by the current needed for electrolysis. Carbon rods and a carbon lining make up the cathode and the anode of what is referred to as an *electrolytic furnace* (Fig. 30–4).

Because of ionization, negative oxygen ions and positive aluminum ions are present in the electrolyte. The aluminum ions are attracted to the cathode. There they obtain electrons and are discharged to form aluminum atoms. The molten aluminum then collects at the bottom of the furnace and is withdrawn.

Several other metals are extracted from their ores by electrolysis. These include magnesium, sodium, potassium, and cesium.

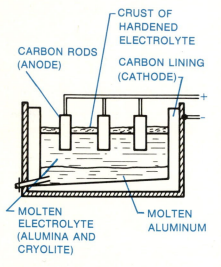

CARBON RODS (ANODE)

CRUST OF HARDENED ELECTROLYTE

CARBON LINING (CATHODE)

MOLTEN ELECTROLYTE (ALUMINA AND CRYOLITE)

MOLTEN ALUMINUM

Fig. 30–4. Basic construction of an electrolytic furnace used for producing aluminum

LEARNING BY DOING

28. Making a Saltwater Rheostat. This kind of rheostat is not used in practical circuits. It does, however, clearly show the effect of ionization in a liquid solution.

*medium-size glass or plastic tumbler filled with water,
preferably distilled*
*2 pieces of sheet copper or aluminum, $\frac{1}{2} \times 2\frac{1}{4}$ in.
(13 × 57 mm)*
1 pilot lamp, no. 46
1 miniature lamp socket, screw-base
common table salt in shaker
glass or plastic stirring rod
*2 6-volt batteries, or a low-voltage power supply with a
current rating of at least 0.4 ampere*
volt-ohm-milliammeter

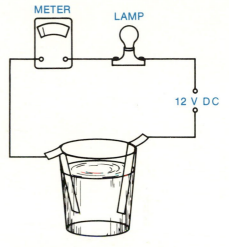

Procedure

1. Bend the pieces of sheet metal. Put them in the tumbler containing the distilled water, as shown in Fig. 30–5.
2. With the volt-ohm-milliammeter adjusted to a range of about 0 to 250 mA, connect the circuit shown in Fig. 30–5. Explain why there is now no current in the circuit.
3. Slowly sprinkle salt into the water and stir. The current should gradually increase to a value that causes the lamp to light dimly.
4. Keep adding salt to the solution until the current has increased to 250 mA. The lamp should now be working normally.
5. Explain the ionization process that changed the nonconducting solution into an electrolyte able to conduct ionic charges.

Fig. 30–5. Tumbler assembly and circuit for Learning by Doing No. 28, "Making a Saltwater Rheostat"

SELF-TEST

Test your knowledge by writing, on a separate sheet of paper, the word or words that most correctly complete the following statements:

1. A solution that will produce positive and negative ions is called an _____ .
2. Pure water is a very poor conductor because it does not _____ easily.
3. As a result of electrolysis, water is broken down into the gases _____ and _____ .
4. Cadmium plating is done to protect a base metal from _____ .
5. Metals extracted from their ores by electrolysis include _____ , _____ , _____ , _____ , and _____ .

FOR REVIEW AND DISCUSSION

1. What is an electrolyte?
2. Why is pure water a very poor conductor?
3. What is meant by the electrolysis of water?
4. Describe an industrial process that uses electrolysis to produce the gases hydrogen and oxygen.
5. Define electroplating.
6. Name two metals used as base metals in electroplating.
7. Give the main reasons for plating base metals with cadmium, with chromium, and with gold.
8. Explain the basic copper-plating process.
9. Name three metals commonly extracted from their ores by electrolysis.

Unit 31　Home Electrical Systems

The electrical system in your home consists of conductors, devices, and various electrical equipment. The purpose of the system is to distribute electric energy throughout the building. Although this unit discusses mostly residential (house) wiring, many of the same materials are found in the electrical systems of other buildings. The main parts of a typical home electrical system are shown in Fig. 31-1.

NATIONAL ELECTRICAL CODE

The *National Electrical Code* is a publication sponsored by the National Fire Protection Association in cooperation with the American National Standards Institute (ANSI). It includes recommendations and guidelines on wiring materials

Fig. 31-1. A typical home electrical system

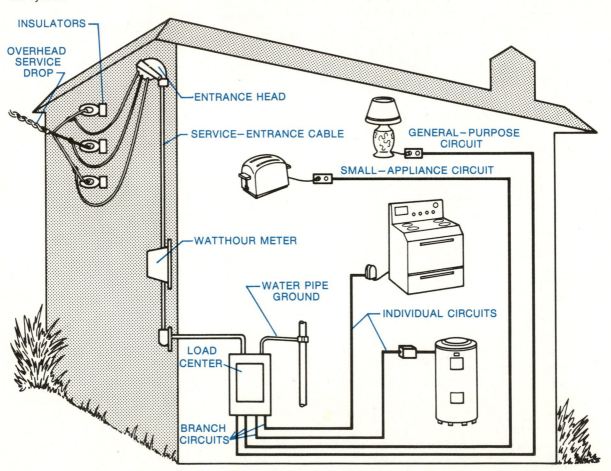

INSULATORS

OVERHEAD SERVICE DROP

ENTRANCE HEAD

SERVICE-ENTRANCE CABLE

GENERAL-PURPOSE CIRCUIT

SMALL-APPLIANCE CIRCUIT

WATTHOUR METER

WATER PIPE GROUND

INDIVIDUAL CIRCUITS

LOAD CENTER

BRANCH CIRCUITS

and methods. Following these guidelines will result in the safe use of electric energy for lighting, heating, communications, and other purposes. The code is reviewed continuously. A revised edition is issued every three years.

The *National Electrical Code* contains the basic minimum requirements needed to protect both persons and property from electrical hazards. Because of this, the *Code* is very useful for developing safety standards throughout the country. These standards, as well as those of local electrical codes, are used by electricians, contractors, and inspectors. Using these standards means that all building wiring will be done in an approved way.

SERVICE DROP

The *service drop* of a house wiring system is the part of the system that extends from a terminal point outside the house to the nearest electrical company distribution line. This line is fed from a distribution transformer. That device steps down the high voltage of a main power line to about the 115 and 230 volts used in the house.

The service drop can be installed overhead or underground. An underground service drop is sometimes called a *service lateral*. The distribution transformer is mounted overhead on a utility pole, at ground level on a concrete pad, or underground in a vault (Fig. 31-2). The distribution transformer may supply electric energy to several houses.

Three-wire Service. In order to supply two different voltages (115 and 230 volts), the service drop is made up of three conductors (Fig. 31-3). One of these, called the *neutral*, is often bare. The other two conductors are insulated.

230-volt Operation. The 230-volt lines brought into a house are used to operate major appliances, such as ranges, hot-water heaters, and air-conditioning units. Comparable voltages used by different electric power companies include 208, 220, and 240. The reason higher voltages are used for these purposes can best be explained by referring to power formulas.

Suppose, for example, that a 1,725-watt appliance is operated at 115 volts. The amount of current used by the appliance under this condition is $I = P/E$, or $1,725/115 = 15$ amperes. Now, suppose that the same appliance is operated at 230 volts. The current is then $1,725/230$, or 7.5 amperes.

The lower value of current (at 230 volts) has two important advantages. First, the wires leading to the appliance can be

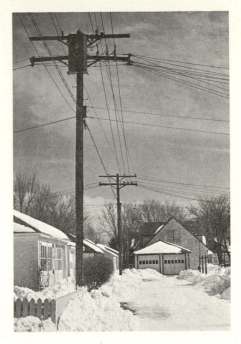

Fig. 31-2. Distribution transformer mounted overhead on a utility pole

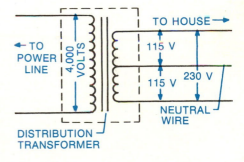

Fig. 31-3. Voltages of a typical three-wire system used in homes

smaller. This reduces the cost of installing them. Second, the lower current results in a smaller power loss in the wires. According to the formula $P = I^2R$, the loss is directly proportional to the square of the current.

SERVICE ENTRANCE

The *service-entrance conductors* extend from the point at which the service-drop conductors are attached at the house to the load center. The complete service entrance is generally thought to include the conductors, the watthour meter, and the load center. Usually the electric power company takes care of the wiring from the power line up to and including the meter.

Service-entrance Cable. The service-entrance cable has three conductors (Fig. 31–4). One of these, the neutral conductor, is bare. It is usually in the form of individual strands of wire. When the cable is to be connected to a terminal point, the strands of the neutral conductor are twisted together to form a single-stranded conductor.

The Watthour Meter. The watthour meter measures and records the amount of electric energy (in kilowatthour units) supplied to a building. The meter is installed by the power company and belongs to it.

The Load Center. The load center, also called a *fuse* or *circuit-breaker panel,* is the unit from which electricity is distributed to different locations in a house (Fig. 31–5). It contains the main-line fuses and the switch. The latter can be used to disconnect all electrical service from a house. The load center also contains the fuses or circuit breakers that protect the branch circuits of the wiring system.

The size of the load center is designated by the number of fuses or circuit breakers that can be placed into it and by its

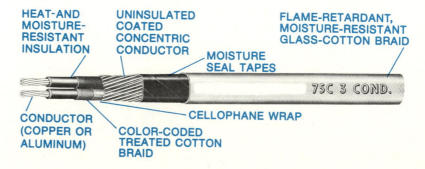

HEAT-AND MOISTURE-RESISTANT INSULATION

UNINSULATED COATED CONCENTRIC CONDUCTOR

MOISTURE SEAL TAPES

FLAME-RETARDANT, MOISTURE-RESISTANT GLASS-COTTON BRAID

75C 3 COND.

CELLOPHANE WRAP

CONDUCTOR (COPPER OR ALUMINUM)

COLOR-CODED TREATED COTTON BRAID

Fig. 31–4. Service-entrance cable (*General Electric Company*)

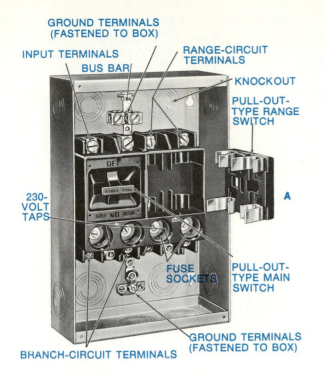

GROUND TERMINALS
(FASTENED TO BOX)

INPUT TERMINALS

RANGE-CIRCUIT
TERMINALS

BUS BAR

KNOCKOUT

PULL-OUT-
TYPE RANGE
SWITCH

A

230-
VOLT
TAPS

OFF

PULL-OUT-
TYPE MAIN
SWITCH

FUSE
SOCKETS

BRANCH-CIRCUIT TERMINALS

GROUND TERMINALS
(FASTENED TO BOX)

B

Fig. 31-5. Load centers: (A) fuse panel; (B) circuit-breaker panel (*Square D Company*)

load, or ampere, rating. The fuse panel shown in Fig. 31–5 at A is rated at 60 amperes. Larger load centers, equipped with more fuse holders, have ampere ratings of 100, 150, 200, or more amperes.

The Service Ground. To reduce the danger of shock and to protect against lightning, the neutral conductor of the service-entrance cable is grounded. This is done by connecting it to the neutral bus bar in the load center. The bus bar is, in turn, grounded by being connected to a ground rod or in older homes, to a water pipe.

BRANCH CIRCUITS

The *branch circuits* of a house wiring system distribute electricity from the load center to the parts of the house. The three common kinds of branch circuits are discussed below.

General-purpose Circuits. General-purpose branch circuits are used for lighting and outlets. The outlets are intended to serve radios, television sets, clocks, and small appliances other than those used for food preparation (such as toasters, broilers, and the like). Lighting circuits are usually wired with

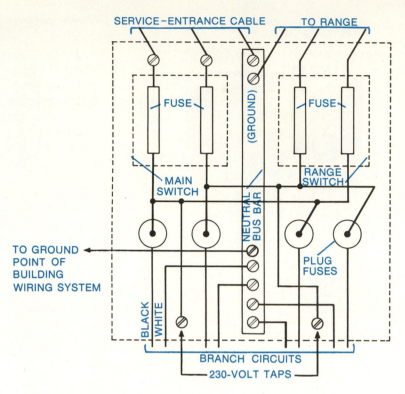

Fig. 31–6. Wiring diagram of a four-pole fuse panel

SERVICE–ENTRANCE CABLE TO RANGE

FUSE (GROUND) FUSE

MAIN SWITCH RANGE SWITCH

NEUTRAL BUS BAR

TO GROUND POINT OF BUILDING WIRING SYSTEM

PLUG FUSES

BLACK WHITE

BRANCH CIRCUITS
230-VOLT TAPS

no. 12 (2.00-mm) or no. 14 (1.60-mm) wire and are protected with 20- or 15-ampere fuses or circuit breakers.

As shown in Fig. 31–6, one wire of each general-purpose circuit is connected to the neutral bus bar of the load center. The grounded neutral wire is color-coded white. The wire connected directly to the fuse or circuit breaker is generally black or red. This circuit wire is often called the *hot* wire.

The grounded white wire of a two-wire, general-purpose circuit is never used as a safety-grounding wire. When the cabinet or case of a product operated from such a circuit is to be grounded, a third grounding wire is used.

Small-appliance Circuits. Small-appliance circuits are used for such appliances as refrigerators, toasters, broilers, coffee makers, and irons. They are wired with no. 12 (2.00-mm) wire and are protected with 20-ampere fuses or circuit breakers.

Individual Circuits. An individual branch circuit is generally used with only one piece of equipment. Examples of individual circuits are those used with electric ranges, dryers, water heaters, heating systems, and air conditioners.

A range circuit is usually a three-wire, 115–230-volt circuit. It has two hot wires and a neutral wire. There is a fuse or circuit breaker in each of the hot wires. A 230-volt individual circuit is a two-wire circuit. Both wires are hot and are protected with a fuse or circuit breaker.

The cabinet of an electric range is grounded by being connected to the neutral wire of the three-wire circuit (Fig. 31–7). However, a separate grounding wire is also connected from a terminal on the frame of the range to a grounding point of the wiring system. The cabinet or frame of an appliance or other equipment operated with a two-wire, 230-volt circuit is grounded with an additional grounding wire. This wire is usually located in the connecting cable. Many electrical appliances and tools operated at 115 volts, such as washing machines, dishwashers, and electric drills, present potential shock hazards. Therefore, they use a separate grounding wire.

230-volt Taps. The 230-volt taps in a panel are for wiring additional individual branch circuits. When they are used for such a circuit, extra fuses or circuit breakers must be installed outside the panel. These are necessary because the circuit is not protected with branch-circuit fuses or circuit breakers in the panel.

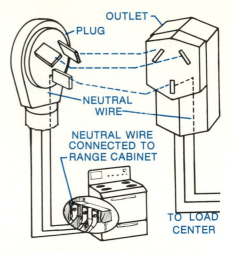

Fig. 31–7. The cabinet of an electric range is grounded using a three-wire range plug and outlet.

BRANCH-CIRCUIT WIRING

Branch circuits are wired with several kinds of approved wiring meant to protect the wire from mechanical damage. The most common of these in house wiring systems are discussed below.

Nonmetallic Sheathed Cable. Nonmetallic sheathed cable usually has two or three insulated wires. It may or may not contain a grounding wire (Fig. 31–8). NM cable is used for all kinds of indoor wiring work. It is never buried in cement or plaster. NM cable has a moisture-resistant overall covering. This allows it to be used for both exposed and concealed work in dry, moist, and damp locations and in outside or inside walls of masonry, block, or tile.

Nonmetallic sheathed cable is made in several sizes, designated as 14/2, 12/2, 12/3, and the like. The number to the left of the slash is the AWG size of the individual conductors. The other number is the number of conductors in the cable other than the bare grounding wire. Therefore, a 12/2 cable has two no. 12 AWG (2.00-mm) insulated wires.

Armored Cable. In armored cable, the insulated wires are in a flexible armor sheath. The sheath is made of a single spiral of galvanized steel tape that is interlocked to form a tubelike

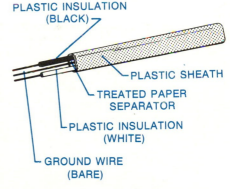

Fig. 31–8. Construction of two-conductor NM nonmetallic sheathed cable with a grounding wire

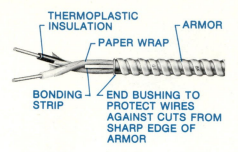

THERMOPLASTIC INSULATION

ARMOR

PAPER WRAP

BONDING STRIP

END BUSHING TO PROTECT WIRES AGAINST CUTS FROM SHARP EDGE OF ARMOR

Fig. 31–9. Armored cable (*General Electric Company*)

enclosure. The cable also contains a bare copper or aluminum bonding strip in contact with the armor along its entire length (Fig. 31–9). The purpose of the bonding strip is to provide continuity of the circuit's ground in case the armor is broken at any point. This is necessary because the armor is usually part of the grounding system.

AC cable is approved for use in dry locations where branch-circuit wiring must be protected against wear or be embedded in plaster. ACL cable is used in damp locations and where oil or other substances may corrode the wire insulation. In this cable, the insulated wires are covered with a lead sheath. The armor is put over this.

Electrical Metallic Tubing. Electrical metallic tubing (EMT) is installed between wiring devices that are to be connected. Individual lengths of *building wire* are then inserted or pulled through the tubing and connected to the devices. For long lengths of tubing, the wires are pulled through with a *fish tape* made of springy steel.

ALUMINUM CONDUCTORS

Some building electrical systems are now being wired with aluminum conductors. Aluminum has a higher resistance than copper. Because of this, conductors made of aluminum must be larger than those made of copper when both could be used for the same purpose. For example, a circuit that could be wired with no. 14 AWG (1.60-mm) copper conductors would be wired with no. 12 AWG (2.00-mm) aluminum conductors.

Aluminum wiring requires special connectors. Wiring devices such as switches and outlets must also be approved for use with aluminum conductors. Aluminum building wires and cables are identified by the word *aluminum* or by its abbreviation, *Al.*

WIRING DEVICES

To complete a home electrical system, a number of wiring devices are used (Fig. 31–10). These are designed and installed to provide the safe, convenient, and reliable use and control of electricity.

Switch Boxes. Most electrical codes require flush switches to be placed in metal enclosures. These enclosures, called *switch boxes,* are also used for mounting outlets. The boxes protect the devices mechanically. They also prevent switch sparking from becoming a fire hazard.

FLUSH-TYPE
SWITCH (SPST)

THREE-WAY
SWITCH

SURFACE-TYPE
SWITCH (SPST)

DUPLEX
CONVENIENCE
OUTLET
(RECEPTACLE)

CLEAT
LAMP HOLDER

CEILING LAMP HOLDER

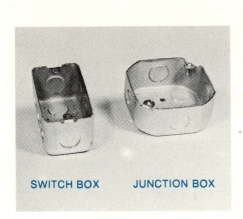

SWITCH BOX   JUNCTION BOX

Fig. 31–10. Common house wiring devices

Junction Boxes. Junction boxes contain the connections in the branch-circuit wiring. The connections are often made with wire nuts or other solderless, pressure connectors (Fig. 31–11). Junction boxes are also used for mounting various kinds of lighting fixtures.

Box Connectors. All cables entering a switch or a junction box must be securely fastened to the box. Some boxes have built-in cable clamps for this purpose. Otherwise, separate box connectors are used (Fig. 31–12).

Surface-wiring Devices. Surface-wiring devices are particularly useful for changing an existing house wiring system (Fig. 31–13). They can be installed without breaking through plaster or wallboard.

Multi-outlet Strips. A multi-outlet strip is a rectangular metal enclosure with wires and outlets preinstalled (Fig. 31–14 at A). Such strips provide a very convenient way of adding to existing wiring. They are very useful for placing outlets in places that could not be reached easily by ordinary cable or conduit wiring. They also can provide several outlets in electrical repair stations (Fig. 31–14 at B).

THREE-WAY AND FOUR-WAY SWITCHES

Three-way switches are used to control a load from two points. Such switches are commonly used in lighting circuits

1. STRIP WIRES AND TWIST TOGETHER

2. PLACE METAL SLEEVE CONNECTOR OVER WIRES AND CRIMP

3. CUT WIRES TO LENGTH

4. PLACE INSULATION OVER SLEEVE

ELECTRICIAN'S PLIERS EQUIPPED WITH CRIMPING DIE

Fig. 31–11. Connecting wires with solderless, pressure connectors (*Ideal Industries, Inc.*)

in hallways, stairways, and doorways. The schematic diagram of a two-point control circuit using two three-way switches is shown in Fig. 31–15 at A.

A combination of three- and four-way switches can be used to control a load from any number of places. The diagram of a three-point control circuit using two three-way switches and one four-way switch is shown in Fig. 31–15 at B.

REMOTE-CONTROL WIRING

Remote-control wiring makes it possible to control lights and appliances in various rooms and areas from a central place. In such systems, local control of lights is still possible. As an example, remote-control wiring makes it possible to turn on a light in the room ahead and to turn off a light in an adjoining room. This can be done with a single switch assembly.

A remote-control wiring system uses relays. These are controlled by a low voltage, usually 24 volts, supplied by a step-down transformer. In some systems, a rectifier diode is connected into the secondary winding or output circuit of the transformer so that the relays can be operated with direct current. The relays may be mounted in outlet boxes or in relay-center boxes or cabinets. A basic circuit showing the connections of the transformer, a relay, and a remote-control switch is shown in Fig. 31–16.

SYMPTOMS OF INADEQUATE WIRING

In recent years, the number of electrical and electronic products used in the home has increased greatly. This has created the problem of overloaded branch circuits. In your own home, too many appliances may be connected to one or more branch circuits. This condition wastes electricity, creates a fire hazard, and often causes appliances to work improperly.

There are several signs of inadequate wiring. The most common of these are listed below:

1. Fuses blow and circuit breakers often open because of branch-circuit overloading.
2. Heating appliances do not warm up as quickly as they should or do not heat to the right temperature.
3. Lights dim when appliances are in use or when motors in appliances such as refrigerators start up.
4. The television picture shrinks when appliances are in use or when appliance motors start up.
5. Appliance motors start with difficulty or fail to start at all.

288

SWITCH

DUPLEX CONVENIENCE OUTLET
(RECEPTACLE)

LAMP HOLDER

Fig. 31–13. Surface-wiring devices (*Bryant Division ot Westing-house*)

LOCK NUT   JUNCTION BOX

EMT
CONNECTOR

ARMORED
CABLE
CONNECTOR

NONMETALLIC
SHEATHED CABLE
CONNECTOR

BOX
KNOCKOUTS

Fig. 31–12. Box connectors (*Midwest Electric Manufacturing Company*)

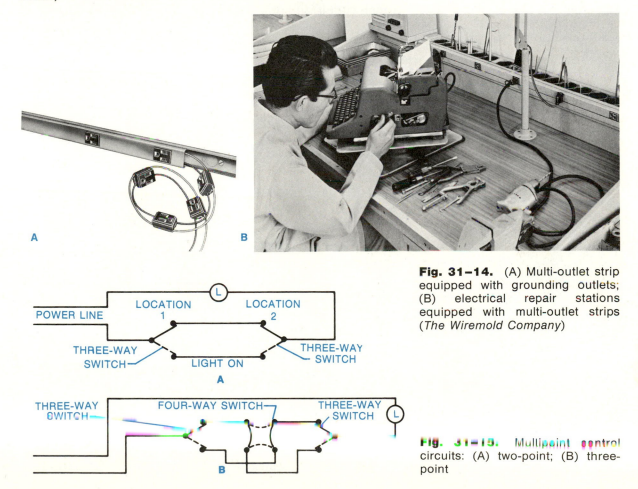

A

B

Fig. 31–14. (A) Multi-outlet strip equipped with grounding outlets; (B) electrical repair stations equipped with multi-outlet strips (*The Wiremold Company*)

POWER LINE

LOCATION
1

LOCATION
2

L

THREE-WAY
SWITCH

LIGHT ON

THREE-WAY
SWITCH

A

THREE-WAY
SWITCH

FOUR-WAY SWITCH

THREE-WAY
SWITCH

L

B

Fig. 31–15. Multipoint control circuits: (A) two-point; (B) three-point

289

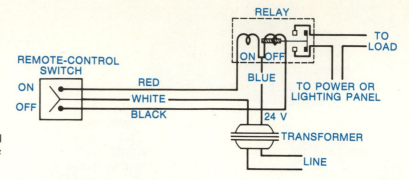

Fig. 31–16. Basic remote-control wiring circuit (*General Electric Company*)

6. Motors operate at a lower-than-normal speed, causing them to overheat.

If any of these signs are present, the wiring system should be thoroughly checked by a qualified electrician. In some cases, an inadequate wiring system can be corrected by simply adding branch circuits or by using wire of a larger size in a branch circuit. In some cases, the entire wiring system may have to be replaced.

Fig. 31–17. Assemblies for Learning by Doing No. 29, "Wiring with Nonmetallic Sheathed Cable"

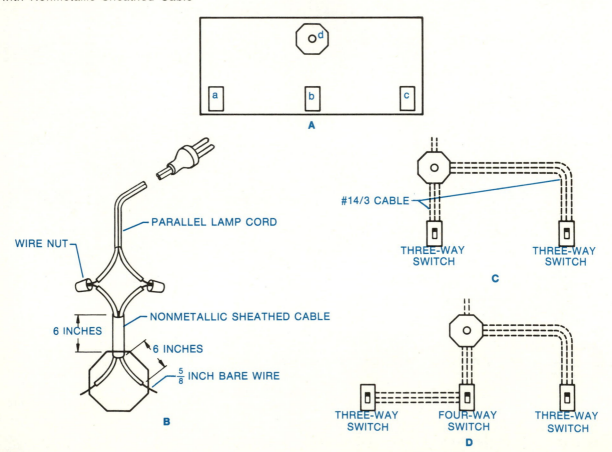

290

29. Wiring with Nonmetallic Sheathed Cable. These wiring jobs will help you learn how several common wiring devices are used in the home. They will also give you experience in connecting a load to three- and four-way switches.

MATERIALS NEEDED

baseboard, $\frac{3}{4}$ × 18 × 30 in. (19 × 457 × 762 mm)
1 junction box, octagonal, 4 × $1\frac{1}{2}$ in. trade size
3 switch boxes, utility type, 4 × $2\frac{1}{8}$ × $1\frac{1}{2}$ in. trade size
1 junction-box cover
3 switch-box covers
1 switch, spst, flush mount
2 three-way switches, flush mount
1 four-way switch, flush mount
2 duplex receptacles
1 lampholder, keyless, medium-base, to fit 4-inch junction box
8 ft (2.4 m) nonmetallic sheathed cable, no. 14/3
5 ft (1.5 m) nonmetallic sheathed cable, no. 12/2
4 ft (1.2 m) parallel lamp cord, no. 18 (1.00-mm)
5 box connectors for nonmetallic sheathed cable
6 wire nuts for joining two no. 14 (1.60-mm) wires
1 rubber-handle attachment plug
1 incandescent lamp, medium-base, 40-, 60-, or 100-watt
straps for nonmetallic sheathed cable

Procedures

1. Mount the three switch boxes (a, b, c) and the junction box (d) on the baseboard as shown in Fig. 31–17 at A.
2. Connect the lamp cord to the junction box as shown in Fig. 31–17 at B. The junction box is also used for mounting the lampholder.
3. **Job 1.** Wire and install the duplex receptacles in switch boxes a and b. Check the wiring by tracing the circuit.
4. **Job 2.** Wire a circuit to control the lamp using the spst flush switch mounted in switch box b.

CAUTION: Never connect switches to the grounded (white) conductor of a building wiring system.

5. **Job 3.** Wire a two-point control circuit (Fig. 31–17 at C).
6. **Job 4.** Wire a three-point control circuit (Fig. 31–17 at D).

Test your knowledge by writing, on a separate sheet of paper, the word or words that most correctly complete the following statements:

1. Information about wiring materials and methods that will result in the safe use of electric energy is given in the _____.
2. The bare wire of a service drop is called the _____ wire or conductor.
3. A three-wire service drop is used to supply two different _____.
4. An appliance operated at 230 volts needs less _____ for its operation than that same appliance would need when operated at 115 volts.
5. Electricity is distributed to different locations in a house from the _____. This unit is also commonly called a _____.
6. The load center contains the _____ or _____ that protect the branch circuits of a wiring system.
7. The neutral conductor of the service-entrance cable is grounded to reduce the danger of _____ and to protect against _____.
8. Three common kinds of branch circuits are _____ circuits, _____ circuits, and _____ circuits.
9. One wire of each general-purpose circuit is connected to the neutral _____ of the load center. The grounded neutral wire is colored _____.
10. The grounded wire of a two-wire, general-purpose circuit is never used as a _____ wire.
11. An electric-range circuit is an example of an _____ branch circuit.
12. _____ nonmetallic sheathed cable is never used outdoors or placed within _____ or _____.
13. A no. 14/3 nonmetallic sheathed cable contains _____ no. 14 AWG insulated wires.
14. Box _____ are used to securely fasten cables to switch and junction boxes.

1. What is the purpose of the *National Electrical Code?*
2. Why is a three-wire service drop used to supply a home electrical system?
3. Give two advantages of operating heavy-duty loads at 230 volts as compared to 115 volts.
4. Define the complete service entrance of a house wiring system.
5. What is the purpose of a load center? Describe two kinds of load centers.
6. How is the house wiring system grounded at the load center?
7. Name and define three kinds of branch circuits in the home.
8. Describe nonmetallic sheathed cable and armored cable.
9. What is the purpose of the bonding strip in armored cable?
10. Describe wiring with electrical metallic tubing.
11. Why are switches and outlets installed in switch boxes?
12. What is the purpose of junction boxes and of box connectors?
13. Describe a multi-outlet strip.
14. Draw diagrams of two-point and three-point control circuits. Explain how these circuits work.
15. What is meant by remote-control wiring?
16. Name five signs of inadequate wiring in a house wiring system.

1. Show your class a copy of the *National Electrical Code.* Describe its contents.
2. Visit a local electrical inspector, building contractor, or electrician and ask for information on the building wiring codes used in your community. Tell your class about what you have learned.
3. Collect and show to your class wiring devices, wires, and cables used in house wiring. Describe the purpose of each.

Unit 32   Electric Motors

People are always looking for ways to do work faster, more efficiently, and more easily. Thus, many machines have been invented to replace manual labor. One of these is the electric motor. This machine changes electric energy into mechanical energy.

MOTOR ACTION

A basic example of how electric energy can be changed into mechanical motion is shown in Fig. 32–1. As current passes through the large wire from left to right, a magnetic field is produced around the wire. The polarity of this field causes it to be repelled by the magnetic field of the permanent magnet. The wire thus moves away from the magnet (Fig. 32–1 at A).

If the direction of the current through the wire is reversed, the polarity of the magnetic field around it also reverses. As a result, the wire moves toward the permanent magnet because of magnetic attraction (Fig. 32–1 at B.)

An example of how circular motion can be produced by simple motor action is shown in Fig. 32–2. A permanent magnet is mounted on a pivot so that it is free to turn between the poles of two electromagnet coils. These coils are connected in series with a battery and a reversing, or double-pole–double-throw (dpdt), switch (Fig. 32–2 at A). The reversing switch is used to change the polarity of the voltage applied to the coils.

When the reversing switch is closed in one direction, the current through the electromagnet coils produces magnetic poles of the polarity shown in Fig. 32–2 at B. These magnetic poles attract the poles of the permanent magnet. It begins to turn in a clockwise direction. As the magnet nears the poles, the switch is opened and the electromagnet is demagnetized. Because the permanent magnet is turning, it will continue to rotate beyond the coils.

The reversing switch is then closed in the opposite direction. The direction of current through the coils is thereby reversed. This, of course, changes the polarity of the electromagnets. Now the permanent magnet is repelled. It continues to turn in the clockwise direction (Fig. 32–2 at C).

COMMUTATOR ACTION

Using a dpdt switch would not be a practical way of changing the magnetic polarity in a motor. In a practical motor, a de-

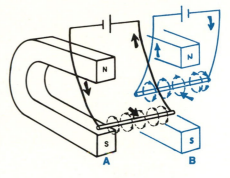

Fig. 32–1. Basic motor action

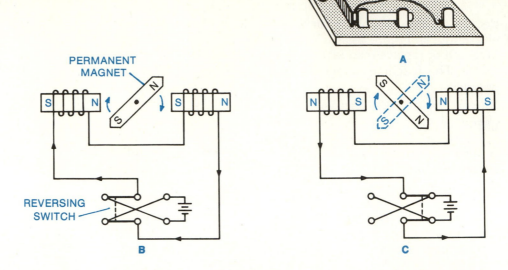

Fig. 32–2. Producing circular motion by simple motor action

PERMANENT MAGNET

REVERSING SWITCH

A

B

C

vice called a *commutator* is used. The action of a commutator in a simple motor is shown in Fig. 32–3.

In this motor, an electromagnet coil, called the *armature,* is mounted on a shaft. It is placed between the poles of a permanent magnet. Since the armature rotates, it is also called a *rotor.* Each end of the rotor coil is connected to a segment of the commutator. The commutator is mounted on the motor shaft. The commutator has segments that are insulated from the shaft and from one another. As the rotor turns, the segments rub against stationary contacts called *brushes* at regular intervals.

In Fig. 32–3 at A, commutator segment 1 is connected to the negative terminal of the battery through brush 1. Segment 2 is connected to the positive terminal of the battery through brush 2. Under this condition, the rotor electromagnet is repelled by the poles of the permanent magnet, and the rotor begins to turn in a clockwise direction. The magnetic polarity of the rotor does not change as it continues to turn to the position shown in Fig. 32–3 at B. Now its poles are attracted by the poles of the permanent magnet.

When the rotor reaches the position shown in Fig. 32–3 at C, the commutator reverses its connections to the battery. Segment 1 is connected to the positive terminal of the battery. Segment 2 is connected to the negative terminal. This changes the direction of the current through the rotor coil and reverses its magnetic polarity. Like magnetic poles are once again near each other. Magnetic repulsion causes the rotor to continue to turn in a clockwise direction.

When the rotor reaches the position shown in Fig. 32–3 at D, the magnetic polarity of its poles does not change. They are attracted by the poles of the permanent magnet. This causes the rotor to continue to turn in a clockwise direction.

MOTOR SIZE

The size of an electric motor is given in terms of a mechanical unit of power called *horsepower* (hp). One horsepower is equal to 746 watts of electric power. One horsepower is defined as the amount of work needed to raise a weight of 550 pounds (250 kg) a distance of 1 foot (305 mm) in 1 second of time. In the metric system, motor sizes are given in watts.

Most of the motors discussed in the following paragraphs are known as *fractional-horsepower motors*. This means that their size is less than 1 horsepower. Other motors used for heavy-duty work may have power ratings of from 1 to 1,000 horsepower (0.746 to 746 kilowatts) or more.

Fig. 32–3. Motor commutator action

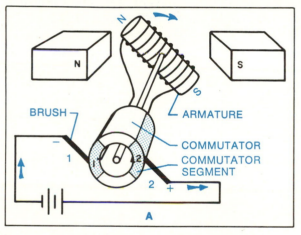

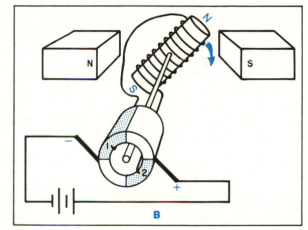

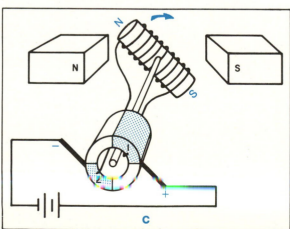

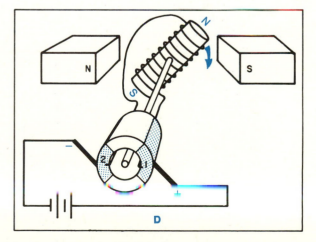

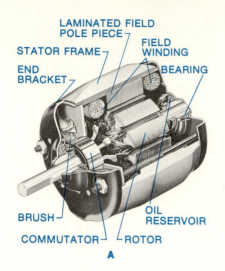

LAMINATED FIELD
POLE PIECE

FIELD
WINDING

STATOR FRAME

BEARING

END
BRACKET

BRUSH

OIL
RESERVOIR

COMMUTATOR

ROTOR

A

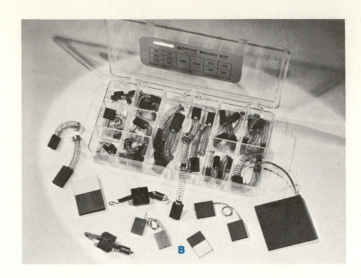

B

Fig. 32–4. (A) Main parts of a dc motor (*Motor Division, Controls Company of America*); (B) carbon brushes used in dc motors and generators (*Helwig Carbon Products, Inc.*)

DIRECT-CURRENT MOTORS

Direct-current motors and direct-current generators are very similar. The magnetic fields of dc motors are produced by stationary windings called the *field* and by rotating windings in the rotor, or armature (Fig. 32–4 at A). The circuit through the rotor windings in the typical dc motor is completed through stationary carbon brushes. The brushes are in contact with commutator segments that are connected to the rotor windings. Several different kinds of motor brushes are shown in Fig. 32–4 at B.

The Series Motor. In a series dc motor, the field and the rotor windings are connected in series (Fig. 32–5 at A). The

Fig. 32–5. Schematic diagrams of dc motors: (A) series motor; (B) shunt motor; (C) long-shunt compound motor; (D) short-shunt compound motor

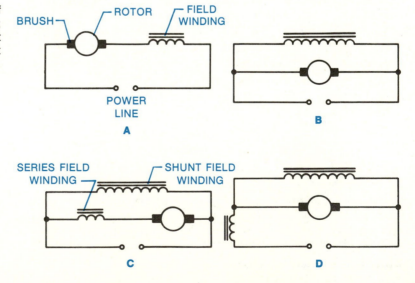

BRUSH

ROTOR

FIELD
WINDING

POWER
LINE

A

B

SERIES FIELD
WINDING

SHUNT FIELD
WINDING

C

D

series motor has a high starting *torque,* or twisting force. This makes it good at starting while connected to heavy loads. The ordinary automobile cranking motor, or starter, is a series motor. Such motors are also used in cranes, hoists, cargo winches, and train engines.

A series dc motor should never be operated without being connected directly to a load or coupled to a load through a gear assembly. Otherwise, the motor will continue to increase its speed to a point where it may be seriously damaged or even destroyed.

The Shunt Motor. In a shunt motor, the field and the rotor windings are connected in parallel (Fig. 32–5 at B). Shunt motors operate at a relatively consant speed when connected to varying loads. They are often used in heavy-duty drill presses, lathes, conveyors, and printing presses.

The Compound Motor. The compound motor is a combination of the series and the shunt types (Fig. 32–5 at C and D). This gives it the high-torque advantage of a series motor and the constant-speed advantage of a shunt motor. Compound motors may be long-shunt or short-shunt, depending on the way the shunt field is connected.

UNIVERSAL MOTORS

Universal motors work on either direct or alternating current. They are similar to the small, series dc motor (Fig. 32–5 at A). They come in sizes ranging from 1/100 to 2 horsepower (7.46 to 1492 watts) with speeds up to 10,000 revolutions per minute (rpm).

Universal motors that work on 120 volts are small but provide a relatively large horsepower output. The direction of rotation of universal motors can be reversed. Their speed can be controlled easily. Because of these features, the motors are used in a wide variety of small appliances and portable tools. These include sewing machines, vacuum cleaners, food mixers, drills, saws, and shears (Fig. 32–6).

REVERSING A MOTOR

The direction of rotation of a dc motor or a universal motor can be reversed by interchanging the connections to the rotor windings or to the field windings. This is sometimes done by using a reversing switch connected to the brushes. An example of such a circuit arrangement using a dpdt switch with a series motor is shown in Fig. 32–7.

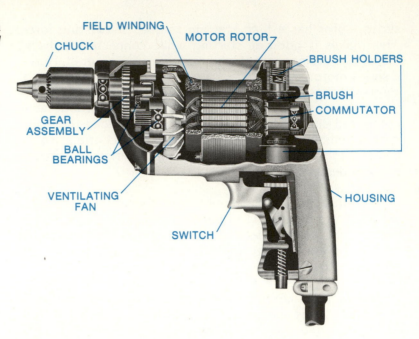

Fig. 32–6. Universal motor in a portable electric drill (*Black and Decker Manufacturing Company*)

CHUCK

FIELD WINDING

MOTOR ROTOR

BRUSH HOLDERS

BRUSH

COMMUTATOR

GEAR ASSEMBLY

BALL BEARINGS

VENTILATING FAN

SWITCH

HOUSING

SPEED CONTROL

The speed of a dc motor can be varied by connecting a rheostat in series with its field or in parallel with its rotor (Fig. 32–8 at A).

The speed of a universal motor connected to an alternating current can be controlled by using a silicon controlled rectifier (Fig. 32–8 at B). The main speed-controlling components of the circuit are potentiometer R1 and capacitor C1. The potentiometer is used to control the charging and the discharging rate of the capacitor. This, in turn, controls the voltage developed across the capacitor and the voltage applied to the gate of the silicon controlled rectifier. By adjusting R1, the amount of current conducted by the silicon controlled rectifier can be varied. As a result, the speed of the motor also changes.

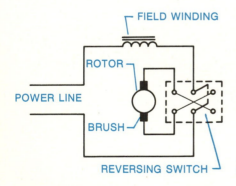

FIELD WINDING

ROTOR

POWER LINE

BRUSH

REVERSING SWITCH

Fig. 32–7. A method of reversing a series motor

ALTERNATING-CURRENT MOTORS

There are several different kinds of ac motors. The two most common are induction motors and synchronous motors. There are four general kinds of small, single-phase induction motors: (1) split-phase motors, (2) capacitor-start motors, (3) shaded-pole motors, and (4) repulsion-start motors. Alternating-current motors that do not have windings in their rotor are known as *squirrel-cage motors.*

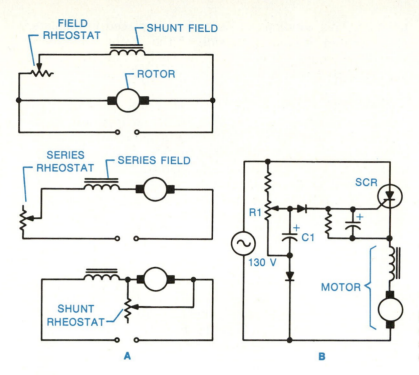

Fig. 32–8. (A) Speed control of dc motors with a rheostat; (B) speed-control circuit using a silicon controlled rectifier

SPLIT-PHASE MOTOR

A typical split-phase motor, a widely used single-phase motor, is shown in Fig. 32–9. Split-phase motors come in sizes ranging from $\frac{1}{20}$ to $\frac{1}{3}$ horsepower (37.3 to 248 watts). Because of their relatively low torque, these motors are used with loads that are easy to start. These include fans, blowers, small pumps, washing machines, clothes dryers, drill presses, grinders, and table saws.

Running Winding. In the split-phase motor, electromagnetic fields are produced by current in the *running*, or *main*,

Fig. 32–9. (A) Main parts of a single-phase split-phase induction motor; (B) schematic diagram of the split-phase field circuit (*Photo courtesy of Delco Appliance Division, General Motors Corporation*)

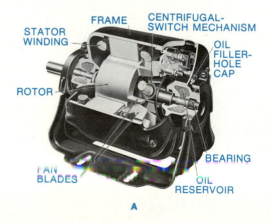

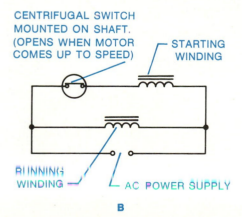

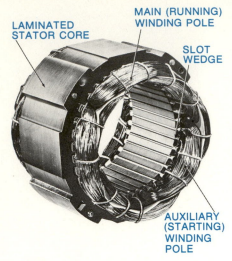

Fig. 32–10. Stator assembly (field) of a split-phase motor (*Wagner Electric Corporation*)

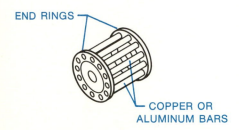

Fig. 32–11. The bars and end rings within the rotor of a split-phase motor. For clarity, the laminated rotor core through which the bars pass is not shown.

winding. This winding consists of coils placed in insulated slots located around the inside of the stator (stationary) core. These coils are connected in series. They form what are called *pole groups* or, simply, *poles* (Fig. 32–10).

Rotor. The rotor of a split-phase motor does not have windings. It is not connected to the power line in any way. The rotor is made up of a laminated iron core that has copper bars connected together by end rings (Fig. 32–11). As the rotor moves through the magnetic fields of the running winding, current is induced in the bars by electromagnetic induction. The induced current produces a secondary (rotor) magnetic field. This is always repelled by the poles of the running winding. This action causes the rotor to turn at a constant speed.

Starting Winding. In order to start a split-phase motor, additional magnetic energy is needed to make the rotor begin turning. This energy is supplied by a second set of stationary coils. These form what is called the *starting winding.* The coils are wound into the insulated slots that also contain the running winding (Fig. 32–10). The starting winding is connected to the motor circuit only during the time needed for the rotor to accelerate to about 80 percent of its full speed. After this, the winding is disconnected from the circuit.

Centrifugal Switch. The starting winding of a split-phase motor is automatically connected into and disconnected from the motor circuit by a centrifugally operated switch. This switch is turned on and off by a governor assembly mounted on the motor shaft (Fig. 32–12). When the motor is turned off or is running at a slow speed, the switch is turned on. The starting winding is now connected into the motor circuit. As the rotor speeds up, centrifugal force causes the

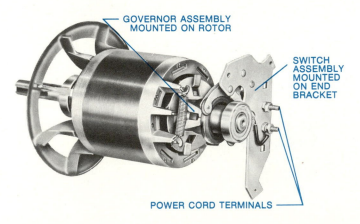

Fig. 32–12. Centrifugal switch used in a split-phase motor (*Robbins and Myers, Inc.*)

governor assembly to move toward the rotor. This turns the switch off, causing the starting winding to be disconnected from the motor circuit. The clicklike sound heard very soon after a split-phase motor is turned on or off is made by the movement of the governor assembly.

Reversing. A split-phase motor is usually reversed by interchanging the connections to the starting winding. A number of different kinds of reversing switches can be used for this purpose. The motor can also be reversed by interchanging the connections to the running winding. If both winding connections are interchanged, the motor will continue to run in its original direction.

Speed Control. The speed of a split-phase motor, unlike that of a universal motor, cannot be varied gradually by reducing the voltage across the field without losing too much torque. For this reason, the speed may be step-controlled by changing the number of running-winding poles. In a three-speed motor, for example, the running winding is arranged so that it can be switched to form two, four, or six poles. The approximate speed that each of these connections produces can be found by using the formula

$$\text{rpm} = \frac{120f}{N}$$

where rpm = revolutions per minute of the shaft
f = frequency of the operating current (expressed in hertz)
N = number of running-winding poles

This formula gives what is called the *synchronous speed* of a motor. In a nonsynchronous motor, the actual speed is always less than the synchronous speed. This is due to what is known as *slip*, which results from losses of electrical and mechanical energy in the motor.

CAPACITOR-START MOTOR

A typical capacitor-start induction motor is similar to a split-phase motor. However, this motor contains a nonpolarized electrolytic capacitor connected in series with the starting winding (Fig. 32–13). The electrical action of the capacitor gives the motor a much greater torque than the starting winding alone can produce. Because of this high torque, capacitor-start motors are used with hard-to-start loads. These include stokers, refrigerators, air conditioners, air compressors, and heavy-duty pumps.

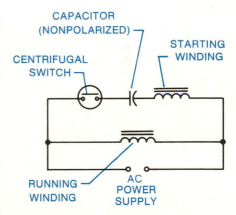

Fig. 32–13. Schematic diagram of a typical capacitor-start motor

SHADED-POLE MOTOR

The shaded-pole motor does not have a starting winding or a capacitor. Instead, the motor has heavy copper *shading coils.* These are wound around each pole piece or part of it (Fig. 32–14). The induced current in these coils produces a magnetic field. This acts with the field-winding magnetic field to provide the rotor with a rotating magnetic field.

Shaded-pole motors are usually less than $1/4$ horsepower (187 watts) in size. Because of their low torque, they are used with light loads. These include small fans, small blowers, record players, tape recorders, and motion-picture projectors.

REPULSION-START MOTOR

The repulsion-start motor has a field winding and a wire-wound rotor similar to the rotor of a dc or a universal motor (Fig. 32–15). The rotor winding, however, is not connected to the power line. The motor also has a commutator and brushes. The brushes are shorted to one another.

When the motor is turned on, the magnetic fields of the field winding induce a voltage in the rotor winding. The secondary magnetic fields produced by the rotor current are then repelled by the electromagnetism of the field winding. This action provides the motor with its starting torque. After the motor reaches about 75 percent of its full speed, a centrifugally operated mechanism short-circuits the rotor commu-

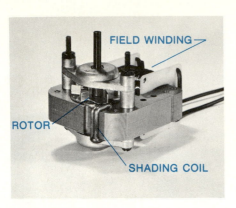

Fig. 32–14. A two-pole shaded-pole induction motor used in a record player (*The General Industries Company*)

FIELD WINDING

ROTOR

SHADING COIL

Fig. 32–15. Cutaway view of a 1/2-horsepower (373-watt) repulsion-start induction motor (*Wagner Electric Corporation*)

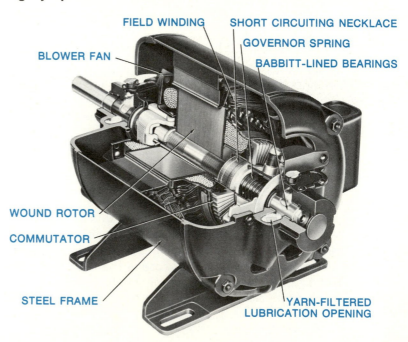

FIELD WINDING   SHORT CIRCUITING NECKLACE

GOVERNOR SPRING

BLOWER FAN   BABBITT-LINED BEARINGS

WOUND ROTOR

COMMUTATOR

STEEL FRAME

YARN-FILTERED LUBRICATION OPENING

tator segments. The motor then continues to work as an ordinary induction motor.

The most common repulsion-start motors range from $\frac{1}{2}$ to 8 horsepower (373 to 5,970 watts) in size. They are operated at 220 volts. Because of their extremely high torque, these motors are used with heavy-duty loads, such as large air-conditioning units, compressors, and pumps.

SYNCHRONOUS MOTOR

A synchronous motor keeps in step with the frequency of the ac power source. For this reason, the motor runs at a more constant speed than other motors. One kind of small synchronous motor is called a *hysteresis motor.*

A common use of synchronous motors is in electric clocks. Such motors are also used to operate time switches, business machines, and several kinds of control mechanisms. A shading coil makes these motors self-starting (Fig. 32–16).

THREE-PHASE INDUCTION MOTOR

In the three-phase induction motor, the field winding consists of three separate sets of coils. Each of these is energized by one phase line, or *leg,* of a three-phase power system (Fig. 32–17 at A). When the motor is connected to a three-phase line, the currents in the coils produce what is called a *revolving,* or *rotating, magnetic field* (Fig. 32–17 at B). This magnetic field acts on the secondary magnetic field produced by the induced rotor current to keep the rotor moving.

A three-phase motor can be reversed by simply interchanging any two of the three power-line conductors. This is usually done with a magnetically operated switch called a *magnetic starter,* or a *motor controller.*

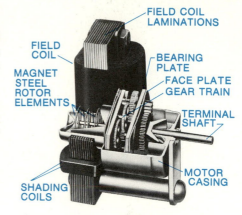

Fig. 32–16. A small synchronous timing motor (*General Electric Company*)

Fig. 32–17. The three-phase induction motor: (A) coil connections; (B) basic operation

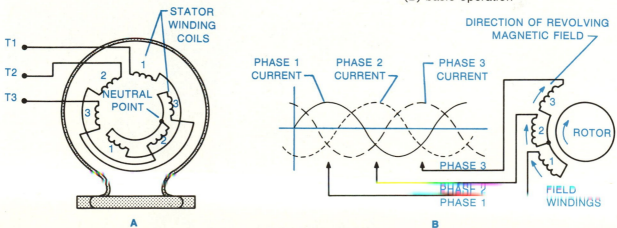

Because of their simple construction, high torque, and high efficiency, three-phase induction motors are found in all kinds of heavy-duty industrial equipment. Most three-phase motors operate at a voltage of 208 volts or higher.

ENERGY-EFFICIENT MOTORS

Electric motors use about 64 percent of all the electricity produced in the United States. Three-fourths of this is used by motors that operate pumps, blowers, fans, and machine tools. These devices are used often in the chemical, metal, paper, food, and petroleum industries.

Because electric motors use such large amounts of electric energy, they also have great potential for energy conservation. The U.S. Department of Energy considers the development and use of energy-efficient motors by industry an important factor in the nation's energy-conservation effort.

Motors rated from 1 through 20 horsepower (0.746 to 14.92 kilowatts) have the greatest conservation potential. Motor efficiency is a measure of mechanical work output compared to the electrical power input. A motor's *power factor* is a measure of how well the motor uses the current it draws (Fig. 32–18).

Energy-efficient motors have several important features. They operate at cooler temperatures. Thus, less electric energy is lost to heat energy. The motors are sized correctly to the load requirements. For example, if the equipment takes a 5-hp (3.7-kW) motor, the motor closest to that requirement should be used. Otherwise, the motor would be overloaded

Fig. 32–18. Making the connections to test a motor before certifying it as an energy-efficient motor (*Gould, Inc., Electric Motor Division*)

or underloaded. Having the right-size motor for the load is important in maintaining a good power factor.

The rotors of energy-efficient motors are made with more aluminum. This reduces losses resulting from current flowing in the rotor bars. More copper is used in the stators. This reduces losses in the motor. More steel and thinner laminations are used to reduce the stator and rotor losses. The air gap between the rotor and stator is less in an energy-efficient motor. A 1-hp (746-W) energy-efficient motor will use 70 watts less under continuous operation than a standard motor of the same wattage.

MOTOR CARE AND MAINTENANCE

An electric motor is a very dependable machine. However, it will provide the safest and most efficient service only if it is used and maintained in the right way. The following suggestions will help:

1. Before installing a new or different motor, read the information on its nameplate. Make sure that the right voltage and system (ac or dc) are being used.
2. It is always safe practice to ground the frame of a motor. This is especially true if the motor is used in a damp location or near any grounded metal object.
3. If water enters the frame of a motor, the motor should be completely dried out and checked for shorts before being put back into operation.
4. Never try to stop a motor by grabbing its shaft.
5. Never overoil a motor. Apply a little oil to the bearings only. Too much oil may damage insulation and cause dirt and dust to collect. Some motors have sealed bearings that do not need oiling or greasing.
6. To prevent an ordinary motor from becoming overheated, keep the air openings in its frame clear.
7. Replace the brushes in a universal motor when they become shorter than $\frac{1}{4}$ in. (6.35 mm). Badly worn brushes will cause too much sparking at the commutator. Too much sparking is also caused by loose brush holders, by loose commutator segments, and by shorts in the rotor winding.
8. Unplug a motor if it seems to be working at a speed slower than normal. Otherwise, the motor could become overheated. The motor may be overloaded or it may have a shorted winding. Check motor speed with a tachometer.
9. Turn a split-phase motor off at once if it hums but fails to start after being turned on. This is very often due to a defective centrifugal-switch mechanism. If the motor is left on, the running winding may burn out quickly.

Table 32–1. Size of Fuses Used with Single-phase, 115-volt Induction Motors

Size of Motor (horsepower)	Equivalent (watts)	Fuse Size (amperes)
$\frac{1}{6}$	124	*
$\frac{1}{4}$	187	*
$\frac{1}{3}$	249	*
$\frac{1}{2}$	373	*
$\frac{3}{4}$	560	*
1	746	*
$1\frac{1}{2}$	1119	25
2	1492	30

*Motors rated at 1 hp (746 watts) or less are considered protected by the branch-circuit fuse in the panel supplying their power. In most cases this will be a 15- or 20-ampere fuse.

Metric Conversion Table for Fig. 32–19

in.	mm
$\frac{1}{8}$	3
$\frac{3}{16}$	5
$\frac{1}{4}$	6.5
$\frac{5}{16}$	8
$\frac{3}{8}$	10
$\frac{1}{2}$	13
$\frac{5}{8}$	16
$\frac{3}{4}$	19
1	25
$1\frac{1}{4}$	31
$1\frac{9}{16}$	39.5
2	51
$2\frac{3}{4}$	70
3	76
$3\frac{1}{8}$	79
$3\frac{1}{2}$	89
$4\frac{1}{2}$	115
6	152
$11\frac{5}{8}$	297

10. An induction motor draws much more than the normal working current as it comes up to full speed. For this reason, such a motor should be protected with a dual-element time-delay fuse. The correct sizes of the fuses to be used with several single-phase, 115-volt induction motors are given in Table 32–1. Small portable motors are adequately protected by the fuse or circuit breaker in the panel supplying their power.

LEARNING BY DOING

30. A Simple Two-pole Universal Motor. As you have learned, electromagnetic energy is used to produce mechanical motion in and electric motor. By making this motor, you will learn more about the basic parts of motors and how they work.

MATERIALS NEEDED

1 piece of wood, $\frac{3}{4} \times 4\frac{1}{2} \times 6$ in.
 $(19 \times 114 \times 152$ mm$)$
1 piece of band iron, $\frac{1}{8} \times \frac{1}{2} \times 11\frac{5}{8}$ in.
 $(3.18 \times 13 \times 295$ mm$)$
1 piece of band iron, $\frac{1}{8} \times \frac{1}{2} \times 6\frac{3}{4}$ in.
 $(3.18 \times 13 \times 170$ mm$)$
2 pieces of sheet brass, $\frac{3}{8} \times 3\frac{1}{8}$ in. $(9.5 \times 79$ mm$)$
2 pieces of tin plate, $\frac{1}{4} \times 1\frac{1}{4}$ in. $(6 \times 32$ mm$)$
1 piece of tin plate, $1\frac{1}{2} \times 3$ in. $(38 \times 76$ mm$)$
60 ft $(18$ m$)$ PE magnet wire, no. 25 $(0.45$ mm$)$; 40 ft $(12$ m$)$ for the field winding and 20 ft $(6$ m$)$ for the armature winding

Procedure

1. Assemble the materials and make the connections as shown in Fig. 32–19.
2. Connect the terminals of the motor to two D-size dry cells connected in series or to a battery-eliminator power supply adjusted to produce an output voltage of about 3 volts.
3. Adjust the brushes so that maximum speed is obtained. The brushes should contact the commutator segments with the same amount of pressure. As brush pressure is increased, the motor speed will decrease.
4. Show how the motor can be reversed by using a dpdt switch wired as a reversing switch.

Fig. 32–19. Assembly and working drawings for Learning by Doing No. 30, "A Simple Two-pole Universal Motor"

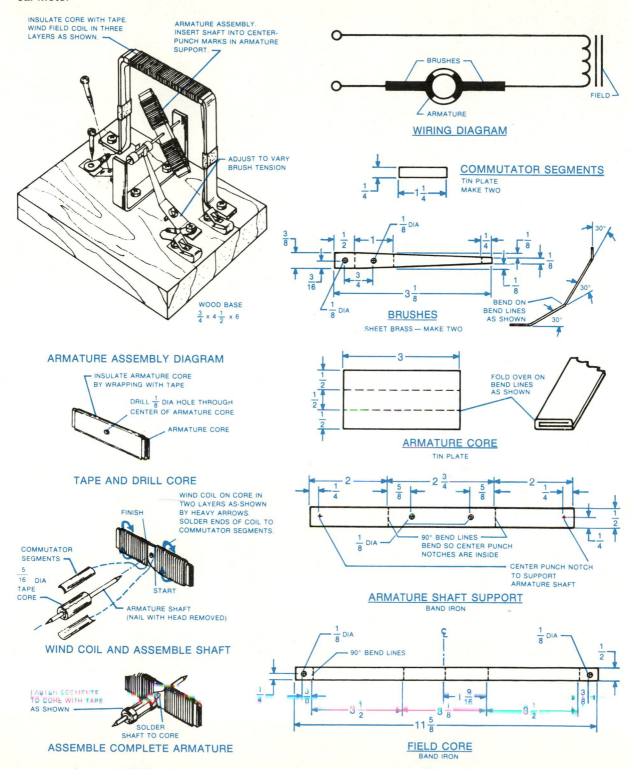

INSULATE CORE WITH TAPE. WIND FIELD COIL IN THREE LAYERS AS SHOWN.

ARMATURE ASSEMBLY. INSERT SHAFT INTO CENTER-PUNCH MARKS IN ARMATURE SUPPORT.

ADJUST TO VARY BRUSH TENSION

WOOD BASE
$\frac{3}{4}$ x $4\frac{1}{2}$ x 6

ARMATURE ASSEMBLY DIAGRAM

INSULATE ARMATURE CORE BY WRAPPING WITH TAPE

DRILL $\frac{1}{8}$ DIA HOLE THROUGH CENTER OF ARMATURE CORE

ARMATURE CORE

TAPE AND DRILL CORE

WIND COIL ON CORE IN TWO LAYERS AS SHOWN BY HEAVY ARROWS. SOLDER ENDS OF COIL TO COMMUTATOR SEGMENTS.

FINISH

COMMUTATOR SEGMENTS

$\frac{5}{16}$ DIA

TAPE CORE

START

ARMATURE SHAFT (NAIL WITH HEAD REMOVED)

WIND COIL AND ASSEMBLE SHAFT

COMMUTATOR SEGMENTS TO CORE WITH TAPE AS SHOWN

SOLDER SHAFT TO CORE

ASSEMBLE COMPLETE ARMATURE

BRUSHES

ARMATURE

FIELD

WIRING DIAGRAM

COMMUTATOR SEGMENTS
TIN PLATE
MAKE TWO

$\frac{1}{4}$ $1\frac{1}{4}$

BRUSHES
SHEET BRASS — MAKE TWO

$\frac{3}{8}$ $\frac{1}{2}$ 1 $\frac{1}{8}$ DIA $\frac{1}{4}$ $\frac{1}{8}$ 30° 30° 30°

$\frac{3}{16}$ $\frac{3}{4}$ $3\frac{1}{8}$ $\frac{1}{8}$ $\frac{1}{8}$ DIA

BEND ON BEND LINES AS SHOWN

ARMATURE CORE
TIN PLATE

3
$\frac{1}{2}$ $\frac{1}{2}$ $\frac{1}{2}$

FOLD OVER ON BEND LINES AS SHOWN

ARMATURE SHAFT SUPPORT
BAND IRON

2 $\frac{1}{4}$ $\frac{5}{8}$ $2\frac{3}{4}$ $\frac{5}{8}$ 2 $\frac{1}{4}$ $\frac{1}{2}$ $\frac{1}{4}$

$\frac{1}{8}$ DIA 90° BEND LINES BEND SO CENTER PUNCH NOTCHES ARE INSIDE

CENTER PUNCH NOTCH TO SUPPORT ARMATURE SHAFT

FIELD CORE
BAND IRON

$\frac{1}{8}$ DIA 90° BEND LINES $\frac{1}{8}$ DIA $\frac{1}{2}$

$\frac{1}{4}$ $\frac{3}{8}$ $3\frac{1}{2}$ $1\frac{9}{16}$ $3\frac{1}{8}$ $3\frac{1}{2}$ $\frac{3}{8}$

$11\frac{5}{8}$

307

Test your knowledge by writing, on a separate sheet of paper, the word or words that most correctly complete the following statements:

1. An electric motor is a machine that changes _____ energy into _____ energy.
2. The size of an electric motor is given in terms of a _____ unit of power called the horsepower. One horsepower is equal to _____ watts of electric power.
3. In a series dc motor, the _____ and the _____ windings are connected in series.
4. The word *torque* means _____ force.
5. In a shunt motor, the field and the rotor windings are connected in _____.
6. The compound motor is a combination of the _____ and the _____ types.
7. Universal motors can be operated with either _____ or _____ current.
8. The direction of rotation in a universal motor can be reversed by interchanging the connections to the _____ windings or to the _____ windings.
9. A circuit using a silicon controlled rectifier provides an efficient way of controlling the _____ of a universal motor.
10. Induction motors and synchronous motors can be operated with _____ current only.
11. The electromagnetic fields necessary for the operation of a split-phase motor are produced by current in the _____ and _____ windings.
12. The rotor of a split-phase motor does not have _____.
13. The starting torque of a capacitor-start motor is increased by connecting a non-polarized electrolytic capacitor in series with the _____ winding.
14. In the shaded-pole motor, heavy copper _____ coils are wound around each pole piece or part of it.
15. The field winding of a three-phase motor consists of three separate sets of _____.

FOR REVIEW AND DISCUSSION

1. Explain how a simple motor works.
2. What does a commutator do in a dc motor?
3. How is the size of a motor given? Define this unit.
4. Describe the series, the shunt, and the compound motors.
5. Describe a universal motor.
6. Tell how the direction of rotation of a universal motor can be reversed.
7. How can the speed of a universal motor be controlled?
8. Name four kinds of single-phase induction motors.
9. Describe a typical split-phase motor. What is the purpose of its starting winding?
10. What is the centrifugal switch in a split-phase motor used for?
11. What is the main advantage of a capacitor-start motor?
12. For what purpose are shading coils used in a shaded-pole motor?
13. Describe a repulsion-start motor.
14. What is the main advantage of a synchronous motor?
15. Describe a three-phase induction motor.
16. Give six procedures that are helpful in using and maintaining electric motors.
17. Why are dual-element time-delay fuses used to protect motors?

INDIVIDUAL-STUDY ACTIVITIES

1. Obtain a commercial universal or split-phase motor. Disassemble the motor and describe its construction to the class. Also explain how the motor works.
2. Different kinds of motors can often be obtained from local appliance dealers, appliance-service technicians, and motor-service technicians. Show some of these motors to the class. Explain the major characteristics of each kind. Discuss also the kinds of equipment that use these motors.

Unit 33   Producing Heat

Electric energy is converted into useful heat energy in a number of ways. In the home, electricity is used for space heating and to produce heat for a variety of uses. Electricity is also used to produce the heat needed for several industrial processes. These include welding, heat treatment, and steelmaking. Special electronic generators are used to produce the heat needed for certain medical treatments.

RESISTANCE HEATING

Resistance heating uses the energy of moving electrons as they collide with particles within a conductor. If enough of these collisions occur, the friction that results will heat the conductor (Fig. 33–1).

The amount of heat produced in any conductor by resistance heating is related to the power formula $P = I^2R$. Thus, the heat produced is directly proportional to the resistance of the conductor and to the square of the current. This means, for example, that if the current is doubled, the amount of heat will be four times the original heat. However, if the current remains constant and the resistance is doubled, only twice as much heat will be produced.

Resistance Alloys. The conductors used for resistance-heating purposes are made of several materials. These materials have the amount of resistance needed to produce heat. The most common of these are alloys made of nickel and chromium; copper and nickel; and nickel, chromium, and iron.

Heating Elements. In many heat-producing appliances and products, the resistance alloy is in the form of a round or flat wire called *resistance wire*. This wire is wrapped around an insulator such as porcelain or mica to form a *heating element* (Fig. 33–2 at A). In other products, the resistance wire is in the shape of a coil or a rod enclosed in a protective metal tube (Fig. 33–2 at B). In this kind of heating element, the resistance wire is usually embedded in a material such as magnesium oxide. This material is both a good electrical insulator and a good conductor of heat. Table 33–1 gives the wattage, resistance, and wire sizes for Nichrome-wire heating elements.

Radiant-heating Cable. Radiant-heating cable consists of resistance wire insulated with a heat-resistant rubber or ther-

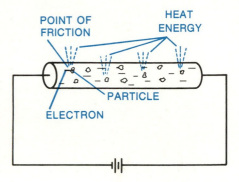

Fig. 33–1.   Resistance heating

309

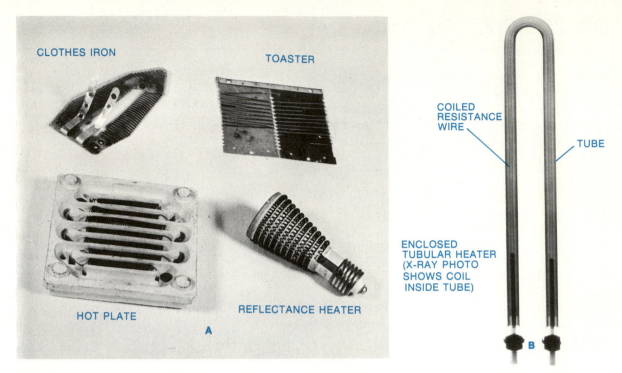

CLOTHES IRON

TOASTER

HOT PLATE

REFLECTANCE HEATER

A

COILED
RESISTANCE
WIRE

TUBE

ENCLOSED
TUBULAR HEATER
(X-RAY PHOTO
SHOWS COIL
INSIDE TUBE)

B

Fig. 33–2. Heating elements: (A) coiled resistance wire wrapped around insulators; (B) coiled resistance wire embedded in insulating material within a metal tube

HEATING CABLE

Fig. 33–3. Installing radiant heating cable (*L. N. Roberson Company*)

moplastic compound. This cable is commonly used for space heating. The cable is placed in a concrete floor or in a plaster ceiling. In a ceiling installation, the cable is first stapled to plasterboard and then covered with plaster (Fig. 33–3).

Radiant-heating cable for space heating comes in several lengths. Each length is controlled by a thermostat. Each has a certain wattage rating, usually in the 400- to 5,000-watt range. The individual lengths of heating elements are color-coded to indicate their wattage rating. The heating elements operate at either 115 or 230 volts.

Resistance Welding. Resistance heating is used for spot welding and for seam welding. In spot welding, the metals to be welded are placed between two welder electrodes. Current passes through the metals from one electrode to the other. The resistance of the metals to the current produces a very high temperature. This welds the metals together at the point of electrode contact (Fig. 33–4).

Seam welding is a special kind of spot welding. On a seam welder, the electrodes are wheels between which the metals to be welded are passed. As with spot welding, current passes through the metals, welding them together along a seam.

Table 33-1.   Resistance-wire Heating Elements

Values for 115-volt Operation			
Wattage Required (watts)	Approximate Resistance at 75°F (24°C) (ohms)	Recommended Sizes of Nichrome Wire (AWG)	
		Minimum	Maximum
100	118.100	30	26
150	78.732	30	26
200	59.050	29	25
250	47.240	28	24
300	39.366	28	24
350	33.742	27	23
400	29.525	26	22
450	26.244	24	20
500	23.620	24	20
550	21.472	23	19
600	19.683	23	19
650	18.170	23	19
700	16.871	22	18
750	15.745	22	18
800	14.762	22	18
850	13.894	21	17
900	13.122	21	17
950	12.431	21	17
1,000	11.810	20	16

Fig. 33-4.   Resistance welding in industry. Sparks from this process show as long white streaks in the photograph. (*Unimation, Inc.*)

ELECTRIC ARC

A simple way of producing an electric arc is shown in Fig. 33-5. To start the arc, the ends of the carbon rods are brought into contact with each other. The heat produced by the current through the point of contact causes a small amount of the carbon to vaporize, or turn to gas. If the contact between the rods is then broken and the ends are separated, current will *arc*, or pass, through the carbon vapor. This produces a very bright light and a very high temperature.

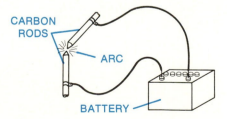

Fig. 33-5.   Producing an electric arc

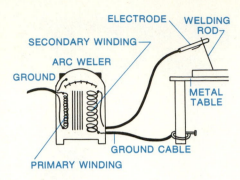

Fig. 33–6. Electric-arc welding

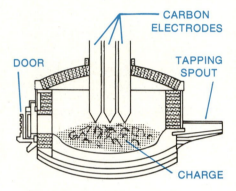

Fig. 33–7. The electric-arc furnace (*American Iron and Steel Institute*)

Fig. 33–8. Infrared lamps: (A) tubular quartz; (B) reflector (*General Electric Company*)

Electric-arc Welding. In electric-arc welding, heat is produced by an arc formed between a welding rod and the metal object being welded. The rod is connected to one terminal of the welder. The metal object is connected to the other terminal (Fig. 33–6).

To strike the arc, the welding rod is touched to the metal. The heat produced as current passes through the point of contact vaporizes some of the rod, thus forming an arc. The heat of the arc is great enough to melt the welding rod and the workpiece. The extra metal needed for the weld is supplied by the rod.

The Electric-arc Furnace. An electric-arc furnace is used to melt metals during the steelmaking process. In such a furnace, carbon electrodes are used to produce heat by means of electric arcs. The electrodes are lowered through the top of the furnace. The furnace contains a *charge,* or mixture, of scrap metals to be melted (Fig. 33–7). Arcs are then produced as current moves from the electrodes to the charge, melting the metal.

INFRARED HEATING

The prefix *infra* means below. *Infrared rays* are a form of electromagnetic radiation that people cannot see. They have a frequency just below that of the lowest-frequency visible light, which is red.

The infrared rays used for heating are usually produced by filament lamps (Fig. 33–8). The filaments of these lamps are operated at a much lower temperature than the filaments of ordinary incandescent lamps. Under this condition, the lamps produce little visible light, but large quantities of infrared rays. When these rays strike a material that absorbs them,

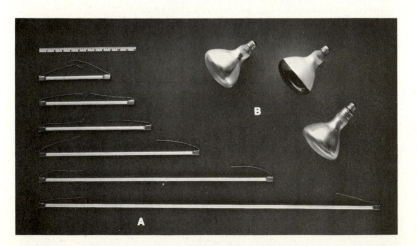

the energy of the rays is changed into heat with a high degree of efficiency.

Infrared lamps are used for drying ink, glue, varnish, and paint (Fig. 33–9). They are also commonly used for defrosting foods and for certain kinds of medical treatments.

INDUCTION HEATING

Induction heating takes place when a current is induced in a material by electromagnetic flux, just as in a transformer. In an induction heater, a high-frequency alternating current of from 5,000 to 500,000 hertz (Hz) is passed through a heater, or work, coil. This coil can be compared to the primary winding of a transformer.

As the magnetic field around the heater coil cuts across a metallic object to be heated, voltages are induced in the object because of electromagnetic induction. These voltages, in turn, produce high-frequency alternating currents that move through the object. The resistance of the metal to these induced currents causes it to become heated very rapidly.

The amount of heat produced and the depth to which this heat will penetrate an object are easily controlled by induction heating. For these reasons, this type of heating is used in a variety of industrial processes. These include forging, heat treatment, and soldering (Fig. 33–10).

DIELECTRIC HEATING

Dielectric heating is used to produce heat in nonmetallic materials. In their simplest form, the working elements of a dielectric heater consist of two metal plates.

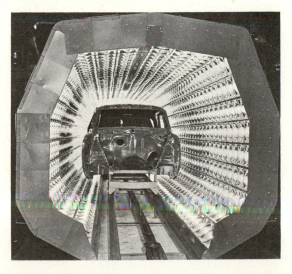

Fig. 33–9. Automobile-body enamel-finish coat being baked in an infrared oven. Banks of infrared lamps provide the necessary amount of heat. (*The Fostoria Pressed Steel Corporation*)

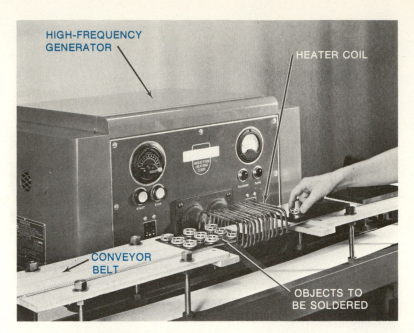

Fig. 33–10. An induction heater being used for soldering (*Induction Heating Corporation*)

The plates are connected to a high-frequency generator. The generator usually has an output with a range of 10 to 30 megahertz (MHz). This produces a high-frequency electrostatic field between the plates and through the material to be heated. This field causes electrons within atoms of the material to be very rapidly drawn or distorted out of their normal paths. The electrons are forced to move first in one direction and then in the other (Fig. 33–11). The friction that results from this back-and-forth movement of the electrons causes the material to be heated very quickly.

Dielectric heaters produce heat that can be easily controlled. In addition, the heat is evenly distributed through all parts of a nonmetallic material. For these reasons, these

Fig. 33–11. Principle of dielectric heating

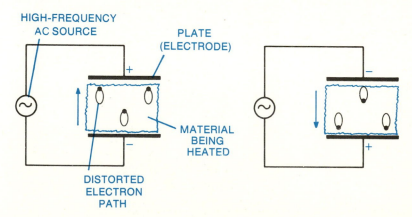

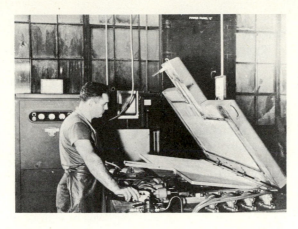

Fig. 33–12. Drying wood glue in a dielectric heater

heaters are widely used for drying glue and for heating wood, plastic, clay, and liquids (Fig. 33–12).

MICROWAVE OVEN

The microwave, or electronic, oven is able to cook foods in a fraction of the time needed for cooking with standard ranges (Fig. 33–13). The oven works similarly to dielectric heating units.

Microwaves are electromagnetic radiations that, because of their ultrahigh frequency (UHF), have a very short wavelength. In the typical microwave oven, the frequency of the radiations is 915 or 2,450 MHz. These radiations are produced by an electron tube known as a *magnetron.*

In some ovens, microwave energy from the magnetron is beamed to a fanlike, motor-driven stirrer through a rectangular, metal, tubelike device called a *waveguide* (Fig. 33–14).

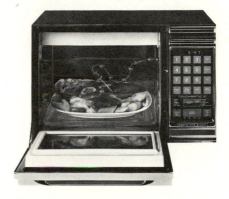

Fig. 33–13. A modern microwave, or electronic, oven (*Amana Refrigeration, Inc.*)

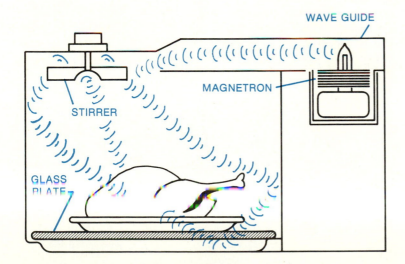

WAVE GUIDE

MAGNETRON

STIRRER

GLASS PLATE

Fig. 33–14. Operation of a microwave oven (*Amana Refrigeration, Inc.*)

The stirrer acts as a reflector, more evenly distributing the energy over the food being cooked.

The molecules in the food in a microwave oven try to keep in step with the ultra-high-frequency polarity reversals of the microwave electrostatic field. This produces heat evenly through all parts of the food by molecular friction.

Materials such as paper, plastics, and ceramics do not absorb microwave energy and so do not become heated. For this reason, food containers made of these materials are excellent for use in a microwave oven. Metal containers should not be used in a microwave oven because they reflect microwave energy and can damage the magnetron.

ELECTRIC RANGE

A typical electric-range heating element is shown in Fig. 33–15. The resistance wire of this unit is placed in a metal tube. It is insulated from the tube by an electrical insulation that is also an excellent conductor of heat.

Temperature Control. Current to each surface heating element is controlled by a heat-selector switch. This switch connects the sections of the heating element to the power line in different ways. For example, three-heat control of a simple two-ring heating element is shown in Fig. 33–16. Some ranges may have six or more heating-element rings that provide up to seven different levels of heat.

The Oven Thermostat. A *thermostat* is a device used to automatically control the operation of heating appliances so that a desired temperature is maintained. It acts as a switch that can turn a circuit on or off. The temperature of the range oven is usually controlled by a bellows thermostat. The main parts of such a thermostat are the bulb, the capillary tube, the bellows, and the switch contacts (Fig. 33–17). The switch contacts are connected in series with the oven power-supply

Fig. 33–15. Construction of an electric-range heating element (*The Tappan Stove Company*)

Fig. 33–16. Switching action for a two-ring, three-heat heating element for a range surface unit

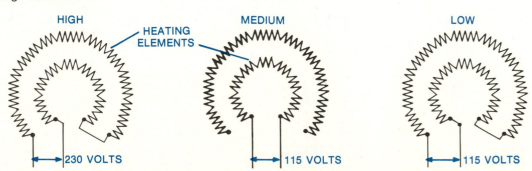

316

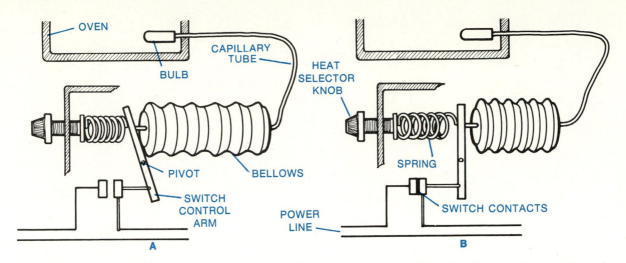

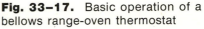

Fig. 33–17. Basic operation of a bellows range-oven thermostat

line. The bulb is filled with a liquid or a gas that expands when heated.

To set the oven temperature, the oven heat-selector dial is turned to the correct position. When the oven temperature increases above this setting, the pressure within the bulb forces the bellows to expand. The movement of the bellows is coupled to an arm that opens the circuit between the switch contacts (Fig. 33–17 at A). As the oven cools, the bellows contract and the switch is turned on (Fig. 33–17 at B). This action of the thermostat is repeated as often as needed to keep the oven at the temperature on the heat-selector dial.

BIMETALLIC THERMOSTAT

The moving part of the common bimetallic thermostat used in many appliances is a strip of two different kinds of metal. These metals, which are rigidly joined together, expand at different rates when heated. As a result, the strip bends as it is heated. This bending is used to make or break a switching contact. The bimetallic strip may be straight or spiral.

Figure 33–18 shows how a simple bimetal thermostat works. In this thermostat, the bimetallic strip is placed near a fixed contact point (Fig. 33–18 at A). The thermostat is now in an open position.

When the thermostat is heated, the brass part of the bimetallic strip expands more than the iron. Because of this, the strip bends upward. It closes that part of the circuit connected to its terminals (Fig. 33–18 at B). When the strip cools, it bends back to its original position. This opens the circuit.

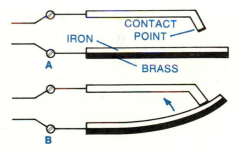

Fig. 33–18. Operation of a bi-metallic thermostat: (A) open; (B) closed

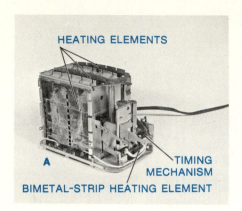

HEATING ELEMENTS

A

TIMING
MECHANISM

BIMETAL-STRIP HEATING ELEMENT

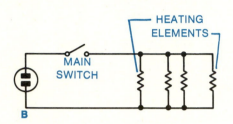

Fig. 33-19. The toaster: (A) heating elements in a typical automatic, pop-up toaster; (B) circuit diagram

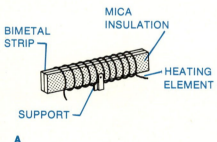

A

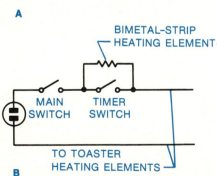

B

Fig. 33-20. Toaster-timer bimetallic strip connected to the toaster circuit

TOASTER

The heating elements of a toaster usually consist of flat, ribbon-shaped resistance wires. These wires are wound on sheets of *mica.* This is a mineral that is an excellent insulator. The elements are mounted in the toaster frame (Fig. 33-19 at A).

The main electric circuit of a typical toaster is shown in Fig. 33-19 at B. The four heating elements are connected in parallel. They have a total wattage rating of about 1,000 watts. The main switch that controls current to the elements is turned on when the toaster bread carriage is pushed down.

The Timer. Automatic pop-up toasters have additional circuits and devices. They are used to time the toasting cycle and to control the carriage-raising mechanism when the cycle is completed. In many toasters, the timer mechanism is controlled by a bimetallic strip. It is heated by a resistance-wire heating element (Fig. 33-20 at A). This element is connected in parallel with a timer switch (Fig. 33-20 at B).

The Toasting Cycle. When a toaster is turned on, current passes through the bimetallic-strip heating element (Fig. 33-21 at A). This heats the strip, and it begins to bend. After the strip reaches a certain position, determined by the setting of the toaster light-dark control, it pushes against a trigger arm. This turns the timer switch on (Fig. 33-21 at B).

The bimetallic-strip heating element is now shorted out of the circuit. This causes it to cool and bend in the opposite direction. It then pushes against a second trigger. This releases the spring-driven carriage, and the carriage moves upward (Fig. 33-21 at C). This action also turns off the main switch and the timer. The toasting cycle is then completed.

CLOTHES IRON

A typical clothes iron is shown in Fig. 33-22. In this iron, two bimetallic thermostats control the operation of the iron and prevent excessive heat.

Temperature Control. Figure 33-23 shows how a simple clothes-iron thermostat works. The thermostat is connected in series with the heating element. When the iron is first turned on, the thermostat contacts are closed. The circuit to the heating element is complete (Fig. 33-23 at A).

As the iron heats, the bimetallic element of the thermostat is also heated. It bends to open the circuit. When the iron has cooled to a certain temperature, the thermostat contacts close once again. This action is repeated to keep the temperature of the iron relatively constant while it is being used.

The temperature of the iron is selected by turning a heat-regulator knob (Fig. 33–23 at B). This increases or decreases the pressure on the bimetallic strip. As the pressure in-

Fig. 33–21. Basic operation of a toasting-cycle control mechanism

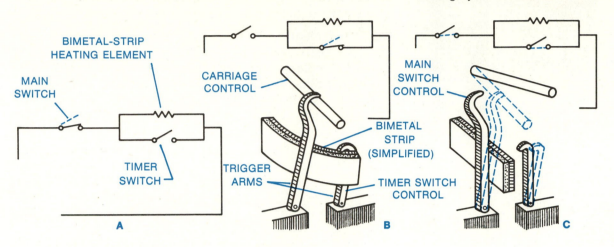

Fig. 33–22. Construction of a clothes iron (*Sunbeam Corporation*)

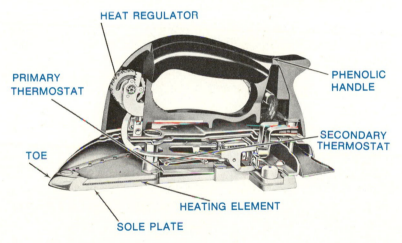

Fig. 33–23. Operation of a simple clothes-iron thermostat

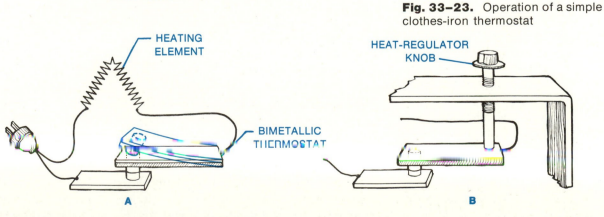

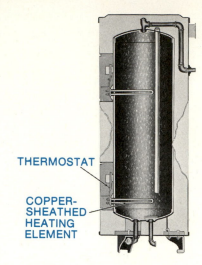

THERMOSTAT

COPPER-
SHEATHED
HEATING
ELEMENT

Fig. 33–24. Construction of an electric water heater (*FauceHot Heater Company*)

creases, the temperature of the iron must also increase before the strip will bend enough to open the heating-element circuit.

WATER HEATER

A typical electric water heater is shown in Fig. 33–24. The heating elements consist of resistance wire insulated from, and enclosed within, copper tubes. As the water in the tank heats, it circulates around the elements. It slowly rises to the top of the tank, where the hot-water outlet pipe is located. The water is automatically kept at the right temperature by bimetallic-strip thermostats. These control the current through the heating elements.

SELF-TEST

Test your knowledge by writing, on a separate sheet of paper, the word or words that most correctly complete the following statements:

1. The amount of heat produced in a conductor by resistance heating is directly proportional to the _____ of the current and to the _____ of the conductor.
2. The most common materials used in making resistance-heating wires are alloys consisting of various combinations of _____, _____, _____, and _____.
3. The two most common kinds of resistance welding are _____ welding and _____ welding.
4. Infrared rays are a form of _____ radiation having a frequency just below that of _____ light.
5. Induction heating is produced by high-frequency induced _____ moving through metallic objects.
6. Dielectric heating is used to produce heat within _____ materials.
7. A microwave oven heats by means of _____.
8. The heat-sensing bulb of a bellows thermostat is filled with a liquid or a gas that _____ when heated.
9. A thermostat acts as a _____, which can be turned on or off by the application of _____.
10. The operation of a bimetallic thermostat depends on the fact that some metals _____ more than others when heated.
11. In a clothes iron, a bimetallic thermostat is connected in _____ with the heating element.

FOR REVIEW AND DISCUSSION

1. Define resistance heating.
2. What is resistance wire?
3. Describe the construction of two common kinds of resistance heating elements.
4. Describe spot and seam welding.
5. Explain how an electric arc is produced between the ends of two carbon rods.
6. Describe arc welding.
7. Define infrared rays.
8. Give several examples of the use of infrared lamps for heating.

9. Explain induction heating. State the purposes for which it is commonly used.
10. Explain dielectric heating. For what purposes is such heating used?
11. What are microwaves?
12. Describe the basic construction of a microwave oven.
13. Describe the construction of a heating element of an electric range.
14. What is a thermostat?
15. Explain how a bellows thermostat works.
16. Describe the construction and operation of a bimetallic thermostat.
17. How is the temperature of a clothes iron controlled?

INDIVIDUAL-STUDY ACTIVITIES

1. Prepare a written or an oral report telling about the industrial applications of resistance heating, infrared heating, induction heating, or dielectric heating.
2. Prepare a written or an oral report telling how the room thermostat in your home operates to control the heating and/or cooling system.

Unit 34   Refrigeration and Air Conditioning

The operation of refrigerators and air conditioners depends on the fact that liquids absorb heat when they *evaporate*, or turn into gas (Fig. 34–1). In cooling appliances, a liquid called the *refrigerant* is circulated through tubes made of thin metal. These tubes are formed into *evaporator coils*.

The refrigerant has a very low boiling point and evaporates in the evaporator coils. As a result, the coils are cooled. They absorb heat from the air around them. After each evaporation, or cooling, cycle, the refrigerant *vapor*, or gas, is compressed and cooled. This changes the vapor back into a liquid. The refrigerant is then ready to be used again.

REFRIGERATOR

Figure 34–2 shows the main parts of an electric refrigerator system. In this and in most other modern electric cooling appliances, the refrigerant used is a substance called *Freon.* Freon is basically a compound of fluorine, carbon, and hydrogen.

When the temperature in the refrigerator cabinet rises above a specific setting, the motor-driven compressor begins to operate. The compressor pumps Freon vapor from the evaporator coils. This reduces the pressure in these coils, allowing liquid Freon to move into them through the capillary tube. A bellows thermostat is used to control the temperature in the cabinet.

As the liquid Freon evaporates, it absorbs heat from the freezer compartment and from the refrigerator cabinet. The

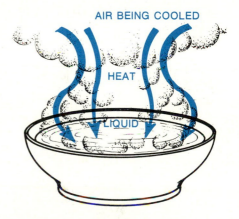

Fig. 34–1. Liquids absorb heat when they evaporate.

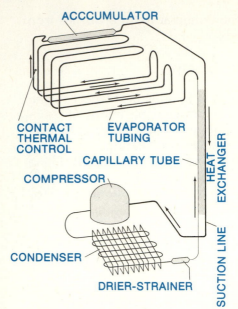

Fig. 34–2. Main parts of an electric refrigerator system

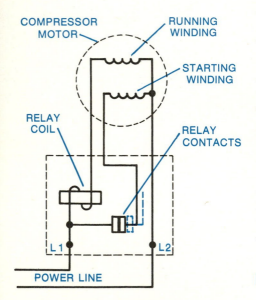

Fig. 34–3. Circuit of a refrigerator motor starting relay

heated vapor is then drawn through the compressor and into the condenser. There it is cooled and changed into a liquid. The liquid Freon is then forced back into the evaporator coils. Here it evaporates once again to complete the cooling cycle. This action is continuous while the refrigerator is in operation.

As more and more heat is absorbed from the cabinet, the temperature within the cabinet drops. This causes the temperature-control-thermostat contacts to open. This, in turn, turns the compressor motor off. The refrigerator remains idle until the cabinet temperature once again rises enough to cause the thermostat contacts to close and start the compressor motor.

The Motor-starting Relay. The refrigerator compressor and the motor that drives it are contained in a sealed unit. The motor-starting winding is connected and disconnected from the motor circuit by means of a relay mounted on the compressor housing.

When the cooling cycle starts, the line voltage is applied to terminals L1 and L2 of the relay (Fig. 34–3). Current now passes through the running winding of the motor and the relay coil. The high starting current in the circuit causes the relay contacts to close. This completes the circuit through the starting winding of the motor.

The current through the running winding decreases in value after the motor has accelerated to its full speed. This causes the starting-relay electromagnet to lose strength. The starting contacts open. The motor then continues to operate with its starting winding disconnected from the circuit.

Overload Protection. The refrigerator motor-starting relay often has a protective device. This device opens the motor circuit when the motor becomes overloaded. An overloaded motor runs slower than usual. It may be seriously damaged by the increased amount of current passing through its running winding.

To prevent damage, a bimetallic thermostat is connected in series with the motor circuit. When an abnormally high current passes through the element, its temperature increases until the thermostat contacts open. The motor is thereby automatically turned off.

ROOM AIR CONDITIONER

A room air conditioner is a refrigeration unit that cools, dehumidifies, cleans, and circulates the air in an enclosed

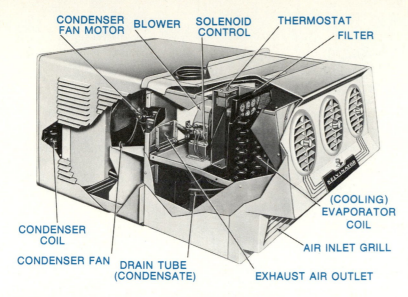

CONDENSER FAN MOTOR BLOWER SOLENOID CONTROL THERMOSTAT FILTER

CONDENSER COIL

CONDENSER FAN DRAIN TUBE (CONDENSATE)

(COOLING) EVAPORATOR COIL

AIR INLET GRILL

EXHAUST AIR OUTLET

Fig. 34—4. A typical room air conditioner (*Kelvinator, Inc.*)

space. In general, the main parts of a typical room air-conditioning unit are similar to those of a refrigerator (Fig. 34–4).

Room air is drawn through the filter to remove dirt, dust, and other particles. The air then passes through an evaporator-coil assembly, where it is cooled. The cool air is then re-circulated through the room by blowers.

As the air moves past the cold evaporator coils, some of the moisture in it *condenses,* or collects, on the coils. As a result, the air leaves the coils in a *dehumidified,* or drier, condition. This makes it more comfortable. The moisture that collects at the coils is discharged from the air conditioner as water or water vapor.

The cooling capacity of a room air conditioner is given in terms of the British thermal unit (Btu). The metric unit is the *joule.* One British thermal unit is the quantity of heat needed to raise the temperature of one pound of water one degree Fahrenheit. The Btu rating of an air conditioner is a measure of how much heat it will remove from a room during a certain period of time, usually 1 hour. The effectiveness of an air conditioner is also indicated by the amount of air it can move in a given period of time. The unit usually used is cubic feet per minute (cfm).

CENTRAL AIR CONDITIONING

A central air-conditioning system cools all areas of a house. In such a system, the condensing unit is located outside the house. When used in a house with a forced-air furnace, the

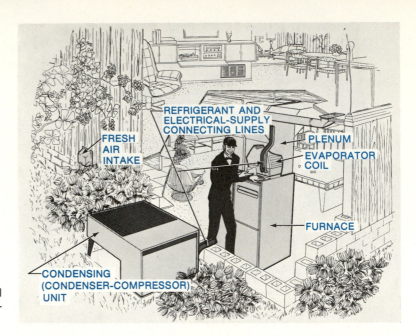

Fig. 34-5. Main parts of a central home-air-conditioning system installation (*Lennox Industries, Inc.*)

Labels in figure: FRESH AIR INTAKE · REFRIGERANT AND ELECTRICAL-SUPPLY CONNECTING LINES · PLENUM · EVAPORATOR COIL · FURNACE · CONDENSING (CONDENSER-COMPRESSOR) UNIT

evaporator-coil assembly is usually located above the furnace in the plenum, or main hot/cool output air duct (Fig. 34-5).

HEAT PUMP

The electrically driven heat pump is a combination heating-and-cooling unit. It transfers heat from one location to another in very much the same way as a refrigerator.

Winter Operation. When the heat pump is used as a heater, a liquid refrigerant is passed through an outdoor coil (Fig. 34-6 at A). As the refrigerant evaporates, it absorbs heat from the outside air. The heat-carrying refrigerant is then compressed. This raises its temperature to 100°F (37.8°C) or more. The refrigerant is then passed through the indoor coil. This acts as a heating unit. Cool air in the rooms is circulated over the warm coils. The warm air is then recirculated through the rooms.

During very cold weather, a heat pump may not be able to absorb enough heat from the outside air to provide proper space heating. In that case, auxiliary resistance-heating units are used. These units are automatically turned on by a thermostat when the temperature drops below a certain point. The units operate until the heat pump alone is once again able to provide enough heat.

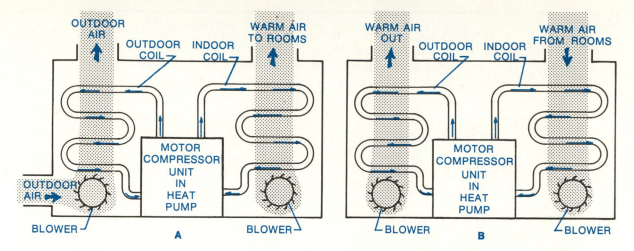

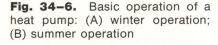

Fig. 34–6. Basic operation of a heat pump: (A) winter operation; (B) summer operation

Summer Operation. During the summer, the operation of the heat pump is automatically reversed by a temperature-control thermostat. Then warm air in the rooms is circulated over the indoor coil, which is cooled by the evaporation of the refrigerant (Fig. 34–6 at B). The refrigerant then carries the heat it has absorbed from the rooms to the outdoor coil. There the heat is released. The refrigerant condenses to begin another cooling cycle.

ELECTRONIC AIR CLEANER

The residential electronic air cleaner removes very small particles that would normally pass through a filter (Fig. 34–7). It is usually installed in the cold-air return duct of a central forced-air heating or cooling system.

In a typical electronic air cleaner, air first passes through a prefilter unit. This traps the larger dust particles. From the prefilter, the air moves past a gridlike assembly of wires. These make up the ionizing section of the cleaner. The wires, suspended between grounded aluminum electrodes, are positively charged at about 6,000 volts. This voltage is supplied by a high-voltage rectifier circuit. This causes particles still in the air to also become positively charged.

The air then passes through the collector section. This consists of aluminum plates negatively charged at about 6,000 volts with respect to the grounded aluminum plates between them. There, the positively charged particles are attracted to the negative collector plates. There they remain until they are removed by washing.

Fig. 34–7. Residential electronic air cleaner (*Whirlpool Corporation*)

Test your knowledge by writing, on a separate sheet of paper, the word or words that most correctly complete the following statements:

1. The operation of refrigerators depends on the fact that all liquids absorb _____ when they _____ .
2. In cooling appliances, a liquid known as a _____ is circulated through thin metal tubes called _____ coils.
3. The most commonly used refrigerant is _____ .
4. The purpose of the condenser in a refrigerator is to _____ the evaporated refrigerant and change it back into a _____ .
5. A room air conditioner _____ , or removes moisture from, the air. This occurs as moisture in the air _____ on cooled evaporator coils.
6. The cooling capacity of an air conditioner is given in terms of the _____ unit.
7. The electrically driven heat pump is a combination _____ and _____ unit.

1. What is the purpose of the refrigerant used in a refrigerator?
2. What is a common refrigerant material?
3. Explain how an electric refrigerator works.
4. Describe how a refrigerator motor-starting relay works.
5. How are refrigerator motors protected against overload?
6. Explain how a typical room air conditioner works.
7. Name and define the unit by which the cooling capacity of an air conditioner is given.
8. What is a heat pump? What is the purpose of the resistance-heating units used with this system?
9. Describe the basic construction and operation of a home electronic air cleaner.

Prepare a written or an oral report on the construction and operation of a refrigerator, room air conditioner, or heat pump.

Unit 35   The Automobile Electrical System

The electrical system of an automobile has a number of electric circuits. These circuits distribute current to the devices and equipment in the automobile (Fig. 35–1). In addition, the modern automobile may contain several or all of the solid-state electronic products and circuit systems shown in Fig. 35–2. The sources of energy for these systems are a battery and an alternator, or ac generator. Fuses, circuit breakers, and relays protect the circuits.

BATTERY

The main purpose of the automobile battery is to furnish the electric energy needed to turn the cranking, or starting,

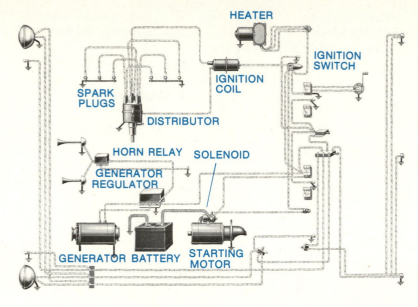

Fig. 35—1. Basic automobile electrical system

motor. The battery is also used to operate other circuit loads when the engine is not running and when the alternator is producing less than a certain amount of voltage. One terminal of the battery, usually the negative terminal, is grounded. It is grounded by being connected to the frame of the auto-

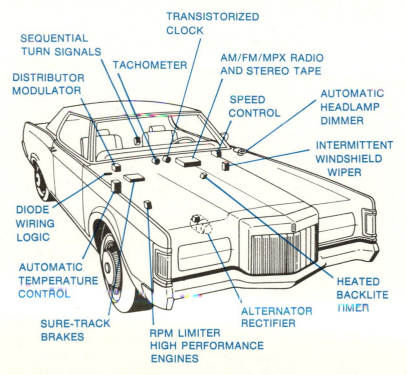

Fig. 35—2. Example of the increasing number of solid-state electronic products and circuit systems found in the modern automobile (*Ford Motor Company*)

Table 35–1. Current Requirements of Common Electrical/Electronic Items in the Automobile

Device or Product	Current (amperes)
air conditioner	10–14
heater	5–7
ignition system	2–3
parking lights	4–8
low-beam headlights	8–14
high beam headlights	12–18
tail lights	1–2
radio	0.4–2
tape player	0.5–2
windshield wiper	2–3
summer starting*	100–225
winter starting*	225–500

*Starting current requirements vary with size of engine and viscosity of oil used.

mobile with a heavy flexible cable. This reduces the number of circuit wires since the frame acts as a return circuit conductor for most of the circuits.

Most modern automobiles use a 12-volt, 6-cell, lead-acid battery. This provides efficient starting and operates an ever-increasing number of electrical/electronic items (Table 35–1).

CRANKING CIRCUIT

When the ignition switch of an automobile is turned to the start position, a circuit is completed from the battery to a relay. The relay connects the battery to the cranking motor. In most automobiles, a series dc motor is used to crank the engine.

In some automobiles, the cranking relay is a separate unit mounted in the engine compartment. A diagram of this kind of cranking circuit is shown in Fig. 35-3.

In other automobiles, the cranking switch completes a circuit from the battery to a solenoid. This is mounted on the cranking motor frame (Fig. 35-4). The solenoid plunger acts as a relay to close the circuit between the battery and the cranking motor. It also operates a shift-lever. The shift-lever couples the motor shaft to the engine flywheel while the engine is being cranked.

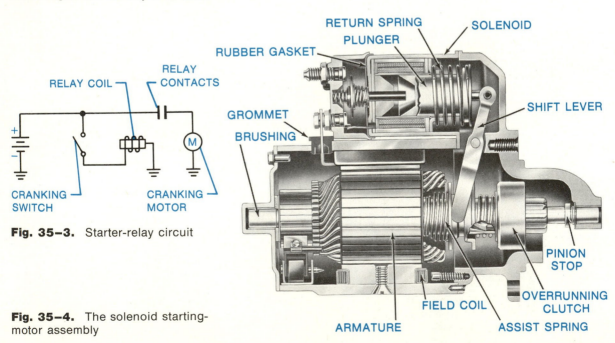

Fig. 35-3. Starter-relay circuit

Fig. 35-4. The solenoid starting-motor assembly

BREAKER-POINT IGNITION SYSTEM

As the crankshaft begins to turn, the cam-driven breaker points in the distributor start to open and close. This makes and breaks the primary circuit of the ignition coil. The ignition coil is actually a step-up transformer (Fig. 35–5). When the primary circuit is broken, the magnetic field collapses quickly. This induces 20,000 or more volts across the secon-

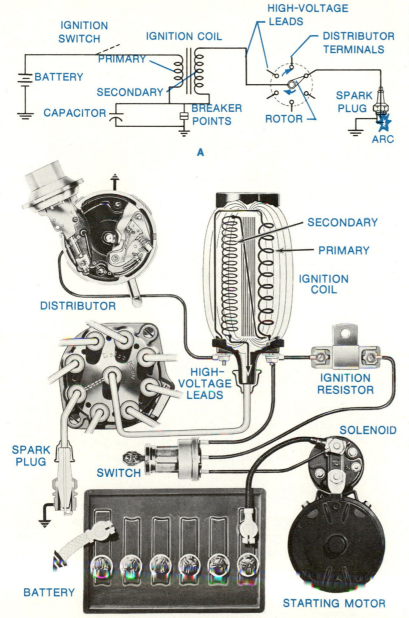

Fig. 35–5. Conventional ignition system: (A) circuit diagram; (B) ignition system and starting-motor connections (*Delco-Remy Division of General Motors Corporation*)

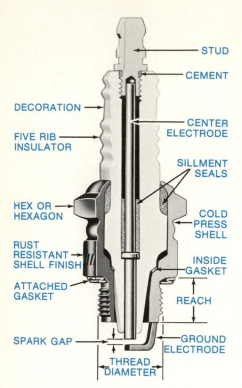

Fig. 35-6. Construction of a spark plug (*Champion Spark Plug Company*)

Labels in figure:
STUD
CEMENT
DECORATION
CENTER ELECTRODE
FIVE RIB INSULATOR
SILLMENT SEALS
HEX OR HEXAGON
COLD PRESS SHELL
RUST RESISTANT SHELL FINISH
INSIDE GASKET
ATTACHED GASKET
REACH
SPARK GAP
GROUND ELECTRODE
THREAD DIAMETER

dary. The high-voltage circuit is completed to the correct spark plug by the rotor under the distributor cap. As the rotor turns, the plugs fire in a desired order. This is called the *firing order*. It is usually found on the intake manifold of the engine. This reference is handy in case the wires should ever become mixed up at the distributor cap.

When the breaker points open, the sudden collapse of the magnetic field also induces a high voltage in the primary circuit of the coil. This voltage would ordinarily cause extreme arcing at the breaker points. However, there is a capacitor across the points. This capacitor absorbs the induced surge and allows the ignition circuit to work properly. A faulty capacitor will cause poor ignition and point burning or no ignition at all.

Spark Plugs. The spark plugs provide the electrical gap in the high-voltage circuit of the ignition system. The gap at the tip of each plug is the site of the spark that ignites the fuel-air mixture in a cylinder of the engine. Spark plugs are exposed to extreme temperatures and pressures. Therefore, they are made of heat-resistant materials. Figure 35–6 shows the details of a typical spark plug.

Cylinders. When the spark plugs fire, the fuel-air mixture in the cylinders is ignited. The rapidly expanding gases produced by this burning push against the pistons. This causes them to move (Fig. 35–7). The movement of the pistons turns the crankshaft. Mechanical energy from the crankshaft is then passed on to the drive wheels through the transmission assembly.

ELECTRONIC IGNITION SYSTEM

In one common electronic ignition system, there are no breaker points in the distributor. A magnetic pickup assembly located over the distributor shaft contains a permanent magnet, a pole piece with internal teeth, and a pickup coil (Fig. 35–8).

As the distributor rotor turns, the teeth of its timer core move past the teeth of the pole piece. As a result, a moving magnetic field cuts across the pickup coil. This induces pulses of voltage across the coil.

The pulses are then applied to an amplifier in the electronic module. The output of the amplifier is connected to the primary winding of the ignition coil. A high voltage is induced across the secondary winding of the coil. This is then applied to the spark plugs through the distributor.

ALTERNATOR

After the engine has started and is running at a certain speed, an *alternator,* or ac generator, replaces the battery as the source of energy.

Construction. The typical alternator is a three-phase generator with a rotating field (Fig. 35–9). The rotor assembly consists primarily of a doughnut-shaped field coil. This assembly is mounted between two iron sections with interlacing fingerlike prongs called *poles.* Field current from the battery is passed to the field windings by two brushes that contact the rotor slip rings.

Most alternators are driven by the fan belt or by a separate belt that couples the rotor shaft to the engine flywheel. As the rotor assembly turns, the magnetic field produced by the field current cuts across the stationary windings of the stator assembly. This action induces an alternating voltage across the stator windings.

Rectification of Output. The voltage produced by the alternator is rectified by six silicon diodes. These are mounted inside the alternator end frame. Rectification is needed because the electrical system of the automobile operates with direct current.

Damage to Diodes. A reversal of the polarity of the automobile-battery voltage may seriously damage the diodes of the alternator or other parts of the electrical system. For this reason, it is important to use the correct polarity while connecting a new battery or a booster battery to the automobile.

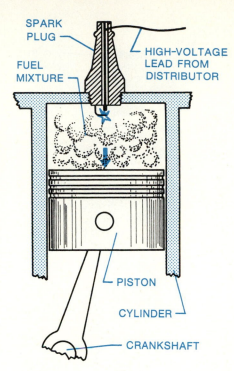

Fig. 35–7. Effect of ignition in an automobile cylinder

ALTERNATOR REGULATOR

The voltage output of the alternator is automatically controlled so that it remains fairly constant at different engine speeds. This is done by the alternator regulator. A typical regulator consists of two relay units, the voltage regulator and the field relay (Fig. 35–10).

The Voltage Regulator. The voltage output of the alternator increases as the engine speed increases. The voltage regulator limits the voltage so that it will not overcharge the battery or damage any part of the electrical system.

When the output of the alternator reaches about 14 volts, the automatic switching action of the voltage regulator puts a resistor in series with the alternator field. This reduces the

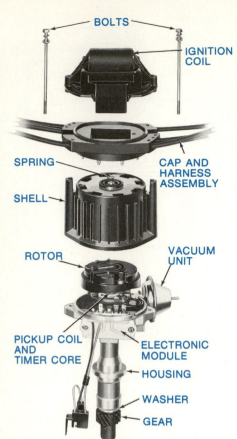

field current. The alternator output voltage is thereby also reduced.

The Field Relay. The field relay acts as a magnetic switch that closes when the relay is energized. This relay provides a low-resistance conductor path between the battery and the voltage-regulator coil. The relay also disconnects the battery from the field windings of the alternator. This system provides complete alternator regulation in a compact form.

LIGHTS

In the automobile, electricity is used to operate several kinds of lamps. Most are single- or double-contact lamps with a bayonet base. For safe driving, the headlights, the tail lights, and the turn-signal lights are most important. Instrument, dome, parking, and backing lights are also common.

Headlamps. Modern automobiles use sealed-beam headlamps. In such a lamp, one or two filaments are sealed into a bulb that also serves as a lens and a reflector (Fig. 35–11). In a two-filament headlamp, the filaments are connected into the circuit so that either one of them can be turned on. This makes it possible to produce high and low beams of light.

After the lamps have been turned on by the main light switch, either the high or the low beam can be selected with the dimmer switch.

Fig. 35–8. Exploded view showing the distributor-assembly components of an electronic ignition system (*Delco-Remy Division of General Motors Corporation*)

Fig. 35–9. The automobile alternator: (A) external view; (B) cutaway view showing main features (*Delco-Remy Division of General Motors Corporation*)

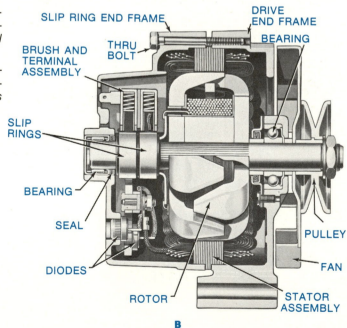

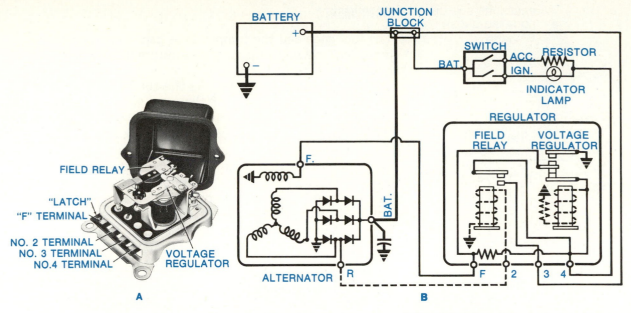

Fig. 35-10. Alternator regulator and regulator circuit

Stop Lights. In many automobiles, two-filament lamps serve as both stop lights and tail lights. In other automobiles, separate lamps are used.

The stop-signal lamps are automatically turned on when the brake pedal is depressed. The pressure of the hydraulic fluid in the master brake cylinder forces a metal diaphragm over switch contacts. When the brake pedal is released, the diaphragm moves away from the contacts, and the lamps go off.

Turn-signal Lights. The flashing of turn-signal lamps is generally controlled by a thermostatically operated flasher unit. This turns the lamps on and off about 90 times per minute.

HORNS

Automobile horns are vibrating devices that work very much like a door bell. In a horn, however, the vibrating arm is connected to a diaphragm. The rapid back-and-forth movement of the diaphragm in an air column produces the sound.

Horns are usually controlled by a relay. When the pressure switch in the steering-wheel assembly is pushed, a circuit is completed from the battery or the alternator through the horn relay (Fig. 35–12). This causes the relay contacts to close. Current then passes through the horn circuit.

Fig. 35–11. Sealed-beam head-lamp (*General Electric Company*)

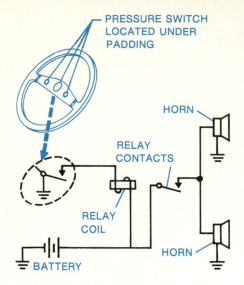

Fig. 35–12. The horn circuit

INSTRUMENTS

The fuel gage and the temperature gage are the most common electrical instruments in the automobile. In one kind of gage, a heating coil is wound around a bimetallic strip (or arm) connected to a pointer. As the coil becomes heated by current, the temperature of the strip also increases. This causes the strip to bend. The amount of bending determines the position of the pointer on the gage scale.

The amount of current in a fuel-gage circuit depends on the position of a float connected to a rheostat (Fig. 35–13). As the fuel supply in the tank decreases, the float causes the resistance of the circuit to increase. This, in turn, causes the pointer to move toward the E (empty) position on the gage scale. In a temperature gage, the current in the circuit is controlled by the temperature of the cylinder block.

ELECTRICAL ACCESSORIES

Common automobile electrical accessories include heaters, air conditioners, fans, windshield wipers, window operators, and seat adjusters. All of these are operated by small electric motors. The accessories are usually controlled by hand-operated switches located on or near the instrument panel.

FUSES AND CIRCUIT BREAKERS

Almost all the lighting and accessory circuits in an automobile are protected by either cartridge fuses or circuit breakers. In many automobiles, circuit breakers are used more widely than fuses.

Common automobile fuses include the 4-, 6-, 9-, 14-, 20-, and 30-ampere sizes. Each size has a different length. The larger the ampere size, the longer the fuse (Fig. 35–14). A wrong-size fuse cannot be inserted into the fuseholder or mounting clip. This is an important safety feature.

THE ELECTRIC CAR

The *electric car* is not new. It first appeared in this country about 1888. It was very popular at the turn of the century. But, by 1930, most electric cars had been replaced with the familiar automobile powered by a gasoline engine.

Today, with the high cost and limited supply of gasoline, there is a renewed interest in the electric car (Fig. 35–15). This car has many advantages. It is practically noiseless, producing only a hum or whine. The electric car does not pollute

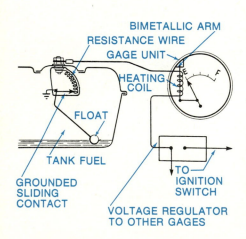

Fig. 35–13. Fuel-gage circuit

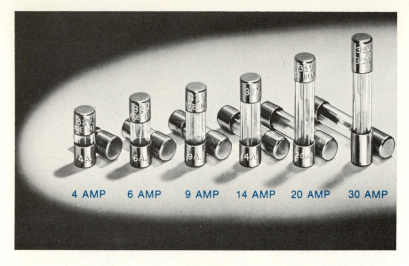

Fig. 35–14. Automobile fuses

the air. It is very maneuverable. It is small and lightweight. The United States has an adequate supply of lead, used in the lead-acid batteries of the electric car.

The electric car does, however, have several disadvantages. It cannot match the gasoline car in range, versatility, performance, or speed. There is no nationwide system for maintenance or for battery recharging.

The electric car has four major systems. These are discussed below.

Power System. Because lead-acid batteries are presently available, some electric-car power systems are designed to operate with them. Eighteen of these batteries are connected to produce 108 volts. Power units made up of lead-acid batteries can be recharged overnight.

Several new power systems are being considered for use in the electric car. Some of these use batteries made from nickel and zinc, nickel and iron, sodium and sulfur, zinc and chlorine, and lithium and a metal sulfide.

Fig. 35–15. Key features of an experimental electric car (*U.S. Department of Energy*)

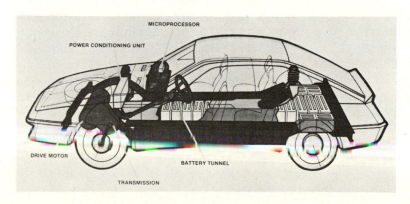

VACUUM CLEANER　　**PANELS OF SOLAR CELLS**

Fig. 35–16. Solar (photovoltaic) cells are used to charge the batteries in an electric vehicle fitted with a vacuum cleaner used for refuse pickup. (*National Aeronautics and Space Administration*)

The objective is to develop a power system that can operate an electric car at a speed of 45 to 55 miles per hour (72 to 89 km/hr) with a range of about 150 miles (240 km) before recharging.

There are two important reasons for using batteries. First, they are compatible with existing electric supply systems. Second, they involve a shift from a petroleum-consuming energy system to a system that uses an abundant or renewable source of energy (Fig. 35–16).

Control System. The control system has the instruments and switches needed to operate the electric car and support systems. Two important meters provide information about the condition of the power system. These are the *voltmeter* and the *ammeter*. The voltmeter reading indicates the charge in the batteries. The ammeter indicates the rate of energy consumption. Other controls include the switches for energizing various units, such as electric brakes, in the car. Because the electronic equipment used in the car is complicated, a central computer unit, called a *microprocessor,* is used to control it. For example, the accelerator uses solid-state devices to control the amount of energy that determines the speed of the car.

Electric Drive System. Some of the experimental electric cars use two dc drive motors, one for each drive wheel. Other electric cars use one motor with a transmission. These dc motors change the electric energy into mechanical energy to run the car.

Support System. The support system includes the accessory equipment, such as windshield wipers and washers, turn signals, heater, and air conditioner.

SELF-TEST

Test your knowledge by writing, on a separate sheet of paper, the word or words that most correctly complete the following statements:

1. Sources of energy for the automobile electrical system are the _____ and the _____.
2. The _____ terminal of the battery is usually grounded by being connected to the _____ of the automobile.
3. In some automobiles, the cranking switch completes a circuit from the battery to a _____ mounted on the _____ frame.
4. The breaker points in an ignition system operate as a _____ switch.
5. High voltage from the secondary winding of the _____ coil is applied to the _____ plugs by means of a _____ inside the distributor.
6. The desired order in which spark plugs fire is called _____.
7. Excessive arcing across the breaker points is prevented by connecting a _____ across the points.
8. The ignition spark is produced in a _____ in the spark plug.
9. The typical alternator is a _____ -phase generator with a _____.
10. The voltage output of the alternator is controlled by the _____.
11. At present, electric-car power systems use _____ batteries.
12. The central computer unit in an electric car is called the _____.

FOR REVIEW AND DISCUSSION

1. What is the main function of the automobile battery?
2. Why is one terminal of the automobile battery grounded to the frame?
3. What is the purpose of the solenoid in a cranking circuit?
4. State the purpose of the breaker points, the ignition coil, and the distributor rotor in a conventional ignition system.
5. Why is a capacitor connected across the breaker points?
6. What is the function of the spark plugs?
7. Briefly explain how one common kind of electronic ignition system works.
8. Why must the voltage output of an alternator be rectified? Explain how this is done.
9. How can alternator diodes be damaged?
10. What does an alternator regulator do?
11. State the purpose of the voltage regulator and the field-relay units of an alternator regulator.
12. Explain how the horn circuit works.
13. What are some major advantages of the electric car?

INDIVIDUAL-STUDY ACTIVITIES

1. Prepare a written or an oral report telling about the recent developments in the use of electronic devices in automobiles.
2. Prepare a demonstration for your class showing some automotive electrical units. Explain their operation.

Communication and Information Systems

Unit 36 Basic Concepts in Communication

People communicate with each other by using one or more of the primary senses—hearing, vision, touch. The basic concepts in communication are shown in Fig. 36–1.

Note that communication is started by an *information initiator.* This is a person who has information to give to someone else, the *information user.* This information can consist of various kinds of messages, including music. The information user is the person who receives and makes use of the information. The information goes through several processes called *encode, transmit, carry, receive,* and *decode* in order to be received by the user. *Feedback* enables the information user to let the initiator know whether the message has been received and understood. The initiator uses feedback to determine if the information needs further clarification. Feedback is essential to an effective communication system. The communication processes for several basic communication systems that use only hearing and vision are summarized in Fig. 36–2. Also shown are communication systems that use electricity and electronics. These systems allow people to communicate over long distances. Common communication systems are the telegraph, teletypewriter, telephone, radio, and television.

VOICE COMMUNICATION

Voice communication is the basic communications process. You engage in this activity daily—in school, at home, or at work. Actually, this process is very complicated. However, you have become so used to it that you do not think about it.

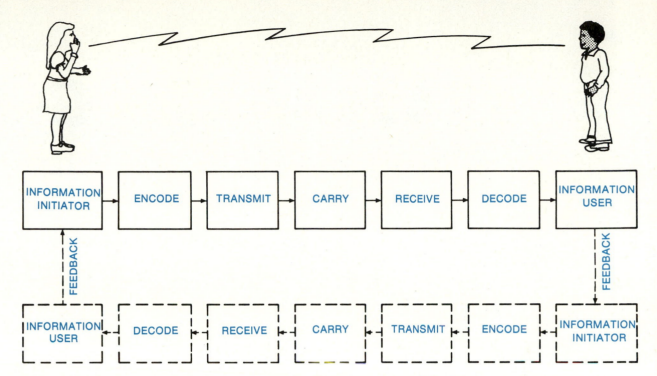

INFORMATION INITIATOR	ENCODE	TRANSMIT	CARRY	RECEIVE	DECODE	INFORMATION USER

FEEDBACK

INFORMATION USER	DECODE	RECEIVE	CARRY	TRANSMIT	ENCODE	INFORMATION INITIATOR

FEEDBACK

Fig. 36–1. Basic concepts in the communication process

You attach meanings to the sounds by the way you say vowels and consonants. This forming of the original language is the *encoding* process. The voice box produces the sound when air is expelled past the vocal cords. These vibrate rapidly and *transmit* an audible sound. The Earth's atmosphere is the medium that *carries* the sound vibrations to the ear— the *receiver.* If the language is heard but not understood by the information user, it must be translated into the language of the user. This process is known as *decoding.* Once the language is decoded, the brain interprets the message. If the language transmitted is the same as the language understood by the user, the brain automatically decodes the message. This makes it possible for the information user to understand the message. Feedback is provided by the reverse process (Fig. 36–1). However, voice communication can carry for only a limited distance.

VISUAL COMMUNICATION

The communication distance can be greatly increased by using visual means, such as flags, smoke signals, and flashing lights. Because of the curvature of the Earth, this distance is limited to a straight line of sight, about 20 miles (32.2 kilometers). However, the distance can be increased to about 40

Fig. 36–2. Types of communication systems

Information initiator makes use of human senses	Processes and elements involved					Information user makes use of human senses
	Message encoding techniques	Transmitting elements	Carrying environment	Receiving elements	Message decoding techniques	
(A) Hearing *Voice communications* 	Sounds spoken with vowels and consonants in the language are understood by person receiving message.	Human voice box (vocal cords, mouth, tongue, etc.) generates sound waves	Earth's atmosphere (distance between communicators: about 275 meters or 300 yards)	Human ear	No need to translate if language is understood by communicator	Hearing Ears receive sound waves. Brain interprets message. 
(B) Vision *Visual communications* 	Light flashes, flag positions, and smoke puffs have assigned alphanumeric meanings (code).	Light flashes, flags, smoke puffs	Visual light (distance between communicators: about 32 kilometers or 20 miles)	Human eye	Translate with same code	Vision Eyes receive visual messages. Brain interprets message. 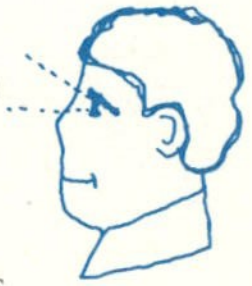
(C) Hearing and touch *Telegraph* 	Hand-operated switch controls spurts of current which have assigned alphanumeric meanings.	Telegraph key	Wire (distance between communicators: worldwide)	Telegraph sounder	Translates electric pulses (current) into sound waves (clicks)	Hearing Ears receive sound waves. Brain interprets message. 
(D) Hearing *Telephone* 	Voice (sound waves) are translated or changed to electric (audio frequency) currents.	Telephone transmitter	Wire (distance between communicators: worldwide)	Telephone earphone	Reproducer (earphone) translates electric currents into sound waves (speech and music)	Hearing Ears receive sound wave. Brain interprets message. 

Sense / System	Encoding	Transmitter	Medium	Receiver	Decoding	Reception
(E) Hearing and touch *Wireless telegraph*	Encoding is the same as in a telegraph,—but uses high-frequency (radio) currents.	Transmitter	Electromagnetic waves (distance between communicators: worldwide)	Receiver	Local oscillator signal mixes with high-frequency currents to produce audio frequencies; reproducer translates audio frequencies into sound waves	Hearing Ears receive sound waves. Brain interprets message.
(F) Hearing *Radio*	Microphone changes voice (sound waves) into electric currents. These currents modulate radio-frequency currents.	Radio transmitter	Electromagnetic waves (distance between communicator: worldwide and interplanetary)	Radio receiver	Detector separates audio frequencies from radio-frequency currents; reproducer (speaker) translates audio-frequency currents into sound waves.	Hearing Ears receive sound waves. Brain interprets message.
(G) Vision and touch *Teletypewriter*	Hand- or tape-controlled keyboard closes selected switches which have assigned alphanumeric meanings. Synchronous distributor rotor develops the train of electric pulses.	Sender (transmitter)	Wire (distance between communicators: worldwide)	Printer (receiver)	Selector magnet translates electric pulses into mechanical action to print alphanumeric characters on paper	Vision Eyes receive visual image from printed paper. Brain interprets message.
(H) Hearing and vision *Television*	TV camera translates images (video) into electric currents. These currents modulate high-frequency currents. Microphone changes sound waves (audio) into electric currents, as in radio.	TV transmitter (video and audio)	Electromagnetic waves (distance between communicators: worldwide and interplanetary)	TV receiver (picture and sound)	Video detector separates video currents from radio currents; picture tube translates video currents into images; audio detector acts same as in radio above	Hearing and Vision Ears receive sound waves from speaker. Eyes receive visual images from TV screen. Brain interprets messages.

miles (64.4 kilometers) by using a high building or a mountain.

As in voice communications, the information initiator starts the process. The nature of the flags (position, color, and so on), the variations in smoke puffs, or the duration of the flashing light are given meanings. This is the *encoding* process. For example, variations in the time a flashing light is on or off can have alphabetical meanings (A, B, C, and so on) and numerical meanings (1, 2, 3, and so on). These meanings are an *alphanumeric code.* Using this code, a flashing light can *transmit* a message. The *carrier* of this message is visible light. The *receiver* of this visual communication is the eye. The difference in time the light is on or off is easily detected by the eye. Therefore, the message can be decoded. Thus, the senses, especially hearing and vision, are essential to the communication process.

ELECTRICAL AND ELECTRONIC COMMUNICATIONS

The use of electricity and electronics makes it possible to communicate over long distances. Figure 36–2 shows how the telegraph, telephone, wireless telegraph, radio, teletypewriter, and television begin and end the process of communication through the primary senses, especially hearing and vision. However, electric energy is used to encode, transmit, carry, receive, and decode the messages.

SELF-TEST

Test your knowledge by writing, on a separate sheet of paper, the word or words that most correctly complete the following statements:

1. The three primary senses used in communication are _____, _____, and _____.

2. The person who starts the communication process is called the information _____.

3. The information _____ is the person who receives the messages.

4. The five basic steps in communication are _____, _____, _____, _____, and _____.

5. _____ is the process by which an information user lets an information _____

know whether the message has been received and understood.

6. _____ communication is the basic communication process.

7. Meanings given to letters and numbers form an _____.

8. In electrical and electronic communications, _____ transmits the messages.

FOR REVIEW AND DISCUSSION

1. Describe the basic concepts in the communication process between two people using voice communications. What are some of the limitations of this process?

2. Compare voice communication with visual communication.

3. List the electrical and electronic communications methods.

Unit 37 Telegraph and Teletypewriter

The telegraph was the first electrical communication system that allowed people to communicate beyond the limited ranges of voice and vision. Figure 37–1 shows a basic telegraph circuit for sending messages in one direction. The electric circuit consists of a battery, a key, a line, and a receiver. The battery is the source of energy. The hand-operated key acts as a switch to open and close the circuit (Fig. 37–2). This causes *pulses* of current to move through the circuit.

Encoding. The operator (the information initiator) controls the pulses by varying lengths of time the key is switched on or off. In the United States, the code used in wire telegraphy is the Morse code. It is named in honor of Samuel F. B. Morse, the American who invented the telegraph. A modified form of the Morse code, called the *international Morse code,* is shown in Fig. 37–3. The international Morse code is often used for amateur radio communications.

Transmitting. The transmitting and encoding processes are done at the same time when the operator's hand first closes the key. This completes the circuit. The pulse ends when the

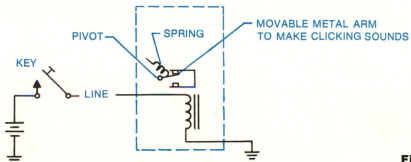

Fig. 37–1. Basic telegraph circuit

Fig. 37–2. A hand-operated telegraph key

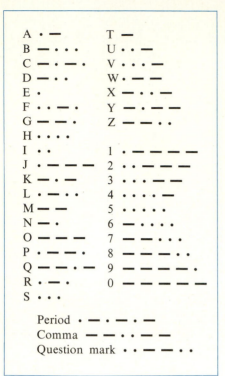

A · —	T —
B — · · ·	U · · —
C — · — ·	V · · · —
D — · ·	W · — —
E ·	X — · · —
F · · — ·	Y — · — —
G — — ·	Z — — · ·
H · · · ·	
I · ·	1 · — — — —
J · — — —	2 · · — — —
K — · —	3 · · · — —
L · — · ·	4 · · · · —
M — —	5 · · · · ·
N — ·	6 — · · · ·
O — — —	7 — — · · ·
P · — — ·	8 — — — · ·
Q — — · —	9 — — — — ·
R · — ·	0 — — — — —
S · · ·	

Period · — · — · —
Comma — — · · — —
Question mark · · — — · ·

Fig. 37–3. The international Morse code consists of combinations of pulses called dots and dashes.

Fig. 37–4. Sounder used in a hand telegraph set

key is released. This opens the circuit. A short pulse is called a *dot.* A long pulse is called a *dash.* The dot is about one-fourth of a second long. The dash is three times longer.

Transmission. The dots and dashes of the code are carried in the form of electric energy by wires. The signals are carried almost instantly through the wires from the transmitter to the receiver.

Receiving. The receiver, called a *sounder,* is a sounding device. It makes clicks loud enough to be heard (Fig. 37–4). The sounder has an electromagnet and a movable arm. A pulse of current is sent through the electromagnet. The electromagnet then attracts the movable arm. The arm moves and makes a click. When the pulse stops, a spring returns the movable arm to its original position. This makes a second click. The time between the clicks represents a dot or a dash.

Decoding. When the electric energy is translated into clicks by the sounder, the decoding process begins. The clicks, heard by ears, form the dots and dashes that represent the alphanumeric characters of the international Morse code. Thus, the message can be interpreted by the brain.

OPEN- AND CLOSED-CIRCUIT TELEGRAPH

Two basic telegraph systems are discussed below:

The Open-circuit System. In this system, current or electrical pulses pass only while a message is being sent. When the telegraph is idle, no current flows (Fig. 37–5). Most countries in Europe use this open-circuit system of telegraphy.

The Closed-circuit System. This system, shown in Fig. 37–6, is used mostly in the United States. It remains closed

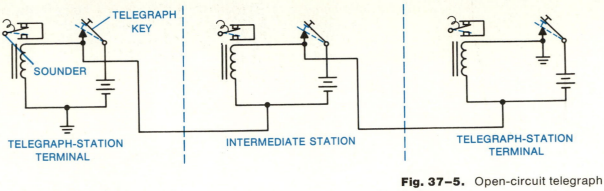

TELEGRAPH
KEY

SOUNDER

TELEGRAPH-STATION
TERMINAL

INTERMEDIATE STATION

TELEGRAPH-STATION
TERMINAL

Fig. 37–5. Open-circuit telegraph system

KEY SWITCH

TERMINAL STATION
(SENDING)

INTERMEDIATE STATION
(RECEIVING)

TERMINAL STATION
(RECEIVING)

Fig. 37–6. Closed-circuit telegraph system

until the operator wants to send a message. First the operator opens a switch connected across the key. This switch opens the circuit. The circuit can then be closed and opened rapidly by depressing and releasing the key. After the message has been transmitted, the key is closed. This completes the circuit so that other stations may use the circuit.

MODERN TELEGRAPH SYSTEMS

Modern telegraph systems use a variety of electrical and electronic equipment. The slow hand key and sounder cannot handle a large volume of messages. Instead, teletypewriters and automatic switching centers are used to send messages with great speed.

TELETYPEWRITERS

The teletypewriter combines the basic elements of a telegraph and a typewriter so as to create coded electrical pulses (Fig. 37–7). A transmitting unit, called a *sender*, can be operated like a typewriter keyboard or by a coded, punched tape. The receiver unit, called a *printer*, can produce regular printed messages on rolled paper or on paper tape. The

Fig. 37–7. A modern, automatic send-receive (ASR) Teletype (*Teletype Corporation*)

printer can also produce messages punched as holes through paper tape. The punched tape is called *perforated tape*. Devices that can both send and receive messages are called *terminal units*. In addition to the keyboard, elements of the teletypewriter include a cabinet, electric-power distribution panels, a transmitter distributor, and a page printer.

Messages are encoded into what is called the *teletypewriter* or *teleprinter code*.

The teletypewriter code works on a *binary system*. This is a system that has two different states—current *on* or current *off*. Using this code, each character—letter, number, or symbol—consists of seven electrical pulses within a time period, for example, 163 milliseconds. When an electrical impulse is *on*, it is called a *mark*. When it is *off*, it is called a *space*. The standard five-unit, start-stop teletypewriter code is shown in Fig. 37–8. Notice that each pulse is rectangular. A rapid rise and/or fall of current forms the marks and spaces. When a paper tape is punched with this code, the punched holes on

Fig. 37–8. Standard five-unit, start-stop teletypewriter code

| CHARACTERS | | WEATHER SYMBOLS | CODE SIGNAL | | | | | | |
LOWER CASE	UPPER CASE	UPPER CASE	START	1	2	3	4	5	STOP
A	—	↑		●	●				●
B	?	⊕		●			●	●	●
C	:	○			●	●	●		●
D	$	↗		●			●		●
E	3	3		●					●
F	!	→		●		●	●		●
G	&	↘			●		●	●	●
H	£	↓				●		●	●
I	8	8			●	●			●
J	'	↙		●	●		●		●
K	(	←		●	●	●	●		●
L	)	↖			●			●	●
M	.	.				●	●	●	●
N	,	⊕				●	●		●
O	9	9					●	●	●
P	0	0			●	●		●	●
Q	1	1		●	●	●		●	●
R	4	4			●		●		●
S	BELL			●		●			●
T	5	5						●	●
U	7	7		●	●	●			●
V	,	∅			●	●	●	●	●
W	2	2		●	●			●	●
X	/	/		●		●	●	●	●
Y	6	6		●		●		●	●
Z	''	+		●				●	●
BLANK	—								●
SPACE						●			●
CARRIAGE RETURN							●		●
LINE FEED					●				●
FIGURES				●	●		●	●	●
LETTERS				●	●	●	●	●	●

☐ SPACE
▨ MARK

346

the tape indicate current *on* (mark). The lack of holes indicates current *off* (space). Start and stop pulses do not appear on the tape. The five information pulses that are on the tape are called *bits*. One of the main advantages of perforated tape is that messages can be punched on the tapes and stored. They can later be sent at high speed.

Encoding Process. The mechanical linkage in the teletypewriter produces the right opening and closing of switches for each character. Figure 37–9 shows how the open or closed switches are sensed. When the key lever is pressed down by an operator, it acts on five crossbars. These, in turn, move locking levels to open or close five different switches. This is done according to the code for the key lever pressed down.

Each switch is matched with a rotating cam. The cams are placed around a distributor. This turns at a set speed. The no. 1 cam senses the no. 1 switch for a specified time, for example, 22 milliseconds. If the switch is closed, current instantly flows through the cam. This indicates a mark. If the switch is open, no current flows. This indicates a space. Next the no. 2 cam senses the no. 2 switch for another 22 milliseconds. This process is repeated for the other cams and switches until the rotation is complete. Thus, a character is encoded electrically by a mechanical motion. The pulses generated, including a 22-millisecond space for start and a 31-millisecond mark for stop, make up one character. These can be sent at a rate, for example, of 60 words per minute. Figure 37–10 illustrates the basic series circuit of a teletypewriter. The pulses generated

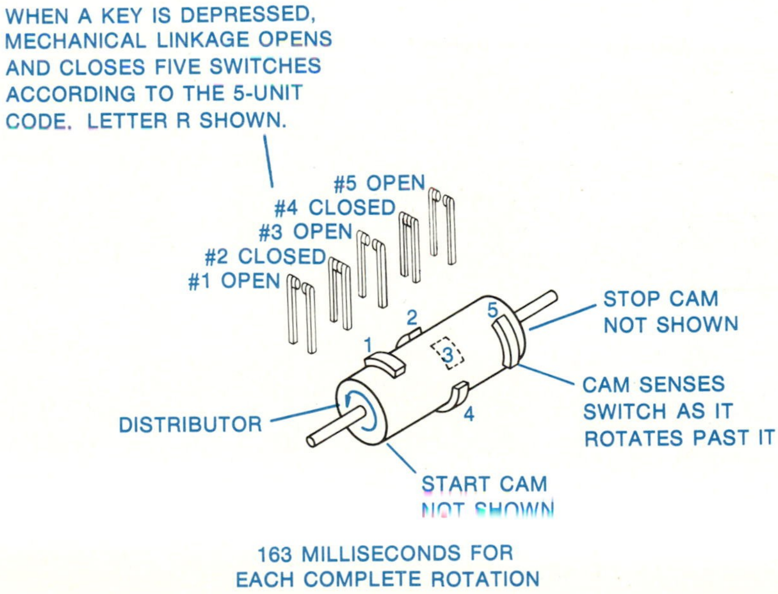

WHEN A KEY IS DEPRESSED, MECHANICAL LINKAGE OPENS AND CLOSES FIVE SWITCHES ACCORDING TO THE 5-UNIT CODE. LETTER R SHOWN.

Fig. 37–9. Distributor for sensing the open or closed switches when the character *R* is depressed

#5 OPEN
#4 CLOSED
#3 OPEN
#2 CLOSED
#1 OPEN

STOP CAM NOT SHOWN

CAM SENSES SWITCH AS IT ROTATES PAST IT

DISTRIBUTOR

START CAM NOT SHOWN

163 MILLISECONDS FOR EACH COMPLETE ROTATION

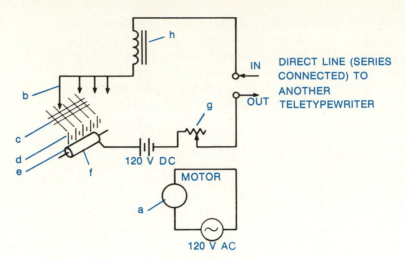

Fig. 37–10. Basic circuit of a teletypewriter: (a) ac synchronous motor; (b) key lever; (c) mechanical linkage; (d) switches; (e) distributor; (f) cam; (g) variable resistance; (h) selector magnet

by the transmitter, or sender, are carried over wires to the receiver, or printer. In radio telegraphy, the pulses are changed into high-frequency radio currents. These are carried by electromagnetic waves to receivers for decoding.

Decoding. Decoding is done by the *selector magnet* in the printer or receiver. The selector magnet is connected in series in the circuit. The pulses from the sending and/or receiving teletypewriter pass through it. The off-on pulses fed into the selector magnet are changed into mechanical action. The message is printed on paper in front of the operator. It can also be printed or punched on tape.

COMMUNICATION NETWORK

Figure 37–11 shows the interconnecting networks between the sender and receiver of information, and vice versa. In a complicated system, a teletypewriter send-receive (S/R) terminal can transmit a customer's order to the manufacturing plant, warehouse, and retail store all at once. In this way, identical information is received by all concerned with the business transaction. If there is a question, any S/R terminal can send a message to clear it up.

Note in Fig. 37–11 that the loop from the weather station teletypewriters goes by land line to a *relay center*. This is a switching center that can be a separate facility. It can also be located, as shown, in a *data-control center* that also has equipment for transmitting and receiving information by radio.

A relay center can receive a message and redirect it to distant points by either radio or wires. This process can be re-

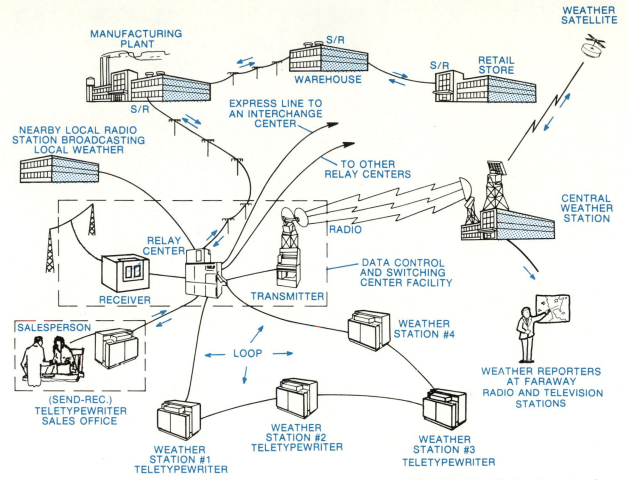

Fig. 37-11. Basic elements of an interconnected data-communication system

peated over and over until the message reaches its destination. An automatic switching system is shown in Fig. 37-12. The messages that come into this center have address codes that automatically direct them to certain places.

SELF-TEST

Test your knowledge by writing, on a separate sheet of paper, the word or words that most correctly complete the following statements:

1. A basic telegraph circuit consists of a _____, a _____, a _____, and a _____.

2. The telegraph key causes _____ of current to move through the circuit.

3. The telegraph code is carried as electric energy by _____.

4. The _____ is the receiving element in a telegraph system.

5. In a _____ telegraph system, a switch is connected across the telegraph key.

6. The teletypewriter's transmitting unit is called a _____.

7. The _____ is the receiving unit in a teletypewriter system.

Fig. 37–12. Installing an automatic (electronic) switching system (*American Telephone and Telegraph Co.*)

8. A teletypewriter's message punched as holes on a strip of paper is called _____.
9. The teletypewriter code works on a _____.
10. A teletypewriter character consists of _____ pulses.
11. The term _____ indicates current on. The term _____ indicates current off.
12. The teletypewriter uses a _____ start-stop system.
13. The five information pulses in a teletypewriter system are called _____.
14. Encoding in a teletypewriter is done by a distributor that has rotating _____ matched with switches.
15. A central point in a communication network that is a switching center and can transmit and receive information by radio is called a _____.

FOR REVIEW AND DISCUSSION

1. What are the basic parts of a telegraph circuit?
2. Describe the encoding process in a telegraph system.
3. Explain how a telegraph sounder works.
4. Describe the open- and closed-circuit telegraph systems.
5. What is the basic function of a teletypewriter?
6. Describe three kinds of teletypewriter terminal units.
7. Describe the pulses that make up the letter R on the teletypewriter.
8. Describe the functions of a data-control center in a communication network.
9. Describe the encoding process in a teletypewriter machine.

Unit 38   The Telephone System

A telephone system changes sound energy, usually speech, into electric energy at the transmitting, or sending, end. At the receiving end, the electric energy is used to reproduce the original sounds (Fig. 38–1 at A).

TELEPHONE TRANSMITTER

The part of the telephone system that changes sound energy into electric energy is the *telephone transmitter*, a microphone. Sound waves are actually a series of varying air pres-

Fig. 38–1. The telephone transmitter changes sound energy into electric energy: (A) basic telephone electric circuit; (B) steady dc current flowing in wires; (C) comparison of steady direct current and audio-frequency currents.

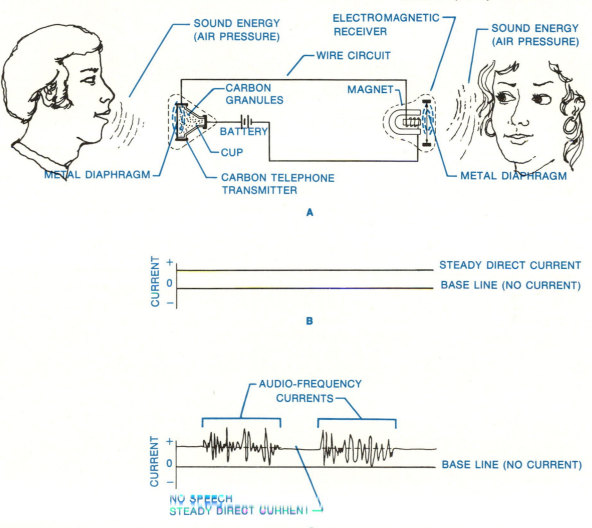

Fig. 38-2. Cross-sectional view of a telephone handset

TRANSMITTER

RECEIVER

PERMANENT MAGNET

CARBON GRANULES

ELECTRO-
MAGNETIC COIL

DIAPHRAGM

DIAPHRAGM

sures. These exert a force on the metal diaphragm of the transmitter. An increase of pressure pushes the diaphragm against a small container of carbon granules. The electrical resistance of the carbon is thereby decreased. This causes the current in the line to increase. When there is no pressure on the diaphragm, the electrical resistance of the carbon granules is high. This causes the current to decrease. The diaphragm vibrates against the carbon granules in step with the sound waves. This vibrating action changes the steady direct current in the telephone circuit into *pulsating,* or *varying, direct current.* The frequencies of the pulsating currents that result are, when changed to sound waves, within the hearing range of humans. Therefore, they are called *audiofrequency* (af) *currents* (Fig. 38-1 at C).

TELEPHONE RECEIVER

The af currents produced in the transmitter are conducted through the circuit to a device called the *telephone receiver,* or *earphone.* In the receiver, the af currents pass through an electromagnetic coil. The resulting electromagnetism causes a metal diaphragm in the receiver to move back and forth. It does this in step with the varying value and frequency of the pulsating current. The pushing of the diaphragm against the air produces sound waves. Thus, the message is first changed from sound energy to electric energy in the transmitter. It is then transmitted to a distant point. There the electric energy is changed back to sound energy by the receiver. A modern telephone has both the transmitter and receiver in one unit, called a *handset.* This is designed so that the transmitter will be near a person's mouth and the receiver near an ear (Fig. 38-2).

Volume and Frequency. You can hear better from the receiver if a person speaks louder or closer to the transmitter. This produces stronger pressures on the transmitter diaphragm. Greater changes in current are thereby created. As a result, the receiver's metal diaphragm vibrates more, and more air is set in motion. This produces a greater volume of sound. An increase or a decrease in the volume of a sound does not usually cause a change in the frequency of the sound.

Bandwidth. Humans can hear sounds that range from about 16 to 20,000 Hz. However, most of the significant content of speech falls in the frequency range below 3,000 Hz. For this reason, voice-communication receivers *attenuate,* or reduce, the high audio frequencies and boost, or increase, the lower frequencies. For example, the *bandwidth,* or frequency range, of a normal telephone channel is about 4,000 Hz.

DIAL TELEPHONE

When you dial a telephone, you are actually encoding a message. For example, when you dial the number 555-1212, pulses of energy are produced by the dial mechanism according to a code. Each telephone number has its own code. After the dial is turned and released, spring tension returns it to its original position. The return speed of the dial is constant no matter what digit is dialed. But the length of time of the return is different for each digit. For example, it takes the dial longer to return after dialing the number 9 than after dialing the number 2. The dial mechanism acts as a switch to open and close the circuit 9 times during the return cycle when the number 9 is dialed. It does this only two times when the number 2 is dialed. These pulses have a meaning used by the switching equipment in the telephone company's central office.

HOW TWO TELEPHONES ARE CONNECTED

There are several ways to connect telephones. Early methods used switches operated by hand. Most systems now have automatic switching equipment.

Automatic switching is done by electromagnetic action or by solid-state devices. In electromagnetic equipment, relays open and close circuits. One such switching system uses a *step-by-step connector switch* (Fig. 38–3). The switch has a set of connecting arms with fingers known as *wipers.* Each contact bank has 100 sets of metal contacts stacked in 10 curved rows. The upper bank is used to determine whether the called line is busy or idle. Each row has 10 sets of contacts. Each set of contacts represents the terminals of one telephone line. Electrical pulses from the dialing mechanism cause the connecting arm in the switch to step up or down to

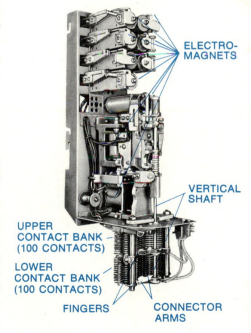

Fig. 38–3. A step-by-step connector switch (*Automatic Electric Co.*)

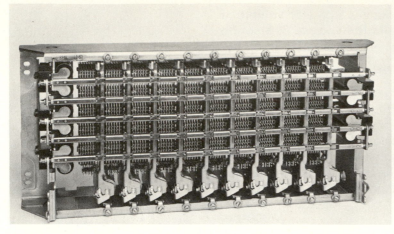

A

B

Fig. 38–4. (A) A typical crossbar switch (*Bell Telephone Laboratories, Inc.*); (B) the ferreed switch (*American Telephone and Telegraph Co.*)

the proper level. Then a connector arm moves across terminals of the switch and stops on the proper contact point. This process is repeated until the entire dialing sequence is performed step by step.

Another kind of switching system uses the *crossbar switch* (Fig. 38-4 at A). It works much like the way railroad cars are switched on tracks in a railroad yard. Only the track or tracks that are empty or not in use can be used. The *ferreed switch* is newer than the crossbar switch and one-fourth of its size (Fig. 38-4 at B). It works faster and uses little power. It is used mostly to establish connections between electronic switching centers.

TOUCH-TONE TELEPHONE

In 1964, the Touch-Tone telephone was introduced. This kind of telephone has push buttons instead of a dial (Fig. 38-5). When a button is pressed, two musical tones are generated at the same time. These tones are transmitted to the central switching office. They are then translated into a series of pulses similar to those made by the dial telephone. One of the main advantages of a Touch-Tone telephone is that a number can be punched more quickly than it can be dialed. The buttons marked * and # are for special services, such as routing calls to follow you when you are not at home. They can also be used to interconnect more than one phone at a time or to signal a busy phone.

Fig. 38–5. A modern Touch-Tone telephone with two special buttons marked * and # (*American Telephone and Telegraph Co.*)

LIGHTWAVE COMMUNICATIONS

Because the number of telephone calls is increasing, telephone wires may take the form of hair-thin glass fibers in the future (Fig. 38–6). Making telephone wires from glass fibers is a new technology developed by the Bell Telephone Laboratories. This new technology involves the use of a laser beam and glass fibers. For example, a small laser transmits pulses of electric energy at audio frequencies in the form of a light beam. The light beam is carried to its destination through glass fibers called a *lightguide* (Fig. 38–7). The popular term for this technology is *fiber optics*.

There are several advantages in using glass fibers in place of copper wires and cables. Glass fibers are much lighter than copper wires. A single pair of hair-sized fibers can carry 672 conversations at the same time. Thus, much less space is used by glass fibers than by conventional wires and cables. The glass fibers are coated with a special organic compound that makes them stronger than steel. Laser beams are not affected by electromagnetic interference. Therefore, they can be used near electric power lines. The world's first lightwave communication system was operated in Chicago, Illinois, in 1977.

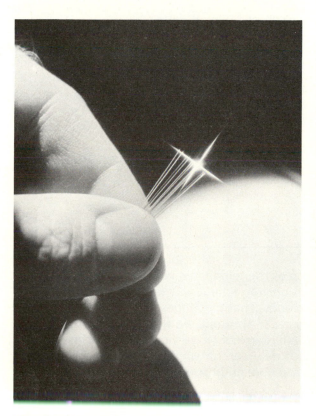

Fig. 38–6. Hair-thin glass fibers (lightguides) (*American Telephone and Telegraph Co.*)

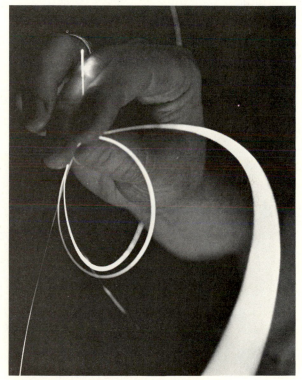

Fig. 38–7. The telephone "wire" of the future ultrapure glass fibers (*American Telephone and Telegraph Co.*)

Test your knowledge by writing, on a separate sheet of paper, the word or words that most correctly complete the following statements:

1. A telephone transmitter changes sound energy into _____ energy.
2. When the carbon granules in a telephone transmitter are compressed by the diaphragm, their resistance is _____.
3. Pulsating electric currents generated in a telephone transmitter by the voice are called _____ -frequency currents.
4. Humans can hear sounds that have a range of about _____ to _____ Hz.
5. In a dial telephone, you actually _____ a message during the dialing process.
6. A _____ switch is a device that automatically makes the connections for the number dialed by step-by-step action.
7. A telephone switch that works like the switching system for railroad cars is called a _____ switch.
8. When a button is pressed on a Touch-Tone phone, two _____ are generated and transmitted to the central switching office.
9. In glass-fiber technology, the laser beam is carried to its destination by a _____.

FOR REVIEW AND DISCUSSION

1. Describe how sound waves are translated into electric currents.
2. Describe how a dialing mechanism works.
3. Explain how a connector switch uses the numbers dialed on a telephone.
4. Give two major advantages of the Touch-Tone telephone.
5. What are three advantages of using glass fibers in place of copper wires for telephone commmunciations?

INDIVIDUAL-STUDY ACTIVITY

Demonstrate how one phone is connected with another by dialing a number. Contact your local telephone company for illustrations and literature on the subject.

Unit 39   Audio Systems

In the discussion of the telephone, you learned that a carbon microphone can change sound waves into *audio-frequency voltages* or *currents*. These voltages or currents are called *audio,* or *af, signals.* In this unit, we will discuss some devices that amplify or make use of these signals. Sometimes af signals are amplified as a person speaks or as music is played, for example, in a school auditorium. Audio-frequency signals also can be stored on a *disk* (record) or on *magnetic tape.*

The parts of a basic audio system are shown in Fig. 39–1. They include a microphone, connecting cables, audio amplifier, and loudspeaker.

MONAURAL SYSTEM

The kind of audio-amplifier system shown in Fig. 39–1 at A is called a *monaural* system. This is because it uses only one *channel*, or path, for input, amplification, and output. Fig. 39–1 at B shows several different input devices for use with a music system. Often a single amplifier is connected to two loudspeakers. If the amplifier has only one channel for amplification, it is a monaural system (Fig. 39–1 at C).

STEREO SYSTEM

Figure 39–2 shows a *stereo* sound system. This system has two amplifiers or two channels for the sound to be amplified. This makes it possible to amplify the sound from each microphone separately. The microphones can be put in different places. They can be used to pick up the sounds from the left

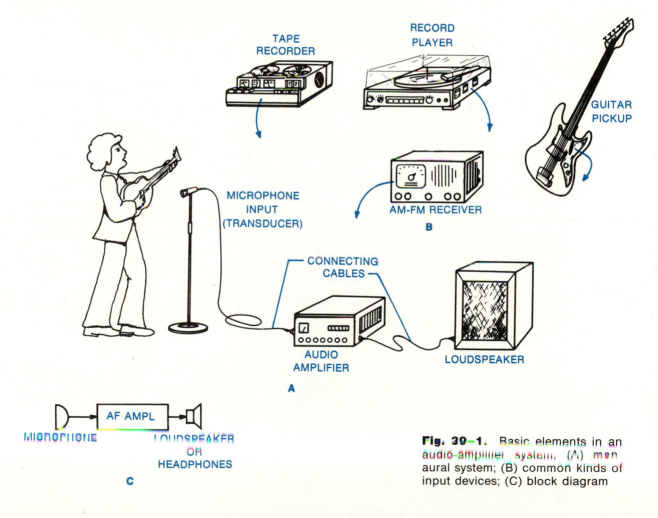

Fig. 39–1. Basic elements in an audio-amplifier system. (A) monaural system; (B) common kinds of input devices; (C) block diagram

357

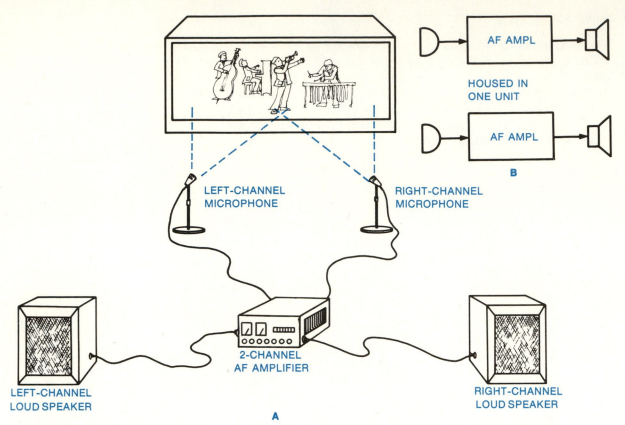

HOUSED IN ONE UNIT

LEFT-CHANNEL MICROPHONE

RIGHT-CHANNEL MICROPHONE

2-CHANNEL AF AMPLIFIER

LEFT-CHANNEL LOUD SPEAKER

RIGHT-CHANNEL LOUD SPEAKER

AF AMPL

AF AMPL

B

A

Fig. 39-2. (A) Stereo sound system (two channels); (B) block diagram

side and the right side of the stage or the sounds from the stage and the echoes from the auditorium. Therefore, this system reproduces more lifelike sounds. Because two-channel sound makes the sound almost three-dimensional, it is called *stereophonic,* or *stereo* for short. The prefix *stereo* means three-dimensional.

FOUR-CHANNEL SYSTEM

Audio systems have been further improved in an effort to obtain as natural a reproduction of sound as possible. The new process involves four-channel stereo systems. These are called quadraphonic, quadrasonic, multichannel, or surround-sound. All these systems may not produce true four-channel sound, but they are attempts to reproduce sound almost as your ears would hear it if it were played "live." This is called *high fidelity.* In a live concert or show, you hear not only what is in front of you but what is to the left, right, up, down, and behind you. A four-channel system tries to duplicate this condition. The basic arrangement for a four-channel sound system is shown in Fig. 39-3.

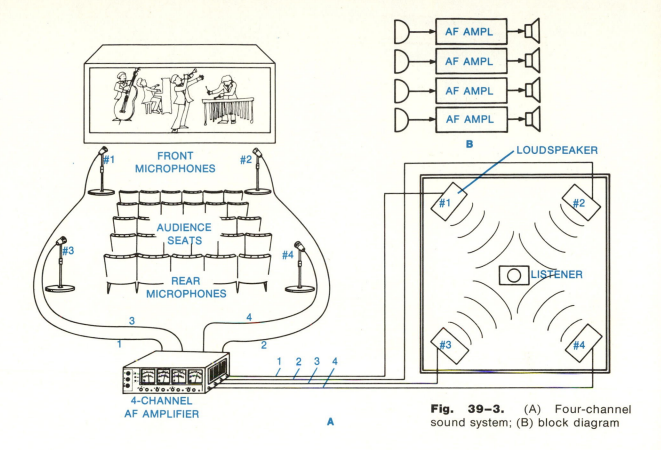

Fig. 39–3. (A) Four-channel sound system; (B) block diagram

COMMON ELEMENTS

Several elements are common to the monaural, stereo, and four-channel amplifier systems. These include microphones, phonograph cartridges, connecting cables, audio amplifiers, loudspeakers, and headphones. Microphones and phonograph cartridges are input devices. Other common inputs include the outputs of tape recorders and radios. Any of these can be monaural, stereo, or four-channel.

MICROPHONE

A microphone changes the energy of sound waves into electric energy. The mechanical energy of the moving air striking the microphone's diaphragm is changed into electric voltages and currents. These represent the audio frequencies and volumes of the original sound wave. There are several kinds of microphones. Two of the more common types are discussed in the following paragraphs:

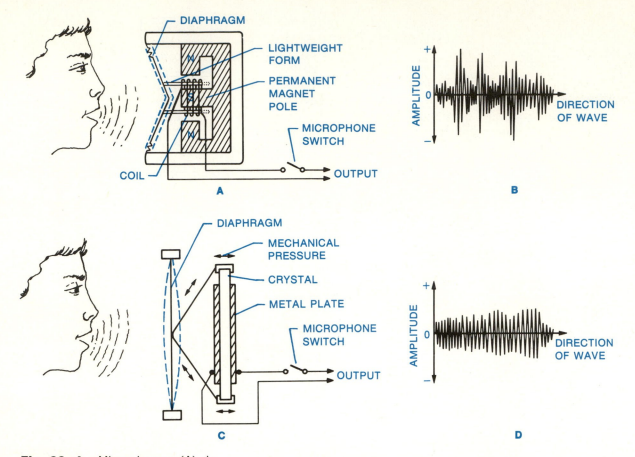

Fig. 39-4. Microphones: (A) dynamic; (B) AF voltage from dynamic microphone; (C) crystal; (D) AF voltage from crystal microphone

Dynamic Microphone. The dynamic microphone has a diaphragm connected to a lightweight form with a small coil of wire wrapped around it. The form and, therefore, the coil are free to move over one pole of a permanent magnet (Fig. 39-4 at A). Sound waves striking the diaphragm cause the form and coil to move.

As the coil moves, it cuts across the magnetic field of the permanent magnet. A voltage is induced across it. The amount and frequency of this voltage vary according to the intensity and frequency of the sound wave. Therefore, the audio-frequency voltage developed across the coil is an electrical representation of the sound wave (Fig. 39-4 at B). Note that the af voltage varies in value, first above the base line (+) and then below it (−).

High-fidelity dynamic microphones can respond to frequencies ranging from about 40 to 20,000 Hz. Since the moving coil of a dynamic microphone has only a few ohms of impedance, it generally has a built-in step-up transformer. This makes it possible to match the impedance of the microphone with that of the input circuit of the amplifier. The matching of the impedances is important. This makes it possible to transfer the maximum amount of energy from the microphone to an amplifier.

Crystal Microphone. The heart of the crystal microphone consists of Rochelle salt crystals or of a ceramic element. The crystals have properties similar to those of quartz. When the crystal is vibrated mechanically, an alternating voltage is developed between its two opposite faces (Fig. 39–4 at C). This is known as the *piezoelectric effect*. The af voltages from dynamic and crystal microphones are similar. However, the output impedance of the crystal microphone is much higher.

CARTRIDGES

The two common kinds of cartridges used in pick-up arms for audio systems are discussed below:

Crystal Cartridge. The crystal phonograph cartridge has a Rochelle salt crystal or a ceramic element, to which is fastened a needle, or stylus. The cartridge is mounted in a pick-up arm (Fig. 39–5). As the needle moves in the grooves of a record, it swings from side to side. This mechanical motion is applied to the crystal, and a voltage is produced. This piezoelectric effect is the same as in the crystal microphone.

Magnetic Cartridge. The magnetic phonograph cartridge performs the same function as the crystal cartridge. There are several kinds of magnetic cartridges, but all work by electromagnetic induction.

In one common kind of magnetic cartridge, the needle is mechanically coupled to a soft-iron armature. The armature is located in the air gap of a permanent magnet (Fig. 39–6). As the needle moves in the grooves of a record, the armature is set in motion. This motion causes the air gap to vary. This, in turn, varies the flux density of the magnetic circuit. A mov-

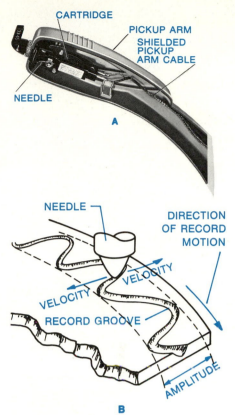

Fig. 39–5. (A) Record-player pick-up arm with crystal cartridge; (B) record groove causes needle to vibrate (*Sonotone Corporation*).

Fig. 39–6. Basic construction features of a magnetic phono cartridge

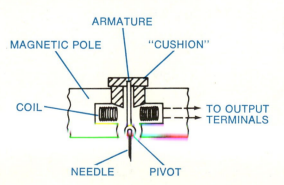

ing magnetic field is produced. When the magnetic field cuts across a stationary coil, it induces a voltage across it. The strength of this voltage depends on how much the needle moves. Thus, the mechanical motion of the needle in the grooves of the record is changed to voltages and currents. These represent sound waves.

In the moving magnetic cartridge, the needle is fastened to a small permanent magnet. As the needle moves, the field of the magnet moves and cuts across stationary coils. This induces a voltage in the coils.

Two important characteristics are often used to compare magnetic phonograph cartridges: frequency response and compliance. *Frequency response* is the range of audio frequencies to which a magnetic cartridge will respond without distortion. In high-fidelity cartridges, this range may be from 20 Hz to 20 kHz. This means that the cartridge is made so its moving parts will vibrate at any frequency in this range without producing a voltage output representing a distorted signal.

Compliance is a measure of the ability of the needle to move easily within the cartridge assembly. The compliance of a cartridge is often indicated in centimeters per dyne units. An example would be 25×10^{-6} centimeters at an applied force of 1 dyne. A high cartridge compliance is desirable. It increases the frequency response of the cartridge and reduces needle and record wear.

AUDIO CABLES

For connecting the parts of an audio-amplifier system, coaxial shielded cable is best. Generally, no more than 20 ft (6.1 m) of cable can be used between a high-impedance microphone and an amplifier without distorting the sound. Connectors, such as cable plugs, phone plugs, and microphone connectors, should be fastened securely to the cable by soldering or by tightening all screws. Many problems in amplifier systems can be traced to loose connections or broken wires in the cables.

Fig. 39-7. AF input voltages are increased by the amplifier.

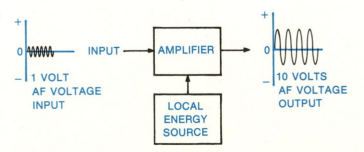

AMPLIFICATION

The main function of an audio amplifier is to increase the volume of the input signal. Figure 39-7 shows a block diagram indicating that 1 volt of input signal is increased tenfold to 10 volts at the output.

THREE-STAGE AUDIO AMPLIFIER

Figures 39-8 and 39-9 show two kinds of three-stage audio amplifiers. One uses transistors and the other uses electron tubes. The following discussion relates mostly to amplifiers containing transistors since none are made with tubes today.

Input Signals. The audio-signal input for the amplifier in Fig. 39-8 could come from a high-impedance microphone or a phono cartridge in a record player. The jack J1 provides the

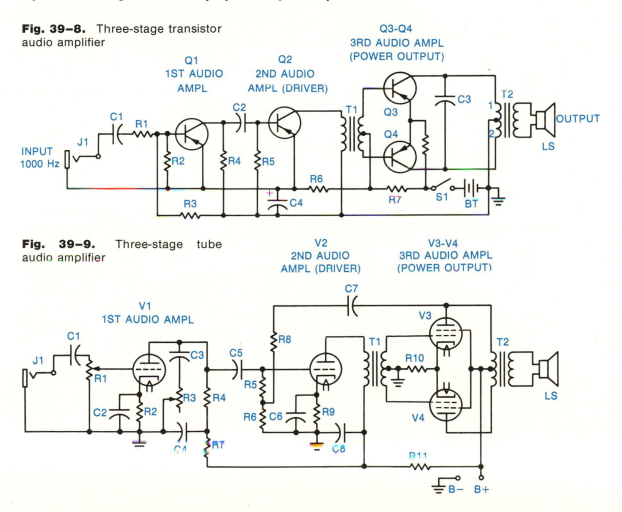

Fig. 39-8. Three-stage transistor audio amplifier

Fig. 39-9. Three-stage tube audio amplifier

cable connection to the amplifier. Capacitor C1 couples the input signals to the amplifier. C1 and resistor R1 working together form an impedance-matching network between the high-impedance transducer and the low-impedance input of transistor Q1.

The first audio-amplifier stage is made up of transistor Q1, resistors R1, R2, R3, and R4, and capacitors C1 and C2. The main current path is from the negative terminal of the battery (also connected to the chassis) through R4, into the collector of Q1, out the emitter, and through resistors R6 and R7 and switch S1 to the positive terminal of the battery. The circuit that establishes the bias for Q1 starts at the negative terminal of the battery and goes through R3, R2, R6, R7, and S1 back to the positive terminal of the battery. The bias for the base of Q1 is developed at the junction of R2 and R3.

Since the collector current of transistor Q1 varies according to the variations of the input signal, a varying voltage drop is produced across Q1. There is a voltage and current gain during this process. Thus, the input signal is amplified by the first stage of this amplifier.

Resistance-capacitance Coupling. The signal output of the first-stage transistor Q1 is coupled to the second-stage transistor Q2 by capacitor C2. Resistor R4 acts as the collector load resistor for Q1. C2 is the dc blocking capacitor that keeps the direct voltage at the collector of Q1 from appearing at the base of Q2. Resistor R5 is a current-limiting resistor through which the necessary base current is applied to Q2.

Transformer Coupling. Interstage coupling between transistor Q2 and transistors Q3 and Q4 (Fig. 39–8) is done by transformer T1. The primary winding of this transformer provides the needed load impedance for the collector (output) of Q2. The center-tapped secondary winding of T1 inductively couples the output signal of Q2 to Q3 and Q4. It acts as the circuit through which the base-emitter voltage is applied to these transistors. This amplifying stage is often called the *driver stage,* and T1, the *driver transformer.*

Transformer coupling is also used between the output stage (transistors Q3 and Q4) and the loudspeaker voice coil. Here, transformer T2 (the output transformer) acts to match the relatively high impedance of the transistor output circuit with the low impedance of the voice coil. Since this is a step-down transformer, its coupling action produces a low voltage across the secondary winding. As a result, the value of the current in the secondary-winding circuit is great enough to energize the voice coil to the level needed to operate the loudspeaker. Note that this transformer-coupling system is similar to the electron-tube circuit (Fig. 39–9).

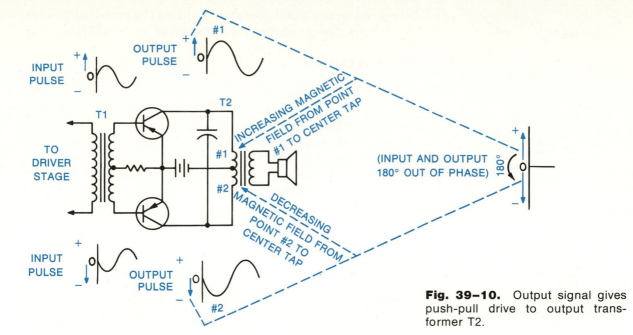

Fig. 39–10. Output signal gives push-pull drive to output transformer T2.

PUSH-PULL OPERATION

The final output, or power-amplification, stage of the three-stage single-channel amplifier circuit (Fig. 39–8) is what is known as a *push-pull stage*. In push-pull operations, two transistors (or tubes) are used. When an amplifier circuit has only one transistor (or tube) in the final stage, the stage is called a *single-ended stage*.

The input signals to a push-pull stage are developed across the ends of the center-tapped secondary winding of inter-stage transformer T1. Therefore, the signal voltages applied to the input circuits of a push-pull stage from the secondary winding of T1 surge first in one direction and then in the other. When this happens, the signals are said to be 180 degrees *out of phase* with each other (Fig. 39–10). The output currents pass through transformer T2.

In this condition, the current from point 1 to the center tap of the primary winding of the output transformer T2 is increasing. The current from point 2 of the winding to the center tap is decreasing. Therefore, the magnetic field around one-half of the primary winding is expanding. The magnetic field around the other half of the winding is collapsing. This action is called a *push-pull effect*. The effects of these magnetic fields combine to induce the voltage across the secondary winding of the output transformer.

One advantage of the push-pull amplifier stage is that it will deliver about twice as much power as a single-ended

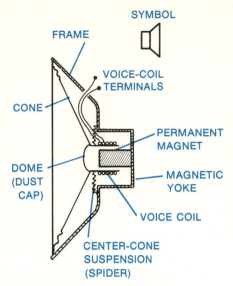

FRAME

SYMBOL

VOICE-COIL
TERMINALS

CONE

PERMANENT
MAGNET

DOME
(DUST
CAP)

MAGNETIC
YOKE

VOICE COIL

CENTER-CONE
SUSPENSION
(SPIDER)

Fig. 39–11. The loudspeaker is one type of reproducer.

Fig. 39–12. Tweeter fitted with horn (*Electro-Voice*)

stage. Also, push-pull operation reduces the hum that power variations may produce in the output of an amplifier.

Power Output. The power output of an amplifier is rated in a number of ways. For example, music, peak, and RMS power output are common ratings. The most meaningful rating is the RMS power output. This is obtained by applying a sine wave at different frequencies to the input. Next the power output is measured across a resistive load. The values of the different frequencies are squared, totaled, and the mean, or average, found. Then the square root is taken to give the RMS value. The RMS rating means the *root mean squared* value. The RMS rating is important for comparing different amplifiers.

LOUDSPEAKER

The loudspeaker changes electric energy into sound. Figure 39–11 shows a cross section of a typical loudspeaker. In this loudspeaker, the voice coil moves freely over, but does not touch, the permanent-magnet pole piece. The voice coil is also connected to the secondary winding of the output transformer T2 (Fig. 39–10). The amplified audio signals are developed across the secondary winding of the output transformer. These signals are applied to the voice coil, thus energizing it. The electromagnetic field of the voice coil is then alternately attracted and repelled by the permanent magnetic field. This causes the voice coil to vibrate. The intensity of this movement varies with the audio signals originally produced by the microphone. Since the voice coil is connected to the loudspeaker's cone, it causes large amounts of air to vibrate. These vibrations produce sound waves. If the microphone, amplifier, and loudspeaker are of high quality, the sounds amplified and reproduced will match the input signals almost exactly.

Kinds of Loudspeakers. A single loudspeaker can produce a good frequency response. However, high-fidelity systems often use three loudspeakers. These are called the *woofer,* the *midrange speaker,* and the *tweeter.* With three speakers, the incoming frequencies are split. The low frequencies (20 Hz to 1,000 Hz) are handled by the woofer. The middle frequencies (800 Hz to 10 kHz) are handled by the midrange loudspeaker. The high frequencies (3.5 kHz to 20 kHz) are handled by the tweeter. High frequencies have little *dispersion,* or scatter. Thus, the tweeter is often fitted with a small horn (Fig. 39–12). A *crossover network* is used to split the input signals into frequency bands for the speakers. These networks, consisting of capacitors and inductors (coils), feed the correct range of frequencies to each of the loudspeakers.

Rating. A loudspeaker is rated by the size of its cone, the impedance and the power capacity of its voice coil, its frequency response, and the weight of its magnet (Fig. 39–13).

The cone diameter indicates whether the loudspeaker is small or large. For example, a 12-inch (304.8-mm) loudspeaker is generally a woofer. A 4-inch (101.6-mm) loudspeaker is a tweeter. Typical impedance ratings for loudspeakers are 4, 8, and 16 ohms. It is important that the impedance "seen" by the output of the amplifier matches that of the loudspeaker. Otherwise, the amplifier can be overloaded. This seriously distorts the signal. In addition, the power-handling capacity should be known for each loudspeaker. If the power amplifier can deliver 30 watts, the loudspeaker should also be able to handle that.

The cabinet that houses the speaker is important for reproducing high-quality sound. For example, a properly designed cabinet keeps the sound waves in front and to the rear of the speaker's cone in phase with each other.

If similar loudspeakers have magnets of the same kind but of different weights, the loudspeaker with the heavier magnet will generally be more efficient. For example, the frequency-

Fig. 39–13. (A) A 15 in. (381 mm) duplex monitor loudspeaker; (B) cross-sectional view (*Altec: Sound Products Division*)

A

B

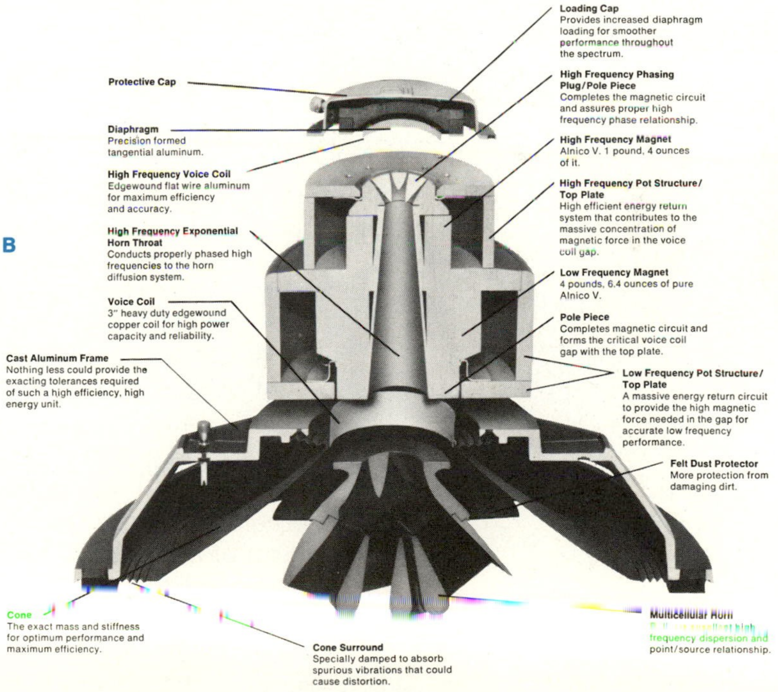

Protective Cap

Diaphragm
Precision formed tangential aluminum.

High Frequency Voice Coil
Edgewound flat wire aluminum for maximum efficiency and accuracy.

High Frequency Exponential Horn Throat
Conducts properly phased high frequencies to the horn diffusion system.

Voice Coil
3" heavy duty edgewound copper coil for high power capacity and reliability.

Cast Aluminum Frame
Nothing less could provide the exacting tolerances required of such a high efficiency, high energy unit.

Cone
The exact mass and stiffness for optimum performance and maximum efficiency.

Cone Surround
Specially damped to absorb spurious vibrations that could cause distortion.

Loading Cap
Provides increased diaphragm loading for smoother performance throughout the spectrum.

High Frequency Phasing Plug/Pole Piece
Completes the magnetic circuit and assures proper high frequency phase relationship.

High Frequency Magnet
Alnico V. 1 pound, 4 ounces of it.

High Frequency Pot Structure/ Top Plate
High efficient energy return system that contributes to the massive concentration of magnetic force in the voice coil gap.

Low Frequency Magnet
4 pounds, 6.4 ounces of pure Alnico V.

Pole Piece
Completes magnetic circuit and forms the critical voice coil gap with the top plate.

Low Frequency Pot Structure/ Top Plate
A massive energy return circuit to provide the high magnetic force needed in the gap for accurate low frequency performance.

Felt Dust Protector
More protection from damaging dirt.

Multicellular Horn
provides excellent high frequency dispersion and point/source relationship.

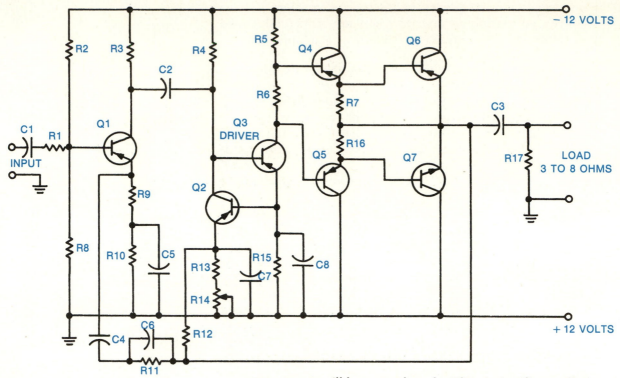

Fig. 39-14. Complementary-symmetry audio-amplifier circuit [*from* Electronics (*magazine*), *McGraw-Hill, Inc.*]

response range will be reproduced without significant distortion by a heavier magnet. The frequency-response range runs from about 50 Hz to 8 kHz in a low-cost loudspeaker. In more expensive loudspeakers designed for high-fidelity systems, the range is from about 20 Hz to 20 kHz.

COMPLEMENTARY SYMMETRY

Some audio amplifiers do not have an output transformer (Fig. 39-14). Instead, the circuit uses p-n-p and n-p-n transistors. Because of their polarity characteristics, these transistors provide the push-pull effect. This kind of circuit is called a *complementary-symmetry audio-amplifier circuit*. The amplifier has an input stage Q1, a driver Q3, and a power-output stage—transistors Q4, Q5, Q6, and Q7.

VOLUME CONTROL

The *volume* is the output sound level of a circuit. The volume control of an amplifier circuit is most often a potentiometer variable resistor. This is usually put into the input circuit of an amplifier stage (Fig. 39-15). The volume control acts as a voltage divider. As the shaft of the potentiometer is adjusted, the sliding arm moves toward either point 1 or point 2. When

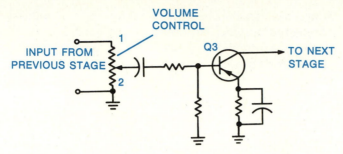

Fig. 39–15. Volume-control circuit

it moves toward point 1, a greater part of the voltage developed between points 1 and 2 of the control is applied to the input circuit of transistor Q3. As a result, the volume of the amplifier circuit is increased. Moving the sliding arm to point 2 decreases the volume.

TONE CONTROL

Tone refers to the balance of low frequencies, or bass, and high frequencies, or treble, in the sound output. A tone control lets you adjust the frequency response of an audio amplifier so as to affect this balance.

A basic tone-control circuit consists of a capacitor and a potentiometer variable resistor connected in series across the output of an amplifier stage. In Fig. 39–16, C2 has a capacitance value that causes it to present a low capacitive reactance to the higher-frequency signals in the output circuit of Q2. The capacitor then tends to bypass those high-frequency signals to ground. This prevents them from reaching the loudspeaker. With the high frequencies weakened, the amplifier produces relatively more bass response. The degree to which this happens is determined by the setting of the potentiometer.

PREAMPLIFIER

A *preamplifier* increases the output voltage of the signal source so that the main audio amplifier can reach full power.

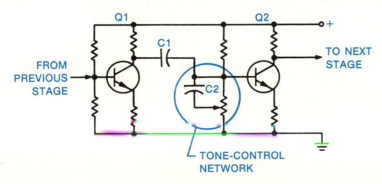

Fig. 39–16. Tone-control circuit

A preamplifier can be either single-stage or multistage. It is connected between an input device with a low output level, such as a magnetic phono cartridge or a dynamic microphone, and the main amplifier. The preamplifier may be a separate unit, or it may be contained in the main amplifier.

STORING SOUND

There are two common devices for storing sounds—the record, or disk, and magnetic tape. The record player reproduces sound stored in the record, and the tape recorder does the same for tape.

RECORD (DISK)

A record is made, indirectly, by a recording machine. This machine changes the original sound into electrical vibrations. The machine's needle-sharp cutting head records these vibrations in a groove it cuts on a master record. During the cutting, the head moves slowly toward the center of the record. This forms an unbroken spiral groove. Inexpensive copies of the master are made from vinyl blanks.

The grooves on monaural records are irregularly wavy. The short waves are high tones. The longer ones are lower tones. The grooves on typical stereo records consist of "walls" cut at right angles (90°) to each other and at an angle of 45° to the record surface (Fig. 39–17). When the stereo master is

Fig. 39–17. A stereo-record groove

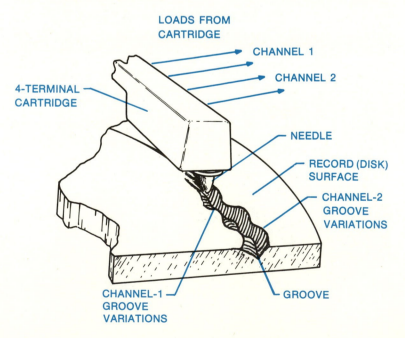

370

being cut, the vibrations from one channel are recorded on one wall of the groove. The vibrations from the other channel are recorded on the other wall. When a record is played with a stereo cartridge (Fig. 39–18), the vibrations from the two walls are picked up on the other wall separately. Thus, the sounds from the two channels are reproduced.

MAGNETIC TAPE

Magnetic tape has a special coating of iron oxide. Sounds are induced on this coating electromagnetically. Two-track, four-track, and eight-track tapes are available today. On a two-track tape, one side is recorded or played on track 1. Then the tape is turned over, and the other track is used (Fig. 39–19). For stereo, four-track tapes are used. Two tracks are used at the same time, one for each channel. Then the tape is turned over, and the other two tracks are used.

TAPE HEADS

Tape heads record sounds on a tape, pick up sounds for play-back, or erase the tape. The tape head is held closely against the tape during recording and playback. During recording, the head changes the audio signals into an electromagnetic field. This field magnetizes the oxide coating on the tape. During playback, the head detects the electromagnetic field on the tape. On some recorders, an erase head demagnetizes the tape by means of an ultrasonic-frequency electromagnetic field. An *ultrasonic frequency* is one that is above the

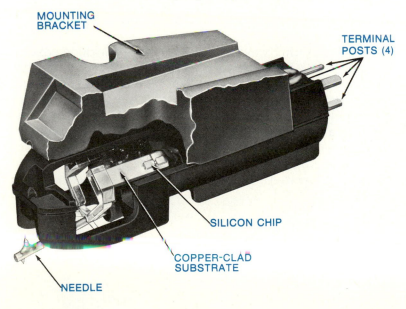

MOUNTING
BRACKET

TERMINAL
POSTS (4)

SILICON CHIP

COPPER-CLAD
SUBSTRATE

NEEDLE

Fig. 39–18. A stereo cartridge (*Sonotone Corporation*)

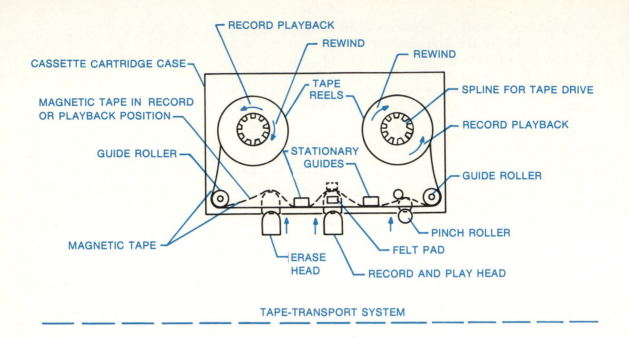

RECORD PLAYBACK
REWIND
REWIND
CASSETTE CARTRIDGE CASE
TAPE REELS
SPLINE FOR TAPE DRIVE
MAGNETIC TAPE IN RECORD OR PLAYBACK POSITION
RECORD PLAYBACK
GUIDE ROLLER
STATIONARY GUIDES
GUIDE ROLLER
MAGNETIC TAPE
PINCH ROLLER
ERASE HEAD
FELT PAD
RECORD AND PLAY HEAD

TAPE-TRANSPORT SYSTEM

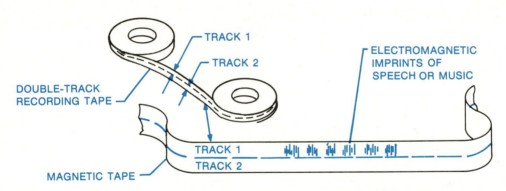

TRACK 1
TRACK 2
ELECTROMAGNETIC IMPRINTS OF SPEECH OR MUSIC
DOUBLE-TRACK RECORDING TAPE
TRACK 1
TRACK 2
MAGNETIC TAPE

Fig. 39–19. Cassette tape cartridge

audio range. Other ways of erasing the tape include passing the tape over a permanent magnet or over an electromagnet energized by a direct current. The tape can be erased and reused many times before it is worn out. A motor and mechanisms pull the tape from a feed reel past one or more tape heads to a take-up reel called the *tape transport,* or *deck.* On some tape recorders, the tape can be set to run at any one of three standard speeds, $1\frac{7}{8}$, $3\frac{3}{4}$, or $7\frac{1}{2}$ in. per second. In general, tapes operated at $7\frac{1}{2}$ in. per second have superior sound reproduction.

Test your knowledge by writing, on a separate sheet of paper, the word or words that most correctly complete the following statements:

1. The parts of a basic audio system are _____, _____, _____, and _____.
2. Four common input devices used with audio amplifiers are _____, _____, _____, and _____.
3. A _____ microphone has a diaphragm connected to a lightweight form with a small coil of wire wrapped around it.
4. High-fidelity dynamic microphones can respond to frequencies ranging from _____ to _____ Hz.
5. Two important characteristics often used to compare magnetic phonograph cartridges are _____ and _____.
6. _____ is best for connecting the parts of an audio-amplifier system.
7. Audio-amplifier stages are often coupled by _____ coupling and/or _____ coupling.
8. A _____ amplifier stage is used when the final amplifier circuit has only one transistor or tube.
9. The most meaningful rating for an amplifier is _____ power output.
10. Three kinds of loudspeakers found in a high-fidelity audio system are the _____, _____, and _____.
11. A _____ network is used to split the input signals into frequency bands for the speakers in a high-fidelity audio system.
12. One kind of transformerless audio amplifier is called the _____ amplifier.
13. A _____ and a _____ are two basic components used to form a tone control for an audio amplifier.
14. The two common means of storing sound are the _____ and magnetic _____.
15. The motor and mechanisms that pull the magnetic tape from a feed reel past one or more tape heads to a take-up reel are called the _____.

FOR REVIEW AND DISCUSSION

1. Distinguish between monaural and stereo sound systems.
2. What is the main function of a microphone?
3. Explain the difference between a dynamic microphone and a crystal microphone.
4. Explain the difference between the crystal and the magnetic phonograph cartridge.
5. What is compliance?
6. What is the main function of an output transformer?
7. Describe how an audio input signal is amplified.
8. Describe the advantages of a push-pull amplifier.
9. Identify three terms used to rate the power output of audio amplifiers. Which term is most meaningful?
10. Explain how a voice coil in a loudspeaker works.
11. What is the purpose of the horn attached to a tweeter loudspeaker?
12. Why is the weight of a magnet in a loudspeaker so important?
13. What is the basic function of a preamplifier?
14. Describe the grooves of a monaural record and of a stereo record. How does the cartridge pick up the sound from the record?
15. What is the function of an erase head on a magnetic tape recorder?

INDIVIDUAL-STUDY ACTIVITY

Prepare a block diagram of a home audio system that would include such inputs as a microphone and phonograph and such reproducers as loudspeakers and headphones. The system might also include a tape recorder and a radio.

Unit 40   Fundamentals of Radio

You have already learned how telephone wires are used to send messages from one place to another. Now you will see how speech and music can be transmitted *without wires*. This is done with a radio system. The radio system, like the telephone, has a transmitter and a receiver (Fig. 40–1). It does not, however, have connecting wires.

TRANSMITTER

The radio transmitter makes it possible to send music and speech by means of *radiated*, or emitted, energy. The radiated energy is transmitted through the air. Understanding how your voice works will help you understand how this is possible. Taking a deep breath is like opening an air bellows. You are forming the energy source for the air that will pass through your vocal cords, throat, and mouth when you talk. The *radio station*, where the equipment is located for transmitting speech and music, also has an energy source and a *high-frequency oscillator*. The energy source produces a high voltage. The oscillator causes a stream of electrons to vibrate back and forth at high frequencies (Fig. 40–2).

Fig. 40–1. Two students operating a basic radio system using two transceivers (transmitter and receiver combined in one unit)

RADIO WAVES

Sound waves are produced by your vocal cords. They are transmitted through the air as varying air pressures. In a radio transmitter, a circuit is used to create waves that, in some respects, are like sound waves. This circuit, the high-frequency oscillator, contains a coil and a capacitor. The oscillator causes a stream of electrons to vibrate, or *oscillate*. Similarly, vocal cords in a person's throat cause a stream of air to vibrate. This rapidly oscillating movement of electrons causes energy to be *radiated*, or sent through the air, in the form of *electromagnetic waves*.

Electromagnetic waves are a form of radiated energy that can do work. When a stone is thrown into the water, water waves radiate out from the point where the stone entered the water (Fig. 40–3). Electromagnetic waves can be compared to these water waves. Figure 40–4 shows this waveform. The higher the frequency, the shorter the wavelength. *Wavelength* is the distance between a point on one wave to a similar point on the next wave. Sound waves, water waves, and electromagnetic waves are similar. Radio waves, however, have a much higher frequency than sound waves do. A radio

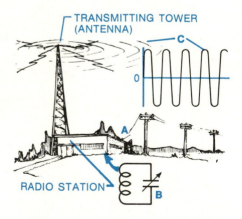

Fig. 40–2. The vibrator (oscillator) in a transmitting station: (A) power lines (energy source); (B) vibrator or oscillator (coil and capacitor); (C) carrier (electromagnetic wave)

wave must have a frequency of at least about 10,000 Hz before it can be used for radio communication.

Ground and Sky Waves. Radio waves travel through space from the antenna of a transmitting station at the speed of light. Some of these waves move along the surface of the Earth. These are called *ground waves*. Ground waves are soaked up rapidly by the Earth. They do not travel very far. Other radio waves move skyward. These are called *sky waves* (Fig. 40–5). Some sky waves are reflected back to Earth by a layer of ionized gas from 60 to 200 miles (97 to 322 kilometers) above the Earth. This is known as the *Kennelly-Heaviside layer* (or *ionosphere*) in honor of the two men who discovered it. Some of the sky waves go through this Kennelly-Heaviside layer and never return to Earth. However, the sky waves reflected back to Earth make long-range radio communication around the Earth's surface possible.

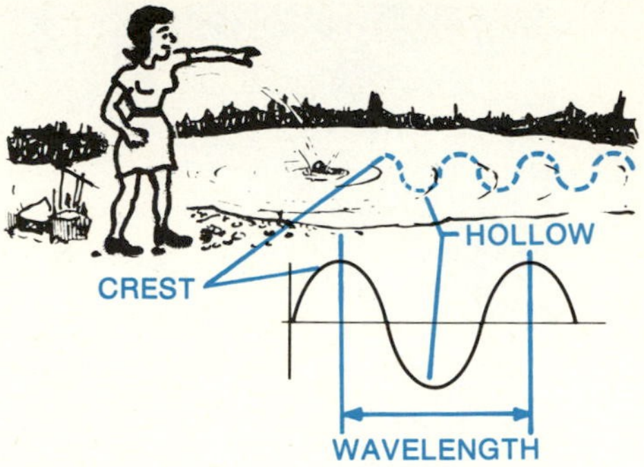

Carrier Waves. A radio wave of a certain frequency is assigned to a radio station. It is called a *carrier wave* because it "carries" speech and music from the radio-station transmitter to radio receivers. The frequencies of the carrier waves used for different kinds of radio broadcasting are assigned by the Federal Communications Commission (FCC).

Fig. 40–3. Throwing a stone into water causes water waves to radiate out from the point where the stone entered.

MODULATION

When you speak, the vowels and the consonants of your speech are molded on the vibrating air stream. This is done by your tongue, lips, teeth, and the walls of your mouth and

Fig. 40–4. Comparison of ac graph (sine wave) to a radio-wave form: (A) alternating current (low frequency); (B) radio wave (high frequency)

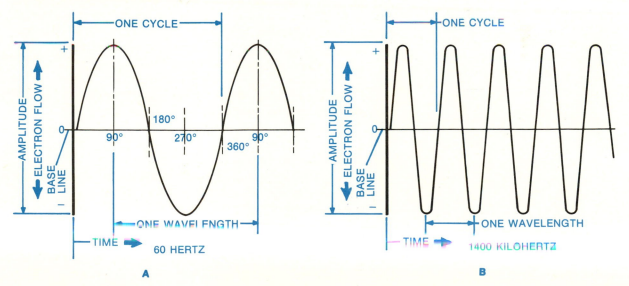

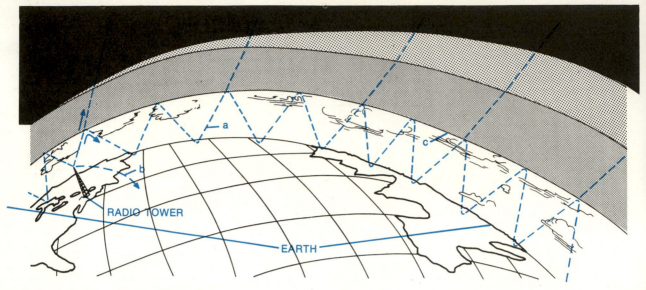

Fig. 40–5. Transmitted radio waves: (a) sky waves; (b) ground waves; (c) Kennelly-Heaviside layer of gas

throat. This molding lets you send meaningful sound waves to another person.

In a radio broadcasting system, the process of molding the electron stream for speech or music is called *modulation*. A *modulator* is a circuit that combines speech or music information with the carrier wave.

A varying current is produced when sound waves strike a microphone (Fig. 40-6). The microphone output is then fed into the oscillator. There the audio and carrier waves are combined. Modulation allows the carrier wave to carry information from one place to another by electromagnetic energy (Fig. 40-7). At the receiving end, this information is then taken from the carrier wave by special circuits in a radio receiver.

Amplitude and Frequency Modulation. The modulated radio wave is developed by combining the carrier wave and the alternating current from a microphone. Note that the frequency of the carrier wave does not change after modulation (Fig. 40-7). The alternating current from the microphone modulates the carrier wave by causing its *amplitude,* or strength, to rise and fall. This process is called *amplitude modulation* (AM).

The carrier wave can also be modulated by changing its frequency. This process is called *frequency modulation* (FM) (Fig. 40-8). The changes in frequency are controlled by the signal from a microphone. Note that the amplitude does not change in frequency modulation.

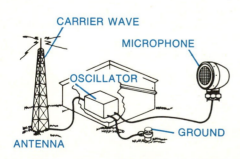

Fig. 40–6. Modulating a radio wave with a microphone

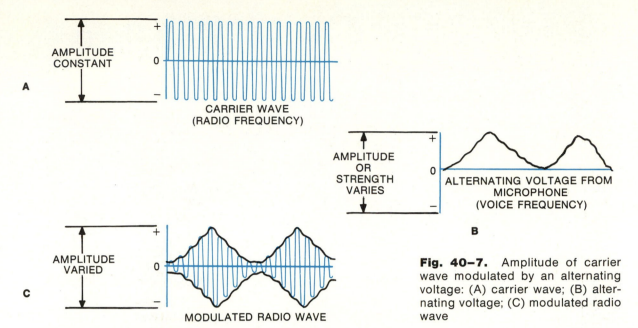

CARRIER WAVE
(RADIO FREQUENCY)

A AMPLITUDE CONSTANT

B AMPLITUDE OR STRENGTH VARIES — ALTERNATING VOLTAGE FROM MICROPHONE (VOICE FREQUENCY)

C AMPLITUDE VARIED — MODULATED RADIO WAVE

Fig. 40-7. Amplitude of carrier wave modulated by an alternating voltage: (A) carrier wave; (B) alternating voltage; (C) modulated radio wave

Broadcast Bands. Commercial AM broadcast stations operate within a carrier frequency band that extends from 535 to 1,605 kHz. The carrier frequency of commercial FM broadcasting extends from 88 to 108 MHz.

AMATEUR RADIO

A person interested in amateur radio can apply to the Federal Communications Commission (FCC) for a license to operate a station (Fig. 40–9). This activity can improve both your communication skills and your technical knowledge of radio.

The Federal Communications Commission provides the rules and regulations for the amateur radio service. At present, there are six classes of operator's licenses: novice, technician, conditional, general, advanced, and amateur extra. The novice class gives the beginner an opportunity to get on-the-air experience quickly. A novice must pass an International Morse code test of five words per minute. The examination also covers FCC rules and regulations and elementary radio theory. The amateur extra class is the most advanced class. This class requires passing an examination on advanced radio theory and operation, including rules and regulations affecting amateur stations and operators and a code test of twenty words per minute. With each rise in class, certain privileges are gained. For example, specific frequencies, such as 3,500 to 3,525 kHz, are reserved for use only by amateur-extra-class operators. Common amateur frequency bands are 3,500 to 4,000 kHz (wavelength about 80 meters),

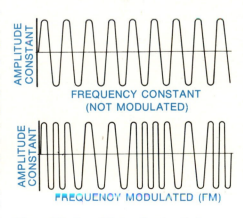

FREQUENCY CONSTANT
(NOT MODULATED)

FREQUENCY MODULATED (FM)

Fig. 40-8. The effect of frequency modulation

Fig. 40–9. An amateur radio station (*photo from* QST, *journal of American Radio Relay League*)

7,000 to 7,300 kHz (wavelength about 40 meters), and 14,000 to 14,350 kHz (wavelength about 20 meters). Higher frequency bands, such as 144 to 148 MHz (wavelength about 2 meters) are also authorized.

PERSONAL RADIO SERVICES

The Federal Communications Commission also publishes rules and regulations governing the operation of *Personal Radio Services*. These services include the stations discussed below:

General Mobile Radio Service Station. This kind of station is intended to be operated between mobile and land stations or between mobile stations. These stations are licensed to be operated on an assigned frequency in the 460 to 470 MHz band with a transmitter power output of not more than 50 watts.

Radio Control (R/C) Service Station. This kind of station is licensed to be operated on an authorized frequency in the 26.96 to 27.23 MHz band. For remote control of objects by radio, the frequency of 27.255 MHz is authorized. The 72 to 76 MHz band is authorized for the radio control of models used for hobby purposes.

Citizens Band (CB) Radio Service Station. This kind of station is licensed to be operated for *radiotelephony* only. *Radiotelephony* is the transmitting of voice messages. This

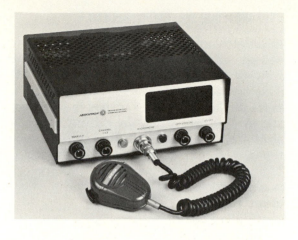

Fig. 40–10. A modern transceiver (*Aerotron, Inc.*)

station provides short-distance mobile radio communications for business or personal activities.

The CB authorized frequencies are in the 26.96 to 27.41 MHz band. Specific frequencies in this band are commonly called *channels*. There are 40 channels. For example, the frequency of 26.965 MHz is channel 1. The frequency of 27.065 MHz, channel 9, is used only for emergency communications. Channel 19, with an assigned frequency of 27.185 MHz, is very popular with truckers and other travelers. The CB equipment, called a *rig*, is authorized to operate with a maximum carrier power output of 4 watts.

Any citizen 18 years of age or older can apply for a license to operate a CB station. Licenses are normally issued for 5 years. The latest rules and regulations of the FCC should be consulted for further details.

TRANSCEIVERS

A *transceiver* is a small unit that combines a transmitter and receiver. A small hand-held unit is popularly called a *walkie-talkie*. A modern transceiver often used in business for instant communication is shown in Fig. 40–10. Several small hand models are available. These are controlled by a crystal and operate at low power (100 milliwatts) in the 49.82 to 49.90 MHz range. A pair of walkie-talkies is needed to carry on a two-way conversation (see Fig. 40–1). These units have 3 to 12 transistors each, depending on the quality and electronic circuitry. Since small transceivers are portable, a 9-volt battery is often used as a power supply. The usual controls on these small units are (1) off-on switch with volume control, (2) push-to-talk button, (3) squelch control, and (4) jack for earphones. The squelch control eliminates background noises and unwanted noises between the transmissions.

SUPERHETERODYNE RADIO RECEIVER

In an attempt to improve radio reception, a special circuit called the *superheterodyne circuit* was invented in 1918. Edward H. Armstrong, an American, invented the circuit. This circuit helped make radio receivers more selective and more sensitive. A radio receiver is *selective* when it can select the modulated carrier wave from the desired broadcasting station and reject all the other carrier signals reaching its antenna. A radio receiver is *sensitive* when it can receive weak radio waves or signals. The superheterodyne circuit has made possible radio receivers that are both highly selective and sensitive.

BEAT FREQUENCY

The special features of the superheterodyne radio circuit are a *mixer*, or *converter*, stage and an oscillator stage called the *local oscillator*. The radio frequency signals generated by the local oscillator usually have a higher frequency than that of the carrier wave to which the receiver is tuned. The signals from the tuning circuit of the receiver and the signals from the local-oscillator circuit are combined in the mixer stage. The result is a signal with a frequency equal to the difference between the frequencies of the two signals being mixed. This is called the *beat frequency*. Producing a beat frequency is known as *heterodyning*.

Figure 40–11 shows how a beat frequency is produced in a superheterodyne radio receiver. The carrier signal to which the receiver is tuned has a frequency of 640 kHz. The signal generated by the local oscillator has a frequency of 1,095 kHz. These two signals are combined in the mixer stage. The output of that stage is the difference, or a beat-frequency signal of 455 kHz. This signal is called the *intermediate frequency* (IF). In this country, most AM superheterodyne radio receivers operate at an intermediate frequency of 455 kHz.

In order to maintain a constant intermediate frequency of 455 kHz, the oscillator circuit must be adjusted whenever a different station is tuned in. This is usually done by means of a variable capacitor controlled by the main tuning knob. As a result, the oscillator output frequency is always 455 kHz higher than the frequency of the station selected.

STAGES IN AN AM SUPERHETERODYNE RECEIVER

The block diagram (Fig. 40–12) shows the electronic stages in a modern transistorized AM superheterodyne receiver. Electron-tube receivers also have similar stages.

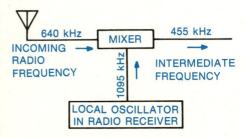

Fig. 40–11. Producing a beat frequency (intermediate frequency)

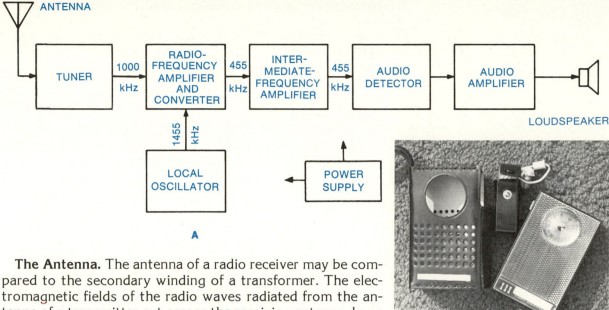

A

Fig. 40–12. (A) Block diagram of a superheterodyne radio receiver; (B) portable radio

The Antenna. The antenna of a radio receiver may be compared to the secondary winding of a transformer. The electromagnetic fields of the radio waves radiated from the antenna of a transmitter cut across the receiving antenna. In so doing, they induce voltages in the antenna circuit. These voltages represent the speech or music information with which the carrier wave has been modulated. The voltages are then applied to the tuning circuit of the radio receiver. Since these voltages "carry" information, they are called radio frequency (RF) signals, or radio signals.

The Tuning Circuit. The tuning circuit for a superheterodyne radio receiver is shown in Fig. 40–13. The circuit consists of variable capacitor C1 and antenna coil L1. The varia-

Fig. 40–13. Tuner and converter (oscillator-mixer) circuits

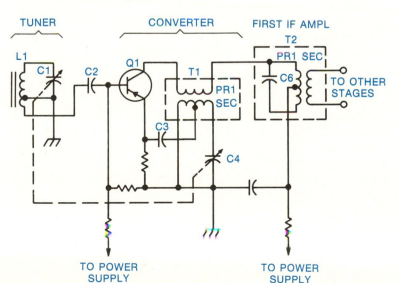

ble capacitor is used to *tune*, or adjust, the circuit to different frequencies.

When the tuning circuit is adjusted to the frequency of a station, the circuit is said to be in *resonance* with the frequency of that station. In this condition, the inductive reactance of coil L1 is equal to the capacitive reactance of capacitor C1. The circuit now offers the least impedance to signals of the station carrier frequency. However, it offers a high impedance to carrier signals of other frequencies. Therefore, all the other carrier signals are rejected by the circuit.

You tune most home radio receivers by adjusting a variable capacitor. However, with other receivers, such as automobile receivers, you tune by varying the inductance of the tuning-circuit coil. This is done by moving a ferrite core into or out of a tube around which the tuning coil is wound. In tuning circuits of this kind, a fixed capacitor is used.

Converter (Oscillator-Mixer). In the converter circuit, incoming RF signals are heterodyned to the intermediate frequency of 455 kHz (Fig. 40–13). Transformer T1, capacitors C3 and C4, and transistor Q1 make up the local oscillator. C4 is *ganged*, or mechanically linked, to capacitor C1. C4 acts with the secondary winding of T1 to establish the frequency of oscillations for Q1. Note that the incoming radio frequencies are also coupled to Q1 through capacitor C2. Q1 heterodynes, or mixes, the two frequencies—a radio signal of 1,000 kHz and a local oscillator signal of 1,455 kHz—to produce the intermediate frequency of 455 kHz.

There are usually several stations broadcasting in an area. For example, three stations, each transmitting with its own frequency of 630 kHz, 980 kHz, and 1,500 kHz, could be tuned in on the receiver. However, no matter what the tuned radio frequency is, the intermediate frequency will always be 455 kHz.

Intermediate Frequency (IF) Amplifier. The IF amplifier amplifies only one frequency, 455 kHz. Thus, its design can be made highly selective and efficient. The IF signals are coupled to their stages by tuned transformers like the first IF transformer T2 in Fig. 40–13. There are usually two IF amplifying stages before the audio signals are detected by a diode. *Detection* involves recovering the modulated audio-frequency signals from the carrier.

Audio Detector and AVC Circuit. The detector CR1, a semiconductor diode, removes the negative-going parts of the IF signals coming from IF transformer T3 (Fig. 40–14). The audio output appears across the volume-control resistor

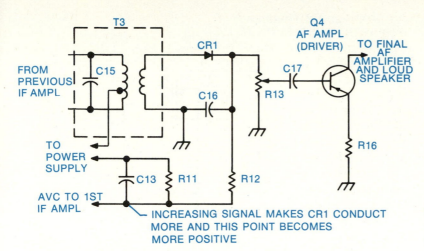

INCREASING SIGNAL MAKES CR1 CONDUCT
MORE AND THIS POINT BECOMES
MORE POSITIVE

Fig. 40–14. Detector, automatic volume control (AVC), and AF amplifier (driver)

R13. It is coupled by capacitor C17 to the audio driver stage, Q4. The amplified audio signals are next fed into a push-pull amplifier and then to the loudspeaker.

The *automatic volume control* (AVC) provides high gain for weak signals at the incoming IF stages. It also reduces the strength of strong signals. For example, when a strong audio signal is detected, the voltage across C13 becomes more positive. The positive-going AVC signal is fed to the incoming IF stages and reduces their gain. The opposite happens when weak signals are received.

ALIGNMENT

Aligning is adjusting the variable trimmer capacitors and the variable cores of transformers and coils in the receiver circuit so that they will work efficiently. This is usually done with a *plastic alignment tool* (Fig. 40–15). Correct alignment of a radio receiver provides the best selectivity and sensitivity. The procedures for aligning different radio receivers are usually given in servicing manuals and handbooks.

FM RECEIVER

The electronic-circuit stages of an FM superheterodyne receiver are shown in the block diagram of Fig. 40–16. These circuits include the antenna, RF amplifier, mixer, local oscillator, power supply, IF amplifier, FM detector, audio amplifiers, and loudspeaker. These electronic circuits have been discussed previously, except for the FM detector. The standard FM broadcasting band in the United States is from 88 to 108 MHz. The intermediate frequency used is 10.7 MHz.

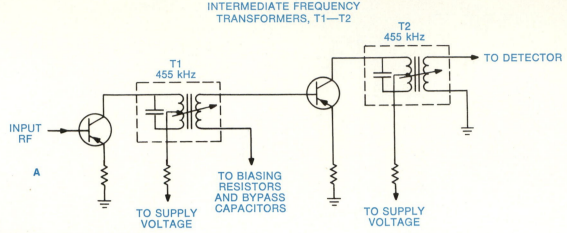

INTERMEDIATE FREQUENCY
TRANSFORMERS, T1—T2

T2
455 kHz

T1
455 kHz

TO DETECTOR

INPUT
RF

A

TO BIASING
RESISTORS
AND BYPASS
CAPACITORS

TO SUPPLY
VOLTAGE

TO SUPPLY
VOLTAGE

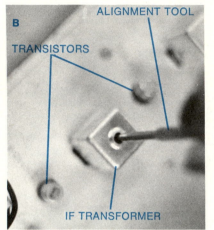

B

ALIGNMENT TOOL

TRANSISTORS

IF TRANSFORMER

Fig. 40–15. (A) Intermediate-frequency amplifier stage; (B) aligning the intermediate-frequency transformer by adjusting the movable iron cores in the coils

FM Detector. Figure 40–17 is a schematic of an FM detector called a *ratio detector*. The input to this circuit is from the previous IF amplifier stages. The ratio detector recaptures the audio signals that are frequency modulated on the carrier signal. In frequency modulation, the audio signals cause the FM carrier signal to change frequency above and below the assigned center frequency.

In FM, the amplitude of the audio signal is proportional to the amount of frequency shift above and below the carrier frequency. Therefore, the ratio detector must be able to produce an output voltage that is proportional to the frequency of the carrier wave at any instant of time.

Advantage of FM over AM. Automobile ignition systems, lightning, and the making and breaking of circuit connections by switching actions create electromagnetic waves. These waves act on amplitude-modulated communications carrier waves and cause them to be disrupted or distorted. As a result, AM radio receivers often produce a noise commonly

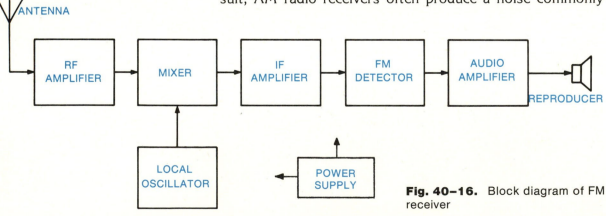

ANTENNA

RF
AMPLIFIER

MIXER

IF
AMPLIFIER

FM
DETECTOR

AUDIO
AMPLIFIER

REPRODUCER

LOCAL
OSCILLATOR

POWER
SUPPLY

Fig. 40–16. Block diagram of FM receiver

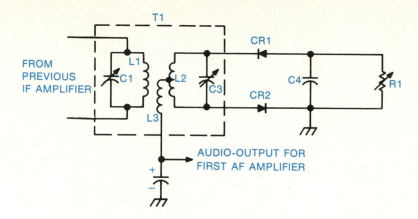

Fig. 40-17. FM ratio detector

called static. Unwanted electromagnetic waves do not cause the frequency of a frequency-modulated carrier wave to change. Because of this, the FM radio receiver is able to reproduce information in the form of speech or music that is not distorted by static.

SELF-TEST

Test your knowledge by writing, on a separate sheet of paper, the word or words that most correctly complete the following statements:

1. A radio system consists basically of a _____ and _____ .
2. In a transmitter, the device that causes a stream of electrons to vibrate back and forth at high frequencies is called a high-frequency _____ .
3. The wavelength of a radio wave depends on its frequency. The higher the frequency, the _____ the wavelength.
4. A radio wave must have a frequency of at least about _____ Hz before it can be used for communication.
5. Radio waves travel at the speed of _____ .
6. _____ waves move along the surface of the Earth. _____ waves move skyward.
7. The radio wave of a certain frequency assigned to a radio station is called a _____ wave.

8. The process of molding the electron stream for speech or music is called _____ .
9. During modulation in an AM system, the _____ of the carrier wave rises and falls according to the audio input.
10. In an FM system, the _____ is varied according to the audio input.
11. Commercial AM broadcast stations operate within a frequency band from _____ to _____ kHz.
12. When a radio receiver is able to select one radio signal and reject all others, it is said to be _____ .
13. A radio receiver is said to be _____ when it can receive especially weak radio signals.
14. The local oscillator of a superheterodyne radio circuit is tuned to a _____ frequency than the carrier frequency to which the receiver is tuned.
15. The standard intermediate frequency used in a superheterodyne radio circuit is _____ kHz.
16. The automatic volume control (AVC) provides _____ gain for weak signals.

385

17. Adjusting the variable trimmer capacitors and the variable cores of transformers in a radio receiver circuit is called _____.

18. The intermediate frequency in FM superheterodyne receivers is _____ MHz.

FOR REVIEW AND DISCUSSION

1. What are the components of a high-frequency oscillator?
2. Compare the frequency of sound waves and radio waves.
3. What effect does the Kennelly-Heaviside layer have on radio waves?
4. Describe the processes of AM modulation and FM modulation.
5. List five common amateur radio bands. Identify their frequencies and wavelengths.
6. Describe how a mixer, or converter, stage works in a superheterodyne radio-receiver circuit.

7. What are the main advantages of the IF amplifier stages?
8. Compare the assigned frequencies for the standard AM and FM broadcasting bands. List three AM- and FM-radio broadcasting frequencies in your local area.

INDIVIDUAL-STUDY ACTIVITIES

1. Obtain a schematic diagram of a complete AM radio. Identify the main components and circuits. Describe the function of each main section, such as the detector, tuner, and so on.
2. Obtain a pair of walkie-talkie transceivers that do not require an FCC license to operate. Read and follow the operating instructions. Find the largest operating range that you and a friend can use to talk to each other. What effects do buildings, hills, and vehicles have on the operating range?

Unit 41   Fundamentals of Television

Two pieces of photographic equipment can be used to show how a television camera works. One is a camera with a ground glass in the back. The other is a light meter. The ground glass of the camera lets you see the image focused through the lens. This image is upside down (Fig. 41–1). Light rays travel in straight lines. If a straight line is drawn from the bottom of an object through a lens, it will meet the ground glass near the top. Thus, the bottom of an object is projected through a lens to the top of the viewing surface.

The light meter measures the *intensity*, or strength, of light striking it. The light meter helps you get the lens opening on the camera. On a dull day, the light meter reads low light intensity. On a bright day, the light meter reads high light intensity.

When light strikes certain substances, an electric current is produced. This *photoelectric effect* was discovered in 1873.

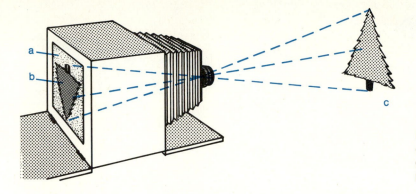

This discovery led to other experiments and eventually, to the first patent on a television system. This was taken out by V. K. Zworykin, an American physicist, in 1928. In the photoelectric effect, electrons are *emitted*, or thrown off, when light or other electromagnetic radiations strike a photosensitive surface. Two common photosensitive materials are cesium oxide and selenium. In a light meter, the emitted electrons from the photosensitive material are conducted by wires through a very sensitive meter. This measures the light intensity.

IMAGE ORTHICON TUBE

The heart of the television camera is the camera tube. The *image orthicon tube* and *vidicon tube* are two kinds in common use today. The basic parts of the image orthicon tube are shown in Fig. 41–2 and discussed below.

Image Section. The image of an object to be televised is focused by lenses onto a light-sensitive material inside the tube. As in a camera, this image is upside down. The image causes electrons to be emitted from the light-sensitive material. The number of electrons emitted depends on the intensity of the light striking the material. A white dress on a performer, for example, would cause more electrons to be emitted than a dark dress.

The emitted electrons are attracted to the *target*, a thin glass plate. This plate is made so that the electron image is transferred to its opposite side, the side facing the scanning beam (Fig. 41–2 at C).

Scanning Beam. The scanning, or electron, beam transfers the picture information to the multiplier section of the tube. Just before the scanning beam strikes the target, it is *decelerated*, or slowed down. This makes the transfer of picture

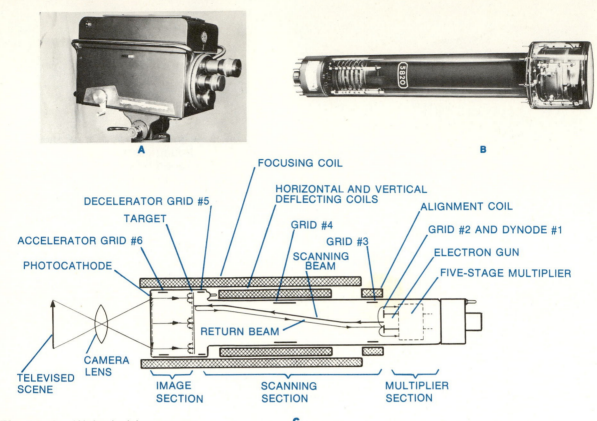

A
B

FOCUSING COIL
HORIZONTAL AND VERTICAL
DEFLECTING COILS
DECELERATOR GRID #5
ALIGNMENT COIL
TARGET
GRID #4
GRID #2 AND DYNODE #1
ACCELERATOR GRID #6
GRID #3
SCANNING
BEAM
ELECTRON GUN
PHOTOCATHODE
FIVE-STAGE MULTIPLIER

RETURN BEAM

TELEVISED
SCENE
CAMERA
LENS
IMAGE
SECTION
SCANNING
SECTION
MULTIPLIER
SECTION

C

Fig. 41–2. (A) A television camera with three lenses; (B) an image orthicon tube (*courtesy of Radio Corporation of America*); (C) schematic arrangement of an image orthicon tube (*courtesy of Radio Corporation of America*)

detail from the target possible. This scanning beam is focused like a light beam into a narrow stream.

The areas of the target where more of the electrons have been emitted are less negative. The scanning beam is repelled less in these areas. The areas where more of the electrons have been retained are more negative. The scanning beam is repelled more in these areas. The result is a return beam of electrons varying in density according to the details of the picture.

Return Beam. The variations in the return beam caused by the light and dark areas of the object being viewed make up the *video*, or *picture, signal.*

The return beam works like a photographic light meter moving across the sensitized image. A meter is more sensitive to light areas than to dark areas. The movement of the needle in the meter would register detail, as does the return beam (Fig. 41–3).

Scanning Section. In order to reproduce an image, the scanning beam must *scan*, or view, the whole image. The

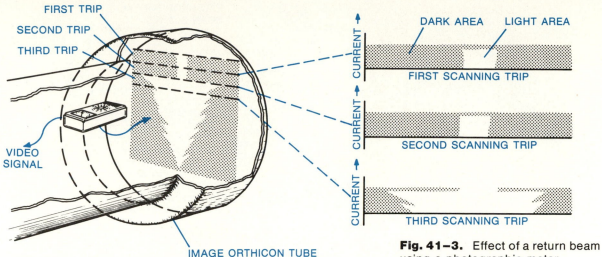

Fig. 41–3. Effect of a return beam using a photographic meter

scanning section causes the scanning beam to scan the image completely.

The scanning beam can be bent left or right by placing certain charges of electricity on horizontal plates like plates 1 and 2 of Fig. 41–4. The beam can be moved up or down by varying the charges on vertical plates 3 and 4. Figure 41–4 shows a greater positive charge on horizontal plate 2. This causes the scanning beam to swerve to the right, since it is attracted to the positively charged plate. The entire target can be scanned by placing various charges on the horizontal or vertical planes. A scanning beam can also be controlled by horizontal and vertical deflecting coils instead of plates (see Fig. 41–2).

The scanning beam scans horizontally and vertically at the same time (Fig. 41–5). Therefore, each horizontal line of scanning slants downward a little. As each horizontal line is completed, the electron beam returns rapidly to the starting

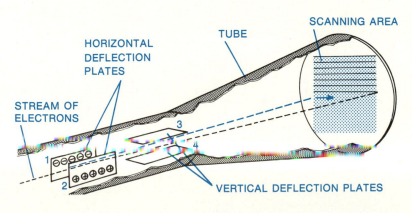

Fig. 41–4. Charges on the plates cause the electron beam to scan in a specified area.

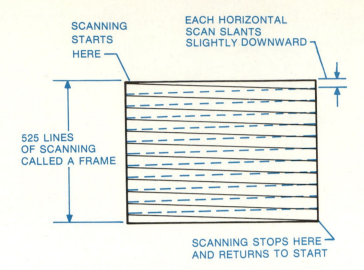

SCANNING STARTS HERE —

EACH HORIZONTAL SCAN SLANTS SLIGHTLY DOWNWARD

525 LINES OF SCANNING CALLED A FRAME

SCANNING STOPS HERE AND RETURNS TO START

Fig. 41–5. Scanning process

point of the next horizontal line. This point is always a little below the previous one.

In television, a series of 30 frames is flashed on the screen every second. Each frame consists of 525 lines. Thus, the scanning rate is 30 × 525, or 15,750 lines, or hertz. A motion-picture film shows 24 frames per second to create the effect of continuous motion. The 30 frames per second of television has the same effect. Figure 41–6 shows the effect of a mechanical connection between scanning systems of the television camera and the television receiver. Actually, they have to be electronically *synchronized*, or locked in step.

VIDICON CAMERA TUBE

Figure 41–7 shows a simplified schematic diagram of a vidicon camera tube. One use of this kind of camera tube is in weather satellites, which takes pictures of the Earth and its

Fig. 41–6. Television camera and receiver scanning systems must be synchronized, or locked in step.

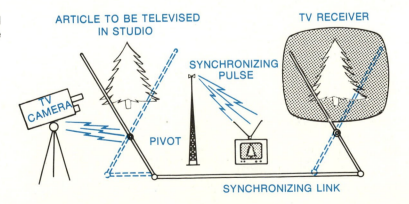

ARTICLE TO BE TELEVISED IN STUDIO

TV RECEIVER

SYNCHRONIZING PULSE

TV CAMERA

PIVOT

SYNCHRONIZING LINK

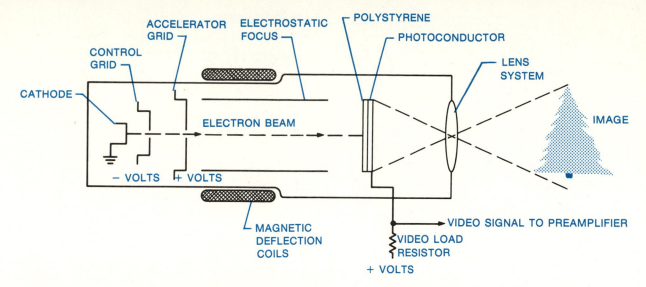

Fig. 41–7. Vidicon camera tube

clouds. The vidicon camera tube does not have a return beam.

When the camera tube is taking a picture, the electrically sensitive material is exposed to the image. The image is converted into an electrical-charge pattern on the surface of this material by the emission of electrons. The electrical-charge pattern is transferred to the polystyrene layer. There it is *read out,* or scanned, by an electron beam moving across and down the surface. The pattern of the electrical charge causes the electron beam, or current, to vary. This variable current forms the video signal. The video-signal voltage appears across the load resistor.

COLOR CAMERA TUBES

To produce a *monochromatic,* or black-and-white, television signal, only one camera tube is needed. However, to transmit color, three or more camera tubes are used. One common kind of color camera has four tubes. One tube produces the *luminance* signal. This is just like a black-and-white signal. All the light reflected from the image is transmitted as the luminance signal. For color, an additional optical system is used. The light is filtered so that only red colors can be seen by the red camera tube, only green colors by the green camera tube, and only blue colors by the blue camera tube. Before transmission, the four video signals (luminance, red, green, and blue) are matrixed, or mixed together. Black-and-white television receivers accept the luminance signal. Color television receivers separate the red, green, and blue signals and reproduce them on their screens (Fig. 41–8).

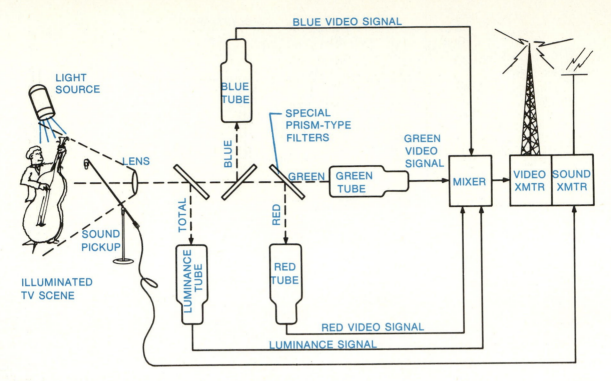

Fig. 41–8. Basic elements in television color transmission using a four-tube color camera

TRANSMISSION OF TELEVISION SIGNALS

The video signal of television is transmitted somewhat like the audio signal of radio. Both video and audio signals are varying direct currents. Each time the electron beam scans the image in the studio camera, this video signal is picked up and transmitted. Sound for a television program is transmitted at the same time as the picture, but on a separate frequency. In a television channel, therefore, there are two carrier frequencies, one for the picture and one for the sound.

Television and frequency-modulation broadcasting stations use high-frequency carrier waves. Television carrier waves are classified as very-high-frequency (vhf) waves (30 to 300 MHz) or as ultra-high-frequency (uhf) waves (300 to 3,000 MHz). Table 41–1 lists the bandwidths and video and sound frequencies for selected television channels. There are 12 vhf channels, numbers 2 to 13. Channel number 1 is not used now because that frequency band has been reassigned by the FCC to another kind of communication service. There are 70 uhf channels, numbers 14 to 83. These uhf channels provide for the expansion of television service, especially through the use of cable television. Cable television, as contrasted to over-the-air television, can provide a wide variety of television channels as well as two-way communication. This is done through a coaxial cable.

Table 41-1. Examples of Video and Audio Frequencies Used in Television Broadcasting

Maximum Bandwidth	Frequency MHz	Nature of Service	Channel Number	vhf or uhf
66–72	67.25 71.75	video sound	} 4	very-high-frequency (vhf) range
180–186	181.25 185.75	video sound	} 8	
204–210	205.25 209.75	video sound	}12	
506–512	507.25 511.75	video sound	}20	ultra-high-frequency (uhf) range
542–548	543.25 547.75	video sound	}26	
818–824	819.25 823.75	video sound	}72	

Because of the high frequencies used in television, the waves take on the characteristics of light waves. As a result, television broadcasting stations are limited to straight-line paths like light paths (Fig. 41–9). To get long-distance transmission, relay towers are often used. Coaxial telephone cables are also used for long-distance transmission. At the local station, the programs are transmitted over the air to local viewers.

TELEVISION RECEIVER

The television receiver (Fig. 41–10) and the radio receiver are alike in many ways. The superheterodyne principle, intermediate frequencies, radio frequencies, amplification, and mixer and detector circuits are used in television as well as radio. Multiterminal modular packages used in a typical color television receiver are shown in Fig. 41–11. Figure 41–12 is a block diagram of a color television receiver. A black-and-white television receiver would be similar. However, it would not have the circuits needed to process the color signals.

TELEVISION ANTENNA

Television antennas pick up both sound and picture signals. One of the simplest antennas for television reception is the

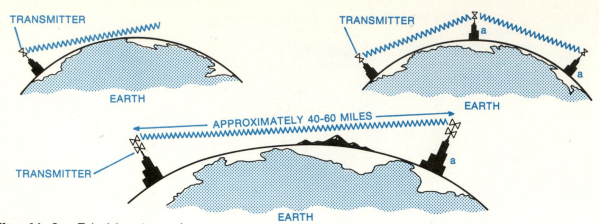

TRANSMITTER

EARTH

TRANSMITTER

EARTH

a

a

APPROXIMATELY 40-60 MILES

TRANSMITTER

EARTH

a

Fig. 41–9. Television transmission: (a) relay towers

dipole, or two-pole, antenna (Fig. 41–13). This is a half-wave antenna opened at the center. Each length of this antenna represents one-fourth of the wavelength of the incoming signal. For example, the length of one arm of a dipole for channel 6 is about 32 inches. Connecting wires leading to the television receiver are fastened to the open ends in the middle of the dipole.

Figure 41–14 shows a five-element antenna. It is used in weak reception areas some distance from the television transmitter.

TUNER

The tuner of a television receiver lets you choose the channel you want. A rotary switching system is often used for this purpose. The rotary switch changes inductors and/or capacitors as needed for tuning in channels. The inductors and capacitors in the tuning circuit of a television receiver perform the same function as in a radio receiver.

Preprogrammed Tuner. A new computerized tuning system is now available on some television receivers. This system has the vhf and uhf channels plus all cable channels preprogrammed and stored in the memory of the *microprocessor.* This is a digital integrated circuit. It is considered the *central processing unit* (CPU) in a *microcomputer.* This microcomputer stores channel frequency information and recalls it on demand (Fig. 41–15). The viewer just presses the channel number and then the "enter" button, and the station is tuned in instantly. A 10-digit keyboard is also available on the television set or in a remote location. This keyboard lets a person choose a channel by touching the proper digits.

Fig. 41–10. Color television receiver (*General Electric Co.*)

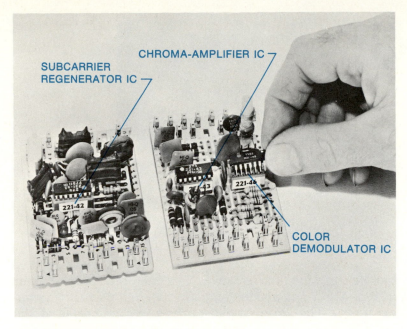

SUBCARRIER REGENERATOR IC

CHROMA-AMPLIFIER IC

221-42

221-43

221-44

COLOR DEMODULATOR IC

Fig. 41–11. An example of an integrated-circuit (IC) color processing circuitry and associated components (*Zenith Radio Corporation*)

AMPLIFIERS

As in a superheterodyne radio receiver, there are radio frequency amplifiers, IF amplifiers, and mixer stages in a television-receiver circuit. However, television IF amplifiers must be sensitive over a range of 6 MHz. This range includes both

Fig. 41–12. Block diagram of a color television receiver

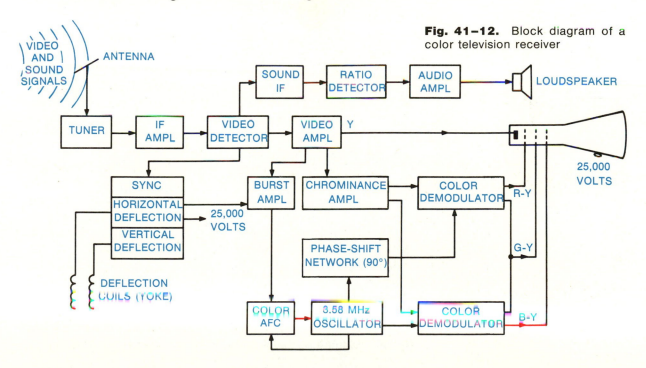

VIDEO AND SOUND SIGNALS

ANTENNA

SOUND IF → RATIO DETECTOR → AUDIO AMPL → LOUDSPEAKER

TUNER → IF AMPL → VIDEO DETECTOR → VIDEO AMPL → Y

25,000 VOLTS

SYNC
HORIZONTAL DEFLECTION
VERTICAL DEFLECTION

DEFLECTION COILS (YOKE)

BURST AMPL

25,000 VOLTS

CHROMINANCE AMPL → COLOR DEMODULATOR → R-Y

G-Y

PHASE-SHIFT NETWORK (90°)

COLOR AFC → 3.58 MHz OSCILLATOR → COLOR DEMODULATOR → B-Y

395

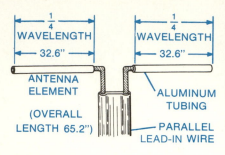

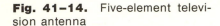

Fig. 41–13. Dipole antenna for channel 6

Fig. 41–14. Five-element television antenna

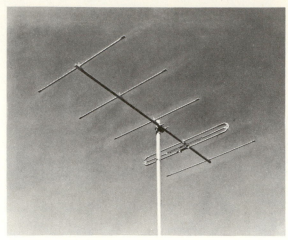

picture and sound carrier waves. The color television receiver in Fig. 41–12 has a video amplifier in addition to the audio amplifiers. The video amplifier increases the power of the video signals. A black-and-white television receiver would not have the circuits for the burst and chrominance amplifiers.

SOUND CIRCUITS

The FM sound circuit contains an intermediate-frequency amplifier, a ratio detector, and an audio amplifier. The efficient *intermediate-frequency amplifier* amplifies the sound signals and increases the sensitivity and selectivity of the receiver toward them. The *ratio detector* separates out the audio component in the frequency-modulation system. The *audio amplifier* increases the audio signal enough to operate a loudspeaker.

SYNCHRONIZING AND SWEEP CIRCUITS

The *synchronizing*, or *sync*, and *sweep*, or *deflection*, circuits receive the sync pulses from the televised picture signal and sweep the television picture-tube beam in synchronization with the camera-tube beam. Like the camera tube, the picture tube has vertical- and horizontal-sweep plates, or coils. These plates sweep the beam across and up and down the chemically coated screen to make the picture.

The horizontal-sweep circuit returns the television picture-tube beam to its starting point. This return sweep, called the horizontal retrace signal, is *blanked*. That is, the retrace signal cannot be seen by the viewer.

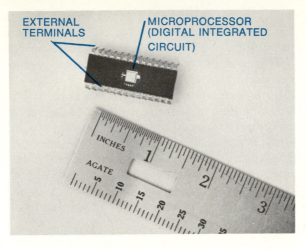

Fig. 41–15. Microprocessor has TV channel frequencies stored in its memory for instant recall. Ruler gives indication of size. (*Zenith Radio Corporation*)

Several controls are used to adjust the operation of the sync and sweep circuits. The horizontal- and vertical-hold controls adjust oscillators within the range of the sync pulses. The height and width controls adjust the amplitude of the sweep. Other controls adjust the horizontal or vertical effect of the sweep to prevent distortion.

HIGH-VOLTAGE SYSTEM

A very high voltage is needed to light a television screen. A system commonly used in television receivers is the *flyback high-voltage* system. This system uses part of the horizontal-deflection circuit to develop a high voltage.

As the horizontal sweep flies back, or returns, to begin another pass across the television screen, its magnetic field induces a high voltage in the coil. This is similar to the way an automobile ignition coil works.

This flyback system does away with bulky transformers and large capacitors. This is because the frequency is high (15,750 Hz) and only a small current is needed to operate the television picture tube. Other electrical elements in the high-voltage system rectify the alternating sweep circuit and smooth any variations in the circuit. A typical color television receiver applies 27,500 volts to the picture tube.

COLOR CIRCUITS

The major color circuits are shown in the lower right-hand section of Fig. 41–12. One interesting feature is the color automatic-frequency-control (AFC) section. This provides the precise signals needed to demodulate the complex color signal. Because of the interference that would result, the color carrier (about 3.58 MHz) cannot be transmitted with the other signals needed for television operation. However, the color carrier is required for color demodulation.

This problem is solved by transmitting short bursts of carrier during the horizontal retrace or flyback. During retrace, interference will not show on the screen. These *color bursts* are amplified and compared with an internally generated color-carrier signal. The comparison stage is called *color AFC*. It corrects the internal 3.58-MHz oscillator as needed. The color signals can then be properly demodulated. The original phase, or hue, relationships of the televised scene at the studio are preserved. The viewer can make small changes in these phase relationships by adjusting the hue, or tint, control.

After demodulation, the R-Y (red minus luminance) and B-Y (blue minus luminance) signals are matrixed, or mixed, to produce a G-Y (green minus luminance) signal. All three signals are then applied to the picture tube. There a positive Y signal from the video amplifier cancels all −Y components. Thus, a full-color picture is produced on the screen.

TELEVISION PICTURE TUBE

The television picture tube reproduces the image on the target of the camera tube in the television broadcasting station. Many of the elements in the television picture tube are similar to elements in the camera tube. The television picture tube produces an electron beam that scans the face of the tube. The scanning system uses either deflection plates or magnetic coils to deflect the beam. Figure 41–16 shows a simplified cross-sectional view of a black-and-white television picture tube.

ELECTRON BEAM

The electron beam is actually a stream of electrons modulated according to the video signals placed on the control grid. The light intensity on the television screen depends on how many electrons strike it and with what velocity, or speed, they do so. The greater the number and the higher the velocity of the electrons, the greater the light intensity. Therefore, the clearness and contrast of the television picture depend on (1) the number of electrons, (2) their velocity, and (3) the kind of screen used.

TUBE SCREEN

On the inside of the face of the television picture tube is a coating of a *phosphor*. This is a material that glows whenever

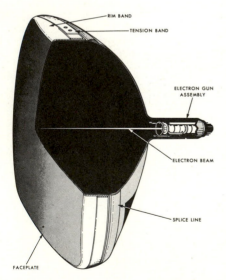

RIM BAND

TENSION BAND

ELECTRON GUN ASSEMBLY

ELECTRON BEAM

SPLICE LINE

FACEPLATE

Fig. 41–16. Black-and-white television picture tube showing essential elements (*Radio Corporation of America*)

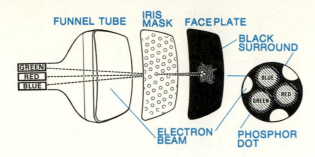

Fig. 41–17. Color television picture tube (*Zenith Radio Corporation*)

a beam of electrons or other radiated energy strikes it. This kind of illumination is sometimes called *cold light.*

Actually, this coating continues to glow after the electron beam has passed. A material that is able to do this is said to be *phosphorescent.* The glow remains for only a short time, but it keeps the picture from appearing to flicker.

In a black-and-white television receiver, the electron beam produces light, gray, and dark spots as it races across and down a screen. If the modulation allows a great many electrons to strike a screen, bright light appears. If only a moderate number are allowed to strike the screen, less light (gray) appears. If no electrons are allowed to strike the screen, no light (black) appears. These various shades make up the picture we see. A lighted television screen that is not receiving a picture is called a *raster.*

COLOR PICTURE TUBE

In one kind of color television picture tube (Fig. 41–17), the video signals (green, red, and blue) are beamed to excite 1,350,000 tiny phosphor dots. When the dots are struck by an electron beam, they glow red, green, or blue.

When this tube is made, 450,000 openings in an iris mask are used to apply the phosphors and a contrast area of jet black on the screen. The iris-mask openings are then enlarged. As a result, the electron beams are larger than the phosphor dots. Thus, the phosphor dots are illuminated completely. The black surround reduces light reflections and allows a clearer faceplate to be used to increase light transmission.

TUNING FOR COLOR

To receive color broadcasts properly, the television set should first be adjusted for black-and-white reception. Then the color controls, color gain and hue, should be adjusted as follows:

Color Gain. The color gain, or intensity, control should be turned clockwise to increase the color intensity. This control is moved to minimum for a black-and-white broadcast. If there is too little gain, the colors will appear pale and weak. If there is too much gain, the colors will appear flushed and poorly focused.

Color Hue. This control adjusts the hue, or tint, of the color. Adjust the control until the facial tones are natural. If the hue is turned too far counterclockwise, the facial tones will appear purple. If the control is turned too far clockwise, facial tones will appear green.

Once the color gain and hue are set properly, they usually do not have to be adjusted again. However, changing channels may require a slight readjustment. When all controls remain at their previous setting, the television receiver will automatically reproduce black and white or color, depending on the broadcast.

TELEVISION PICTURE INTERFERENCE

Many kinds of electrical apparatus can interfere with a television picture. Automobile ignition systems, flashing signs, electric motors, diathermy equipment, and other television and radio receivers and transmitters can all cause interference. Most television sets today reduce this interference to a minimum.

From Automobile Ignition Systems and so on. Interference from automobile ignitions, airplanes, electrical appliances, and so on cause speckled, streaked pictures. This condition is most noticeable in weak-signal areas.

From Diathermy Equipment. This kind of medical equipment produces a herringbone pattern and possibly one or two horizonal bands across the face of the picture. Weaker interference of this kind may cause a general fogging or blurring of the image.

From Radio Signals. Interference from radio signals may create moving ripples or diagonal streaks on a screen. These signals may be produced by nearby citizen, commercial, amateur, or police station radios. Other television or frequency-modulation receivers operating near the set may produce the same effect.

From Rebounding Signals. This interference accounts for a multiple image on the screen. It is caused by the rebounding of the television signal from a nearby building, hill, or mountain. Actually, this rebounding causes two or more signals to be received by the antenna. The rebounding signal is received a little later than the direct signal. The former appears as a "ghost" image on the screen.

From Fringe-area Reception. At the fringes of the transmission area of a television station, the picture may appear as a snowy pattern on the screen. This condition can sometimes be partially corrected by moving the television receiver within a room. An outside antenna and rotator should be used for best reception in fringe areas.

CABLE TELEVISION

In the late 1940s, cable television became a reality. It is called the Community Antenna Television (CATV) system. This system brings television signals into communities located too far from the television station. It is also used in communities in which television signals are blocked by deep valleys or high mountains. The antennas of cable systems are located in areas with good reception. Then television signals are distributed by cable to subscribers for a fee. At the present time, the average monthly fee is about $7.00 for the cable service. The average installation cost is about $15.00. Most cable companies distribute television signals through coaxial cables. These are strung over telephone poles or placed underground.

Since cable systems have more channels than are broadcast, they can provide additional services. Cable systems often have separate channels for weather reports, stock-market reports, wire-service news, FM radio, and movies. Cable operators also originate their own programs. They provide cable channels for use by the public, for education, for government use, and for lease.

New cable systems having more than 3,500 subscribers generally provide 25 television broadcast channels. Some systems have the capacity for two-way communication (Fig. 41-18). This allows subscribers to shop for merchandise, conduct banking business, reply to community surveys, and have utility meters read.

Cable systems often offer pay cable service in addition to their basic service. Pay cable service allows subscribers to view special programs, such as feature films and sports events. Since its beginning in late 1975, pay cable programming has become very popular and economical.

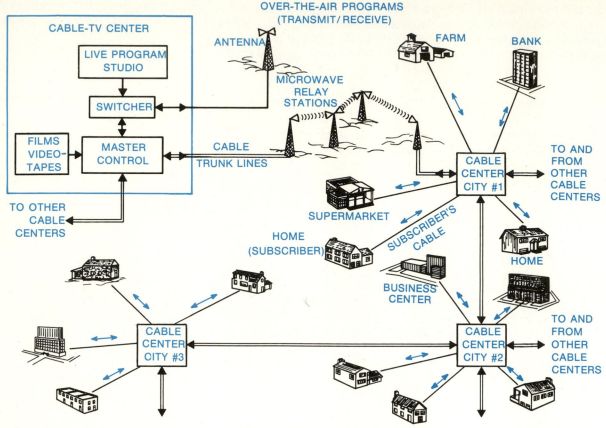

Fig. 41–18. A cable television two-way communication system

Test your knowledge by writing, on a separate sheet of paper, the word or words that most correctly complete the following statements:

1. Two common television camera tubes are the _____ and the _____.
2. Variations in the return beam make up the _____ signal.
3. The scanning beam is bent left and right and up and down by charges on the _____ and _____ deflection plates.
4. Each frame on the television screen consists of _____ horizontal scanning lines.
5. When the television transmitter and receiver are locked in step, they are said to be _____.
6. In the vidicon tube, the video-signal voltage appears across the _____ resistor.
7. All the light reflected from the image is transmitted as the _____ signal.
8. A television channel contains both _____ and _____ signals.
9. Television carrier-wave frequencies are classified in the _____ and _____ ranges.
10. There are _____ vhf and _____ uhf television channels.
11. A simple kind of television antenna with two arms is called a _____.
12. A rotary switching system used in a television has _____ and _____ for tuning.

13. A tuner with a microcomputer is called a
    _____ tuner.
14. The _____ amplifier and the _____
    amplifier are two amplifiers found only in
    color television receivers.
15. The television picture tube, like the camera tube, has an _____ that scans the
    face of the tube.
16. The coating on the inside surface of a television picture tube is a _____ .

FOR REVIEW AND DISCUSSION

1. Discuss the common relationships of the
   principles of a television camera, a photographic camera, and a light meter.
2. Describe the image section, scanning
   beam, return beam, and scanning section
   of an image orthicon tube.
3. Describe the basic difference between the
   image orthicon tube and the vidicon tube.
4. How are the color signals developed in a
   color television camera?
5. Identify the electronic circuits found in a
   color television receiver that would not
   be found in a black-and-white television
   receiver.
6. List and explain the functions of the electronic circuits found in a typical television
   receiver.
7. Describe the function of the burst amplifier in a color television receiver.
8. Explain the purpose of the synchronizing
   sweep circuits in a television receiver.
9. Describe how the electron beam develops
   the picture in a black-and-white television
   receiver.
10. Describe the interference pattern most
    likely to be caused by automobile ignition
    systems, by diathermy equipment, and by
    rebounding television signals.

INDIVIDUAL-STUDY ACTIVITY

Find out if there is a cable television system
available in your area. Describe the system,
how a person may subscribe, and how the
system is installed in the subscriber's home.

Unit 42   Telemetry

Telemetry is making measurements at one place and sending
these measurements for recording and study to another place
far away. The information obtained may be in a form suitable
for display on a television tube, for recording on tape, or for
use in analog or digital computers. In space explorations
without astronauts, telemetry has helped scientists perfect
the design of vehicles by monitoring their performance during flight. In flight with astronauts, telemetric signals add to
the messages sent by the astronauts. Telemetry makes it
possible to compute flight data with heavy equipment on the
ground. Data include such information as facts and figures.
These data are used for study or for making a decision.

POWER REQUIREMENTS

The power needed to send telemetric signals from satellites
and space probes is much less than that needed for commer-

cial radio and television stations. The space explorer, *Mariner IV*, sent data back to Earth from more than 191 million miles (307 million kilometers) away with a 10-watt transmitter. Radio contacts, but no telemetric data, were received from *Mariner IV* when it was 216 million miles (348 million kilometers) away.

ANTENNA SYSTEMS

In part, so little power is needed on a spacecraft because ground radio systems have very sensitive receiving equipment and high-gain antennas. For example, the antenna used with the Mariner spacecraft was 85 ft (26 m) in diameter. It was about 20,000 times more effective than the average rooftop television antenna. Ground stations with these antenna systems can use radio signals from satellites and spacecraft to track them with a high degree of accuracy. These systems also make it possible to communicate and receive data from spacecraft and satellites (Fig. 42–1).

PROGRAMMING AND CONTROL

One of the electronic subsystems used in a weather satellite is for programming and control (Fig. 42–2). This subsystem allows the satellite to be controlled either automatically or by command from ground stations in contact range. The satellite responds to a command-address unit. This unit allows one satellite out of a group to be contacted. For example, first an address tone of a specified frequency and duration is transmitted from the ground station. If the satellite is within range and the address tone is received, the programming and control subsystem is then activated.

Satellites have a number of auxiliary controls. The magnetic attitude control is an example. This control guides the spin axis of the satellite along a predetermined path. Other auxiliary controls are the despin control, spin-up rocket control, and telemetry and beacon controls. This last system has two continuous wave radio signals used to track the satellite. The beacon signals are also used as radio-frequency carrier waves for telemetric signals.

One of the important elements in the satellite programming and control system is the satellite clock. A clock is like an oscillator. It generates periodic signals. The clock must be synchronized with the master clock at the ground station. The master clock at the ground station is calibrated with the standard time signals. These signals are transmitted by the U.S. National Bureau of Standards radio system, such as sta-

Fig. 42–1. An 85-foot-diameter (26-meter-diameter) parabolic antenna used for spacecraft and satellite communications (*General Telephone and Electronics Corporation*)

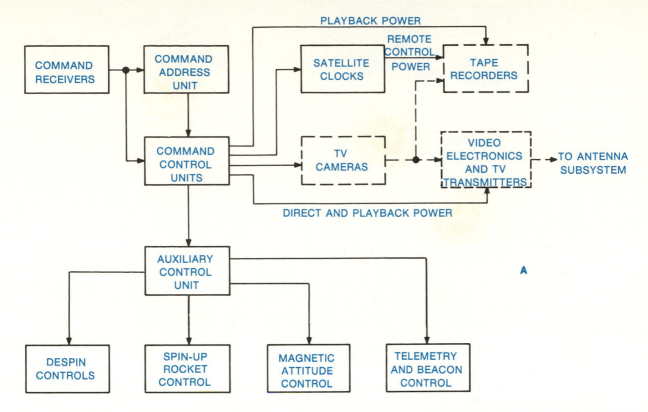

PLAYBACK POWER

COMMAND RECEIVERS → COMMAND ADDRESS UNIT

SATELLITE CLOCKS

REMOTE CONTROL POWER

TAPE RECORDERS

COMMAND CONTROL UNITS

TV CAMERAS

VIDEO ELECTRONICS AND TV TRANSMITTERS → TO ANTENNA SUBSYSTEM

DIRECT AND PLAYBACK POWER

A

AUXILIARY CONTROL UNIT

DESPIN CONTROLS

SPIN-UP ROCKET CONTROL

MAGNETIC ATTITUDE CONTROL

TELEMETRY AND BEACON CONTROL

tion WWV. The satellite clock maintains proper synchronization and programming for the entire satellite system around the world. Different tone signals are used to set different clocks. Once set and working, the satellite clock helps control operating steps.

DATA ACQUISITION FACILITIES

Because of the large amount of data transmitted by communications satellites and other spacecraft, a system of ground stations has been built to receive and pass on this information. This system is called the Space Tracking and Data Acquisition Network (STADAN). *Acquisition* means "the act of obtaining or gaining." The network sends all its data to the Communications and Computing Center at the Goddard Space Flight Center of the National Aeronautics and Space Administration (NASA) in Greenbelt, Maryland. From there, the data are sent on to the users.

Figure 42-3 shows the basic parts of a weather-satellite system. The satellite provides television pictures of the Earth. These pictures show clouds, storms, and pollution. The pictures are relayed by radio to the ground station cen-

B

Fig. 42–2. (A) Block diagram of a programming and control subsystem of a weather satellite (*National Aeronautics and Space Administration*); (B) a weather satellite

405

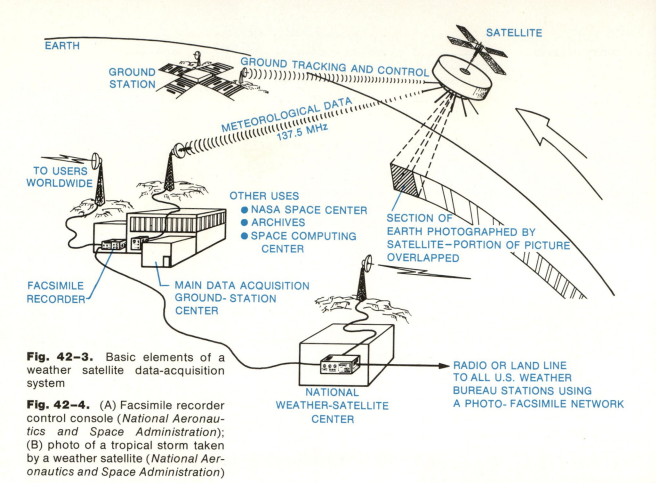

EARTH

SATELLITE

GROUND STATION

GROUND TRACKING AND CONTROL

METEOROLOGICAL DATA
137.5 MHz

TO USERS WORLDWIDE

OTHER USES
● NASA SPACE CENTER
● ARCHIVES
● SPACE COMPUTING CENTER

SECTION OF EARTH PHOTOGRAPHED BY SATELLITE—PORTION OF PICTURE OVERLAPPED

FACSIMILE RECORDER

MAIN DATA ACQUISITION GROUND- STATION CENTER

Fig. 42–3. Basic elements of a weather satellite data-acquisition system

NATIONAL WEATHER-SATELLITE CENTER

RADIO OR LAND LINE TO ALL U.S. WEATHER BUREAU STATIONS USING A PHOTO- FACSIMILE NETWORK

Fig. 42–4. (A) Facsimile recorder control console (*National Aeronautics and Space Administration*); (B) photo of a tropical storm taken by a weather satellite (*National Aeronautics and Space Administration*)

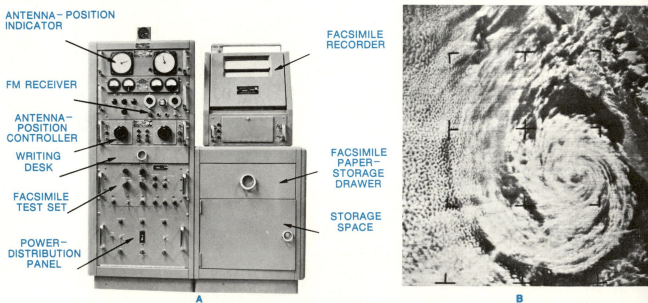

ANTENNA– POSITION INDICATOR

FACSIMILE RECORDER

FM RECEIVER

ANTENNA– POSITION CONTROLLER

WRITING DESK

FACSIMILE PAPER– STORAGE DRAWER

FACSIMILE TEST SET

STORAGE SPACE

POWER– DISTRIBUTION PANEL

A

B

ter. Other signals from the satellite are converted into usable drawings by facsimile recorders (Fig. 42–4 at A). These drawings or graphs are very helpful in locating hurricanes and other weather disturbances (Fig. 42–4 at B).

Other weather data, such as surface temperatures, are also received by the center at the same time. These are sent by radio-teletype or teletypewriter from about 1,500 local weather stations located throughout the United States.

SELF-TEST

Test your knowledge by writing, on a separate sheet of paper, the word or words that most correctly complete the following statements:

1. Little power is needed on a satellite because ground radio systems have very sensitive _____ and _____.
2. The programming and control subsystem in a weather satellite responds to an address tone of a specified _____ and _____.
3. The _____ signals provide the carrier waves for telemetric signals.
4. A satellite clock must be _____ with the master clock at the ground station.
5. Signals from satellites are converted into usable drawings by _____.

FOR REVIEW AND DISCUSSION

1. Define the term *telemetry*.
2. Give two major advantages of telemetry in space exploration.
3. List four kinds of auxiliary controls in a weather satellite.
4. Explain the primary function of a satellite clock.
5. Describe the parts of a satellite data acquisition system.

INDIVIDUAL-STUDY ACTIVITY

Prepare a report on the use of telemetry in the moon-landing, Voyager, Skylab, or Pioneer space programs.

Unit 43   Computers and Data Processing

One basic need of business, industry, and government is to handle large amounts of information quickly with minimum error. Thus, systems are needed for gathering information, for making mathematical computations, and for storing information so that it can be easily obtained.

Today, these and many other related functions are performed by computers and auxiliary units. A *computer* is a programmable electronic device that can store, retrieve, and process data. Computers and auxiliary units are known as *automatic data processing* (ADP) systems. There are two main kinds of computers. These are the *analog computer* and the *digital computer*.

COMPARISON OF BASIC ANALOG AND DIGITAL SYSTEMS

The automobile fuel-level circuit is an example of an analog measuring system. Its purpose is to first "sense" the quantity of fuel in the tank. This quantity is then indicated by a *readout*. This takes the form of the position of a pointer on a scale. Since the pointer can move to many positions on the scale, it continuously indicates fuel quantity. For this reason, an *analog measuring system* can be defined as one that is able to respond instantly to any condition of what may be a wide range of conditions.

An indicator light such as the oil-pressure light in an automobile is an example of a *digital measuring system*. In this system, the light is at any time either on or off. Conditions of this kind are known as *binary conditions*. *Binary* means "made up of two parts."

DIGITAL COMPUTERS

The circuits of a digital computer respond to one of only two conditions. These may be switches that are on or off, voltages that are either positive or negative, or two voltages that are different in value. Although the input and the output information to be processed by a digital computer may have different forms, the circuits that process this information are digital.

Most data processing is done by digital computers. There are many models and varieties of digital computers. They range from small portable computers to large, complicated systems. These systems are made up of many pieces of interconnected equipment. Figure 43–1 shows some of the kinds of equipment commonly found in a large computer center.

BASIC ELEMENTS IN AN ADP SYSTEM

Generally, all computers have the following five basic elements: input, output, storage, arithmetic (logic), and control (Fig. 43–2).

COMPUTER INPUT

The computer input device feeds data into the computer system. The main input devices used are paper tape, punched

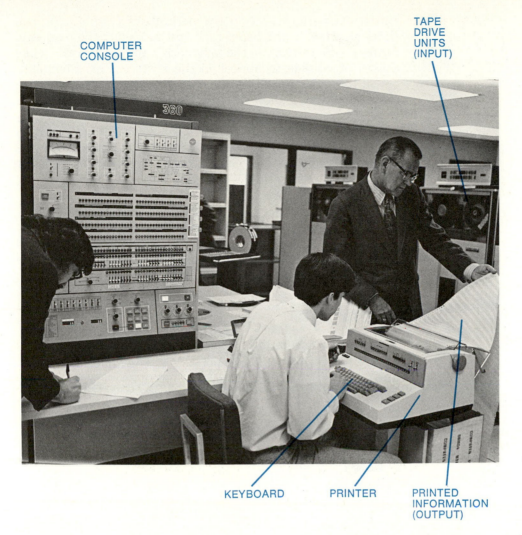

COMPUTER
CONSOLE

TAPE
DRIVE
UNITS
(INPUT)

KEYBOARD   PRINTER   PRINTED
INFORMATION
(OUTPUT)

Fig. 43–1. Examining the printed output from an automatic data-processing computer (*U.S. Department of Housing and Urban Development*)

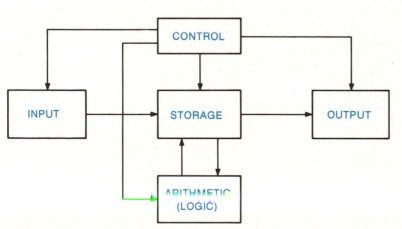

CONTROL

INPUT   STORAGE   OUTPUT

ARITHMETIC
(LOGIC)

Fig. 43–2. Basic elements in an ADP system

PUNCHED TAPE

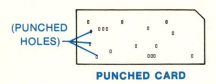

(PUNCHED HOLES)

PUNCHED CARD

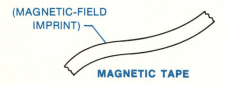

(MAGNETIC-FIELD IMPRINT)

MAGNETIC TAPE

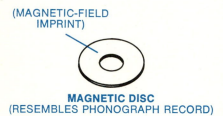

(MAGNETIC-FIELD IMPRINT)

MAGNETIC DISC
(RESEMBLES PHONOGRAPH RECORD)

Fig. 43–3. Computer inputs

cards, magnetic tape, and magnetic discs (Fig. 43–3). The holes punched in the paper tape and cards and the magnetic patterns imprinted on the magnetic tape and discs cause pulses of voltage to appear in the system. For example, a 5-volt pulse may stand for the digit 1. The absence of a pulse may stand for the digit 0. These digits form what are known as *binary codes*. These are codes that are made up of only two parts. In this case, the two parts are the digits 1 and 0. These digits are called *bits*. A group of bits makes up one *alphanumeric character*. Such a character stands for a letter or number. This coding system is similar to that used for teletypewriter communications (see Unit 37).

Magnetic tapes are used to store large amounts of data. Large computer systems use equipment to handle magnetic tapes called *tape drive units* (Fig. 43–4). These units control the movement of the tape and can sense the bits that make up the characters on the tape.

COMPUTER OUTPUT

The output information obtained from a computer is often in the same form as the input: paper tape, punched cards, magnetic tape, or magnetic discs. Information printed directly on

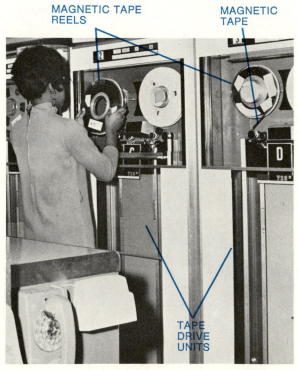

MAGNETIC TAPE REELS   MAGNETIC TAPE

TAPE DRIVE UNITS

Fig. 43–4. Installing a magnetic tape reel on a tape drive unit (*U.S. Department of Labor*)

paper is also a very common form of computer output. In some computer systems, the output information is displayed on a cathode-ray tube. This is similar to a television picture tube.

STORAGE

Data must be stored in the computer before it can be processed. One of the common ways of storing data is to magnetize tiny iron cores with each pulse of electricity. The magnetic fields that surround the cores represent stored data. These fields, and thus the data, can be sensed. Since the tiny iron core can retain its magnetic field, it has the property of *memory* (Fig. 43–5).

Other devices have "memory," too. For example, you have probably used a retractable ball-point pen. When you push down on the top of the pen and release, the point is in place for writing. When you push and release again, the point disappears. If you push up and down on the top of the pen in rapid succession, you will notice that the point appears *once* for every *two* pushes.

The ball-point pen action is similar to an electronic device called a *toggle flip-flop*. The toggle flip-flop is a kind of electronic switching circuit. The toggle flip-flop output is a digital 1 for every other 1 input. If the input signal is suddenly removed, the flip-flop output retains the same value it had after the last input signal. Because of this, you can say it has memory, too.

A flip-flop can exist in many forms. The most common form uses transistors. Modern computers use tunnel diodes, metal oxide silicon (MOS) field-effect transistors, and integrated circuits as memory to store data. Each of these has some different advantage over the others, such as high speed, low cost, or small size.

ARITHMETIC (LOGIC)

The arithmetic element of a digital computer has the ability to do comparisons, make decisions, add, and subtract. The computer does these things by rapidly counting and adding electrical pulses with specially designed *logic circuits*.

CONTROL

In order to perform its operations automatically, a computer must be *programmed*. A *computer program* is a series of instructions for the computer to follow. The program is

Fig. 43–5. A typical core plane containing 1,024 sixteen-bit words. The inset shows an enlarged view of the cores with wires interleaved. Each circular core is one memory bit. (*Hewlett-Packard Co.*)

stored in the computer in the same way as the information it controls. The program can cause the computer to perform operations involving the stored information. These might include adding, dividing, reading a magnetic tape, or printing on paper.

LOGIC CIRCUITS

A computer is made up of a large number of transistors, diodes, resistors, capacitors, and magnetic cores. Many of these are connected together to form what are known as *logic circuits*. These circuits, or *gates* as they are sometimes called, are the main building blocks of a digital computer. They are able to perform mathematical computations and to change these computations into usable information. The three basic kinds of logic circuits are the AND, the OR, and the NOT circuits.

Logic circuits are actually switching circuits. They are used to provide an output that depends on the input conditions. When certain inputs are given to a logic circuit, it has the ability to decide what the output of the circuit is to be.

The output of a logic circuit is one of two conditions, on or off. In computer work, an "on" condition most often stands for the number 1. An "off" condition usually stands for 0.

AND Circuit. The basic AND logic circuit is shown in Fig. 43–6. Here, switches A and B (the input devices) are used with a battery and a lamp (the output device).

Since both switches A and B are in series, both A *and* B must be closed before the lamp C (output) will light. Assume that the open (off) condition of a switch stands for 0 and that the closed (on) condition of a switch stands for 1. Likewise, assume that the unlighted condition of the lamp stands for 0 and that the lighted (output) condition of the lamp stands for 1. You can then, by means of a truth table, describe the operation of the circuit in terms of its input-output relationships (Fig. 43–6). For each set of inputs A-B, there is an output C listed. From the truth table, it can be seen that the AND circuit is completed and the output is 1 if and only if both inputs are 1. If either input or both inputs are 0, then the output is 0.

The relationship between the inputs and output of an AND circuit can be expressed mathematically by $C = A \cdot B$, or $C = AB$. The dot represents AND. When no symbol appears, AND is assumed. A special algebra, known as *Boolean algebra*, is used to analyze computer logic. Boolean algebra is named after George Boole, an English mathematician and

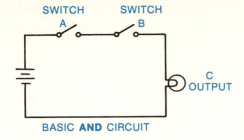

SWITCH A   SWITCH B

C OUTPUT

BASIC **AND** CIRCUIT

Truth Table

Inputs		Outputs
A	B	C
0	0	0
0	1	0
1	1	1
1	0	0

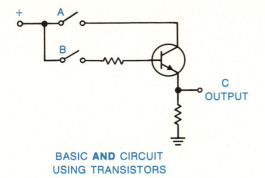

BASIC **AND** CIRCUIT
USING TRANSISTORS

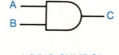

LOGIC SYMBOL
FOR **AND** CIRCUIT

Fig. 43–6. The AND logic circuit

logician. His theories on logic are used in computers where the mathematics is restricted to the two quantities 0 and 1. The logic symbol and basic wiring for the AND circuit are shown in Fig. 43–6.

OR Circuit. The input devices of the basic OR circuit are two or more switches connected in parallel (Fig. 43–7). In this case, the lamp will light if *either* switch A or switch B is closed and if both A *and* B are closed. The truth table shows that the OR circuit has a 1 output if either or both inputs are 1. The output is 0 only if both inputs are 0. This is expressed mathematically as $C = A + B$. The output C is 1 when either A or B is 1. The logic symbol and basic wiring for the OR circuit are shown in Fig. 43–7.

NOT Circuit. The NOT circuit is also called an *inverter*. Its main function is to produce *signal inversion*. This means that it will produce an output signal that is the reverse of the input signal.

Figure 43–8 illustrates the NOT circuit. When switch A is closed, the relay is energized and the normally closed (NC) contacts are opened. Likewise, when switch A is open, the relay contacts are closed. The truth table shows that if the input is 1, the output is 0, and vice versa. When inverted quantities are required, the NOT circuit is used. The NOT circuit is expressed as $B = \bar{A}$. The bar over the symbol indi-

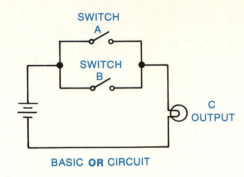

SWITCH
A

SWITCH
B

C
OUTPUT

BASIC **OR** CIRCUIT

Truth Table

Inputs		Outputs
A	**B**	**C**
0	0	0
0	1	1
1	1	1
1	0	1

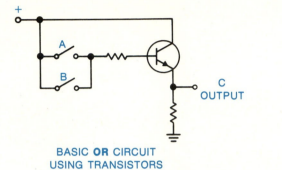

BASIC **OR** CIRCUIT
USING TRANSISTORS

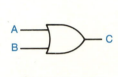

LOGIC SYMBOL
FOR **OR** CIRCUIT

Fig. 43–7. The OR logic circuit

cates an inversion, or complement, of the quantity. The logic symbol and basic wiring for the NOT circuit are shown in Fig. 43–8.

LOGIC CIRCUIT DESIGN

Suppose a circuit with the following characteristics is needed: output C is 1 if inputs A and B are both 1; C is 1 if inputs A and B are both 0; C is 0 for all other inputs. Expressed mathematically, these statements are

(1) $C = 1$ if $A = B = 1$
(2) $C = 1$ if $A = B = 0$
(3) $C = 0$ if otherwise

To design a logic circuit with these characteristics, each requirement is first fulfilled individually. They are then combined to give the output. Requirement (1) can be fulfilled with a simple AND, $C = A \cdot B$, as shown in Fig. 43–9. Requirement (2) takes a circuit that is 1 when both inputs are 0. This can be done with an AND circuit, but with the inputs inverted. If both inputs are 0, then the inputs to the AND circuit are 1, and hence the output is 1.

These two circuits can be fed into the OR circuit so that either circuit gives an output when the right inputs are present. This circuit arrangement is called a *combinatorial circuit.*

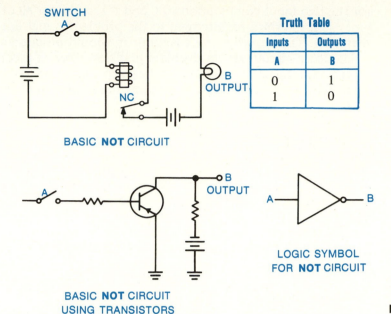

Truth Table

Inputs	Outputs
A	B
0	1
1	0

BASIC **NOT** CIRCUIT

B OUTPUT

A — B

LOGIC SYMBOL
FOR **NOT** CIRCUIT

BASIC **NOT** CIRCUIT
USING TRANSISTORS

Fig. 43–8. The NOT logic circuit

A table for the first part of the problem can be made. First all the possible combinations of inputs are listed. With two inputs, there are four combinations. For each set of inputs, the output C is determined using the rules for the logic circuits used. With a specific set of inputs, intermediate logic values can be determined, and the output value is clear. This output value is listed in the truth table. A new set of inputs is then determined. This is done for all combinations. The final truth table is shown in Fig. 43–9. Note that the output C is 1

Fig. 43–9. Steps for designing a logic circuit: (1) the AND circuit; (2) the AND circuit, with inputs inverted using NOT circuits; (3) combinatorial circuit; (4) truth table for the conditions of the problem

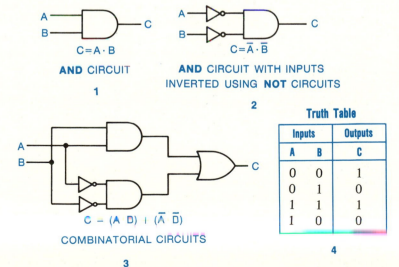

$C = A \cdot B$

AND CIRCUIT

1

$C = \overline{A} \cdot \overline{B}$

AND CIRCUIT WITH INPUTS
INVERTED USING **NOT** CIRCUITS

2

$C = (A \cdot B) + (\overline{A} \cdot \overline{B})$

COMBINATORIAL CIRCUITS

3

Truth Table

Inputs		Outputs
A	B	C
0	0	1
0	1	0
1	1	1
1	0	0

4

for the two input combinations 0-0 and 1-1, and 0 for all others. Also, the third requirement in the problem is met automatically. This is often the case, but it should be checked.

The logic-circuit design problem just described is a common circuit found in computers. It has a special name: *coincidence circuit*. Its output is 1 if its inputs are *coincident*, that is, both 1 or both 0.

BINARY NUMBERING SYSTEM

The logic circuits discussed above are mathematical. It is possible to make logical decisions using these circuits. A machine designed for this purpose, such as a computer, becomes an efficient tool in problem solving. However, a *decimal* number system, one with 10 numbers, is the mathematical system used today. Such a system will not work directly in a digital device such as a computer. Therefore, a binary numbering system is used. This mathematical system is ideal for use in digital machines.

The binary system consists of two digits, 0 and 1. In electronic terms, a *binary condition* is one in which a switching device is on or off, conducting or nonconducting, energized or deenergized. Thus, the logic circuits discussed before are binary as to their output conditions. These conditions lend themselves to the binary numbering system.

Any numbering system can be identified by its *base*, or radix. The base is the fundamental number of a system of numbers. The decimal system, for example, uses a base of 10. This numbering system has 10 different digits, 0 through 9. The binary system uses a base of 2 with two digits, 0 and 1.

In both these numbering systems, the position of the number is important. In the decimal system, the number 10 is quite different from 01. Each position in a multidigit number stands for a power of the base. The powers of 10 are $10^0 = 1$, $10^1 = 10$, $10^2 = 100$, $10^3 = 1,000$, and so on. We commonly call these positions (reading from right to left) *units, tens, hundreds, thousands,* and so on. The powers of 2 are $2^0 = 1$, $2^1 = 2$, $2^2 = 4$, $2^3 = 8$, and so on. A rule to remember is that any number raised to the zero power equals 1. For example, $2^0 = 1$, $8^0 = 1$, $10^0 = 1$. Another rule is that any number raised to the first power is the number itself. For example, $2^1 = 2$, $8^1 = 8$, $10^1 = 10$. The decimal system can express any quantity with ten digits. The binary system can also express any quantity with the digits 0 and 1. For example, the decimal number 2 is expressed in the binary system as 10. This number is read as *binary one-zero*, not ten. The relationship between a decimal number and its binary equivalent number is shown in Fig. 43–10.

Decimal Numbers	Decimal Value of Binary Digit					
	2^5 (32)	2^4 (16)	2^3 (8)	2^2 (4)	2^1 (2)	2^0 (1)
0	0	0	0	0	0	0
1	0	0	0	0	0	1
2	0	0	0	0	1	0
3	0	0	0	0	1	1
4	0	0	0	1	0	0
5	0	0	0	1	0	1
6	0	0	0	1	1	0
7	0	0	0	1	1	1
8	0	0	1	0	0	0
9	0	0	1	0	0	1
10	0	0	1	0	1	0
20	0	1	0	1	0	0
30	0	1	1	1	1	0
40	1	0	1	0	0	0
45	1	0	1	1	0	1
50	1	1	0	0	1	0
57	1	1	1	0	0	1

Fig. 43–10. Relationship of decimal numbers to binary numbers

ADVANTAGES OF THE BINARY CODE

To represent electronically any number using digits of the decimal system would require 10 different circuit conditions, each of which would stand for a certain digit. With the binary code, any number (or letter of the alphabet, when encoded into bits) can be expressed with only two circuit conditions. This, in turn, makes it possible to simplify the logic-circuit systems used in electronic counters and in computers.

COMPUTER APPLICATIONS

We have seen how it is possible to "communicate" with the computer through binary symbols. These symbols form a *computer language*. This language makes it possible to program the computer easily to meet specific needs. A *computer program* is a set of instructions expressed in computer language. Different programs can cause the same computer to do different things.

The ordinary electronic calculator provides a good illustration of this new technology. The electronic circuitry performs operations on given inputs to obtain required outputs. The central processing unit (CPU) is the control circuit for the calculator. This keeps the system working in harmony as it responds to the step-by-step instructions of the computer

Fig. 43–11. An electronic calculator (*Hewlett-Packard Co.*)

program. The CPU is a digital integrated circuit called a *microprocessor*. The keys pressed on the keyboard provide the instructions for the microprocessor. The electronic calculator is a complete digital *microcomputer* (Fig. 43–11). It is so called because of its small size, *micro-* meaning "small."

In the future, computers will control the production of more and more goods in the industries of the United States. Office work is changing rapidly because of the *word processing machine*. This machine makes it possible to type rough drafts onto an easy-to-read televisionlike screen. The typist need not worry about typing errors, erasing, or starting the page over. Corrections can be easily made on the television-like screen. Once the draft is correct, a printer can produce the final copy at more than 600 lines per minute.

Electronic data processing by computer is even used in "fast food" restaurants. No longer do the attendants write your food order on a piece of paper. As you state your order, they just press keys on a typewriterlike machine. When the transaction has been completed, you receive a printed paper tape. At the same time, the machine automatically stores the information electronically for future use. This information can be retrieved easily by managers for reordering or accounting purposes.

SELF-TEST

Test your knowledge by writing, on a separate sheet of paper, the word or words that most correctly complete the following statements:

1. The letters ADP mean _____.
2. The two basic kinds of computers are _____ and _____.
3. A fuel-level circuit is a simple form of an _____ measuring system.
4. Stored data in a computer can be in the form of _____.
5. Three basic logic circuits are the _____, _____, and _____ circuits.
6. A logic circuit can have one of two states, _____ or _____.
7. The _____ logic circuit is similar in function to two parallel switches.
8. A circuit that produces an inverted form of the input signal is called a _____ circuit.

FOR REVIEW AND DISCUSSION

1. Give some reasons for the increased use of computers by business, industry, and government.
2. List at least four different operations that can be performed by a computer from instructions stored in a computer program.
3. Define a logic circuit.
4. Describe the truth table for an AND logic circuit.
5. Draw a schematic diagram of an AND circuit.
6. Draw the symbols for the AND, OR, and NOT logic circuits.
7. Explain why binary numbers are used in computers.

INDIVIDUAL-STUDY ACTIVITY

Visit a local company that makes use of automatic data processing. Tell your class what you have learned.

Electricity and Industry

Unit 44   The Electric Power Industry

The first electric power plant was built in New York City in 1882 under the direction of Thomas Alva Edison. Early electric power plants were small and generated direct current. One of the main reasons that small local plants were used at that time is that a direct-current output cannot be transmitted economically over long distances. When direct current is so transmitted, much energy is lost in the conductors in the form of heat. The use of alternating current, which does not have such a large power loss, made large central power plants economical. Today huge electric power plants have almost completely replaced smaller plants.

Modern power plants are of three main kinds: (1) hydroelectric plants, (2) steam-electric palnts, and (3) nuclear plants. All power plants use large alternating-voltage generators, or alternators, to produce electric energy. This energy is delivered to users by transmission networks. The generators are usually driven by turbines, often called *prime movers.* These turbines are turned by the energy of moving water or by steam pressure. The basic principle of a simple turbine is shown in Fig. 44–1. Smaller generators are usually driven by gasoline or diesel engines.

HYDROELECTRIC POWER PLANTS

In a hydroelectric power plant, water from a river, a lake, or a reservoir created by a dam is directed against the blades of a turbine. The water pressure causes the shaft of the turbine to rotate. This shaft turns the rotor of a generator (Fig. 44–2).

A number of hydroelectric power plants in the United States are operated by the federal government. The dams

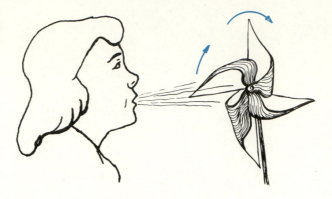

Fig. 44-1. Principle of a turbine. Air pressure forces the vanes (blades) of the pinwheel (rotor) to rotate.

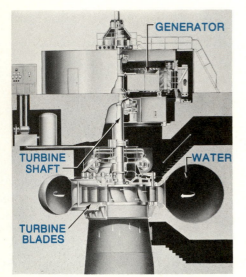

Fig. 44-2. A vertical hydroelectric generator

built in connection with these plants are also important in flood-control and navigation (Fig. 44-3).

STEAM-ELECTRIC POWER PLANTS

Steam-electric plants were the earliest kind of power plant and are still the most common. In these plants, water is heated by coal, oil, or natural gas in a boiler to create steam (Fig. 44-4). In some electric plants, the steam is heated to 1000°F (538°C) and is under a pressure of 3,600 pounds per square inch (24 800 kilopascals). This steam is directed at the turbine rotor (Fig. 44-5). The velocity of the steam is converted into rotary motion by a series of blades or buckets on the turbine rotor. After the energy of the steam is used up in the turbine, the steam is condensed into water and recycled to the boiler. This process is continuous. The generator turns at a constant speed even though the load requirements on

Fig. 44-3. Aerial view of Shasta Dam and Shasta Lake—a large hydroelectric power plant near Redding, California (*U.S. Bureau of Reclamation*)

COAL
PILE

Fig. 44—4. A large steam-electric plant

the generator change. As more electric energy is needed, more fuel is burned. More steam is thus made available to be converted into mechanical energy and then into electric energy. This process is an *energy conversion system.* Some high-efficiency steam-electric plants use as little as ¾ pound (0.34 kilogram) of coal to generate 1 kilowatt hour of electric energy.

NUCLEAR POWER PLANTS

In a nuclear power plant, the fuel used is uranium. An atomic particle, the neutron, strikes the nucleus of a uranium atom and splits it about in half. This process is known as *atomic fission.* Fission causes the release of a large amount of energy and more neutrons. These neutrons split more uranium and more energy and neutrons are released, and so on. The result is a *nuclear chain reaction.* In an atomic bomb, this chain reaction releases tremendous amounts of heat energy almost instantly. However, it can be controlled so as to release energy slowly.

In a nuclear power plant, a controlled amount of heat is produced by nuclear chain reactions in what is called a *reactor.* This heat boils water, and the steam produced is used to operate turbines. These turbines are similar to those in a steam-electric power plant.

The main elements of a modern nuclear power plant are shown in Fig. 44–6. Note that in this kind of atomic electric plant, the water is cycled directly from the main circulating pump to the reactor vessel. The reactor is inside the reactor

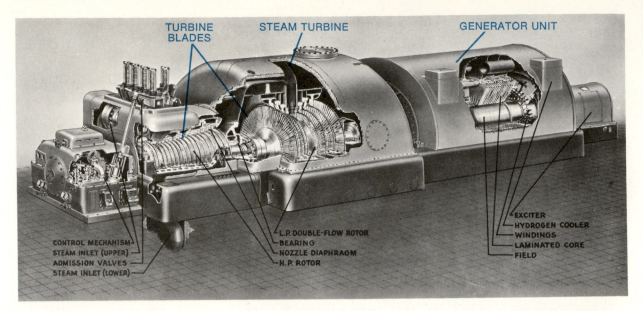

TURBINE BLADES   STEAM TURBINE   GENERATOR UNIT

EXCITER
HYDROGEN COOLER
WINDINGS
LAMINATED CORE
FIELD

L.P. DOUBLE-FLOW ROTOR
BEARING
NOZZLE DIAPHRAGM
H.P. ROTOR

CONTROL MECHANISM
STEAM INLET (UPPER)
ADMISSION VALVES
STEAM INLET (LOWER)

Fig. 44–5. A large steam turbine-generator unit. The turbine blades are mounted on the high-pressure (HP) rotor and the low-pressure (LP) rotor. (*General Electric Co.*)

Fig. 44–6. The main elements of a nuclear-electric plant. This is a direct-cycle, boiling-water plant.

vessel. The water surrounding the reactor is heated to about 600°F (316°C) and is changed into steam at a pressure of 1,000 pounds per square inch (6900 kilopascals). The water used is demineralized. It thus helps slow down the neutrons, making them hit large numbers of atoms. A recirculating pump also helps regulate the fission. It does this by reducing the number of bubbles occurring in the water as the steam is produced. The fewer the bubbles, the greater the rate of the atomic reaction.

Reactor and Reactor Vessel. The reactor vessel in some plants is 65 ft (20 m) long and weighs nearly 700 tons (635

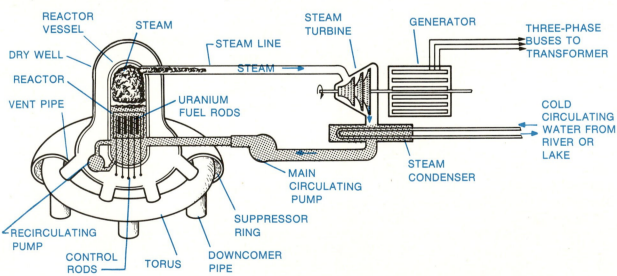

REACTOR VESSEL   STEAM   STEAM TURBINE   GENERATOR   THREE-PHASE BUSES TO TRANSFORMER

DRY WELL   STEAM LINE

REACTOR   STEAM

VENT PIPE   URANIUM FUEL RODS

COLD CIRCULATING WATER FROM RIVER OR LAKE

STEAM CONDENSER

MAIN CIRCULATING PUMP

RECIRCULATING PUMP

SUPPRESSOR RING

CONTROL RODS   TORUS   DOWNCOMER PIPE

metric tons). It is supported in an upright position and is partly filled with water. The reactor contains the uranium fuel rods. The fuel rods throw off neutrons. During this process, some of the atoms split up, giving off heat energy.

Other rods of neutron-absorbing material are used to control the rate of the chain reaction in the reactor. The control rods are placed between the fuel rods. By moving the control rods in and out of the reactor, the amount of heat can be regulated. By controlling the heat from 115 tons (104 metric tons) of uranium in the core of the reactor, a plant can produce electric power equal to that obtained from 6 million tons (5.4 million metric tons) of coal.

Steam Generator. The steam generated at the top of the reactor vessel is directed through pipes at the turbine blades, making the blades spin. The turbine shaft turns the generator. In some modern plants, the generators convert over 800,000 hp (597,000 kW) into electricity. When the electricity is produced, it is transformed and distributed over regular transmission lines.

Steam Condenser. After the steam is used in the turbine, it is condensed to water in a *steam condenser*. There, large amounts of cold water are used to speed the condensation of the steam to water. The water is then recycled through the system by the main circulating pumps. Since this plant converts into steam the water that surrounds the reactor, it is called a *direct-cycle boiling-water plant.* Because of the large amounts of cold water needed by the condenser, plants that use steam turbines are usually found near large rivers or lakes. Some condensers use 250,000 gallons of water per minute (15.8 cubic meters per second) to condense the steam from the turbines.

Safety. Many safeguards are used in nuclear power plants. For example, in Fig. 44–6, a dry well surrounds the entire reactor vessel. Should a break occur in the reactor vessel, the steam and water would rush out into the dry well. The steam and water would then go through the vent pipes into the *torus,* the doughnut-shaped base of the reactor vessel, and the downcomer pipes. These pipes are placed around the torus with their openings under water. The water helps condense the steam, thus reducing the pressure. The large suppressor ring that surrounds the dry well has a chamber that acts as a cushion for any quick change in pressure. This protects the entire system from damage. If water is removed from the reactor vessel, the reactor will automatically shut down.

One of the major advantages of nuclear power plants is that they are operated with nuclear energy. This is available in practically unlimited quantities. The utilization of this energy, together with a wise use of coal, oil, and natural gas, can help meet energy needs in the future. However, safety factors are very important in the use of nuclear energy.

Nuclear power plants are massive and complex structures. They are designed to withstand earthquakes and tornados. They have many special safety devices and systems that help protect the plant, the plant personnel, and the public in the event of serious accident. However, in spite of these precautions, an accident did occur on March 28, 1979, at the Three Mile Island nuclear power plant. This plant is located near Middletown, Pennsylvania, on the Susquehanna River. The accident raised questions in the minds of many people about the use of nuclear energy to generate electricity.

It is important, therefore, that all citizens be fully informed about nuclear power plants. Learn about their safety devices. Discuss the methods used to treat and dispose of the wastes from nuclear plants. Find out what programs are used to train the people who operate the plants. Only when you know and understand the facts and risks involved, can you help make intelligent decisions about the future of nuclear power. Nuclear power plants in the United States are operated under the general direction of the *Nuclear Regulatory Commission* (*NRC*).

PRODUCTION OF ELECTRIC ENERGY

In 1979, coal-fired electric power plants produced 47.8 percent of the total amount of electricity used in the United States. Oil-fired plants produced 13.5 percent; gas-fired plants, 14.7 percent; hydroelectric plants, 12.4 percent; nuclear plants, 11.4 percent; and all other sources, 0.2 percent.

TRANSMISSION SYSTEM

To carry electric energy from power plants to the places where it will be used takes large transmission and distribution systems (Fig. 44–7). The main parts of these systems are discussed below.

Transmission Lines. Electric energy is transmitted from power plants by overhead wires supported by high towers. The center strands of these wires are made of steel to give them strength. The outer strands are made of aluminum because of its lightness and ability to carry current. The wires

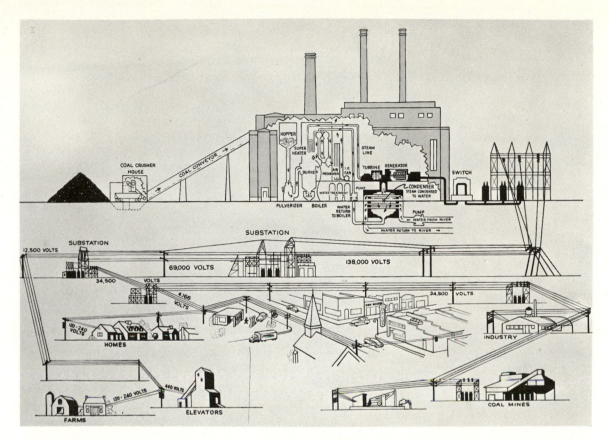

Labels within figure: COAL CRUSHER HOUSE, COAL CONVEYOR, HOPPER, SUPER HEATER, STEAM LINE, BURNER, AIR PREHEATER, I.D. FAN, TURBINE, GENERATOR, SWITCH, CONDENSER STEAM CONDENSED TO WATER, PUMP, PULVERIZER, BOILER, WATER TREATMENT, WATER RETURN TO BOILER, PUMP, WATER FROM RIVER, WATER RETURN TO RIVER, SUBSTATION, SUBSTATION, 12,500 VOLTS, 69,000 VOLTS, 138,000 VOLTS, 34,500 VOLTS, 4,166 VOLTS, 34,500 VOLTS, 120-240 VOLTS, HOMES, DRY GOODS, INDUSTRY, 120-240 VOLTS, 440 VOLTS, FARMS, ELEVATORS, COAL MINES

Fig. 44–7. How electric energy is produced and delivered

are insulated from the towers by porcelain insulators to prevent the loss of electric energy. Figure 44–8 shows line workers stringing wires for a high-voltage (250,000-volt) transmission line. In some cases, especially in cities, wires run underground in ducts.

Transformers. In an alternating-current system energy loss is reduced. This is so because the transformer makes it possible to raise the ac voltage to very high values. Some transmission systems today are using 750,000 volts. A higher voltage allows the same level of electric energy to be made available at a lower current. This results in a lower energy loss and higher efficiency. Transformers make it possible to transmit electric energy economically over distances of 200 to 300 miles (322 to 483 kilometers). In addition to stepping up, or raising, the voltage for long-distance transmission, transformers step down, or lower, the voltage to the requirements of the load.

Figure 44–9 shows two transformers connected in parallel being fed by isolated *buses,* or heavy wires, directly from a three-phase alternator. Each of the three phases of the alter-

Fig. 44–7. How electric energy is produced and delivered

Fig. 44–8. High-voltage transmission tower: (a) high-voltage insulators; (b) conductors

nator is isolated from the other. Each bus is separated by insulators in the center of a pipelike protector. The internal wiring connections are shown in Fig. 44–10. The high-voltage three-phase leads can be seen at the top of Fig. 44–9. These leads go directly to a substation.

Fig. 44–9. Two transformers connected in parallel (*Potomac Electric Power Co.*)

THREE–PHASE
HIGH–VOLTAGE
TRANSMISSION
LINES

BUSES FROM
GENERATOR

TRANSFORMER

Substation and Distribution Networks. Figure 44–11 shows a *substation,* or distribution center. Substations step down the voltage for industrial and home use. They distribute electric energy conveniently to various loads. They make it easy to isolate any trouble in the distribution system.

Circuit Breakers. A circuit breaker is an automatic electric switch. It opens by itself whenever an overload causes excessive current. This protects the transmission lines and other parts of the circuit. Because of the high currents and voltages involved (as high as 34.5 kV and 1,200 A), the circuit breakers are placed in oil. Smaller circuit breakers operate in air. They

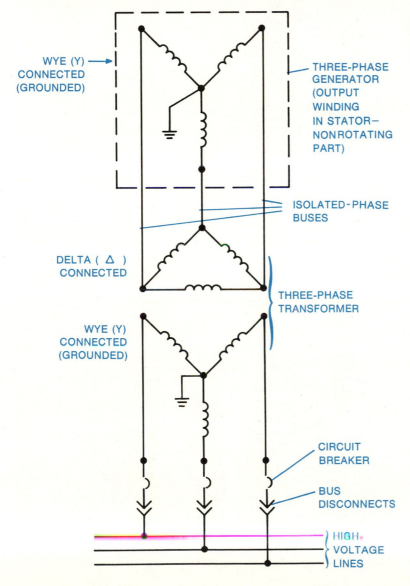

WYE (Y)
CONNECTED
(GROUNDED)

THREE-PHASE
GENERATOR
(OUTPUT
WINDING
IN STATOR—
NONROTATING
PART)

ISOLATED-PHASE
BUSES

DELTA (Δ)
CONNECTED

THREE-PHASE
TRANSFORMER

WYE (Y)
CONNECTED
(GROUNDED)

CIRCUIT
BREAKER

BUS
DISCONNECTS

HIGH
VOLTAGE
LINES

Fig. 44–10. Internal wiring connections from a three-phase generator to a three-phase transformer and transmission lines

Fig. 44–11. Substation (*General Electric Co.*)

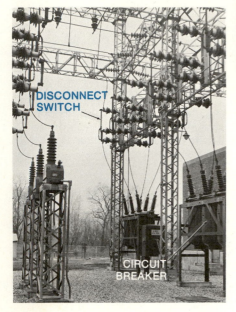

DISCONNECT SWITCH

CIRCUIT BREAKER

Fig. 44–12. Close-up view of sub-station switchgear

are called *air circuit breakers.* Figure 44–12 shows two circuit breakers installed in a substation. Above the circuit breakers is a disconnect switch. Maintenance personnel can open this switch from the ground and thereby completely disconnect the load from the source. This allows work to be done on the system safely.

ORGANIZATION

An electric power company has many departments and working groups. These provide efficient service, maintain facilities, and plan the improvements and expansions needed to satisfy the demand for electric energy.

SELF-TEST

Test your knowledge by writing, on a separate sheet of paper, the word or words that most correctly complete the following statements:

1. The first electric power plant was built in the year _____ .
2. The three main kinds of electric power plants are _____ , _____ , and _____ .
3. The rods that regulate the chain reaction in a nuclear plant are called _____ rods.
4. Voltages as high as _____ volts have been used in transmission systems.

FOR REVIEW AND DISCUSSION

1. Explain why the dc power plants were generally small and provided only local electric service.
2. Explain why turbines are called prime movers.
3. What is meant by the term *energy-conversion system*?
4. Describe the main elements of a nuclear power plant.
5. List the main parts of a transmission and distribution system for electric energy.
6. What are three functions of a substation?

Unit 45   Research and Development in Electronics

The term *research* means "a thorough study or investigation of some topic or topics." In industry, one main purpose of research is the improvement of existing products and processes. The other is the *development* of new ones that are useful. New products and processes are most often made possible because of discoveries made by experimentation during research. Thus, research and development, or R & D, are closely related.

In the field of electronics, many areas are constantly being researched in an effort to improve and expand technology. These include studies of molecular electronics, solid-state physics, bio (life) functions, space exploration, data processing, and magnetics.

RESEARCH AND ENGINEERING

Research and engineering personnel work closely together. The engineers develop the ideas provided by research scientists into usable products and systems. An example of a product system that is the result of close cooperation among scientists, engineers, and technicians is shown in Fig. 45-1.

Research scientists and engineers are also responsible for tests and evaluations done on each piece of equipment or

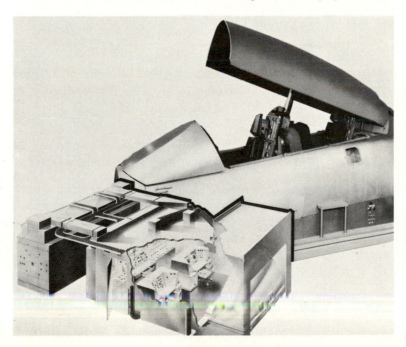

Fig. 45-1. Flight simulator for a jet aircraft (*Melpar*)

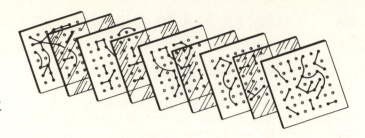

Fig. 45-2. Lay-up of a multilayer, etched, printed-circuit board prior to bonding (*Melpar*)

material developed. Environmental, electrical, mechanical, and physical-chemical tests are made. These provide data on the quality, performance, and dependability of a product.

DEVELOPING A NEW PRODUCT

The multilayer circuit board shows how the research and development process results in a specific product. This circuit board is made by compressing interconnected printed circuit boards together so that they take up as little space as possible (Fig. 45-2).

There are several steps in producing a new and complex product of high quality. First, the materials and processes that will be involved must be determined. Second, their *parameters* (ranges of limits and capabilities) and characteristics must be defined and measured. These steps provide information about the effects of the environment on the material. This information relates to temperature, shock, vibration, humidity, fastening methods, warp and twist, and so on. The third step is the setting up of effective and economical control procedures. The knowledge about the product gained through these steps is used to modify materials and processes as needed. The flow diagram in Fig. 45-3 shows the processes in the development of a new material or product and their interrelationships.

ENGINEERING SERVICES AND MANUFACTURING

The main function of a research and development company is to design and develop new products and/or materials. However, such a company needs a *manufacturing division* to support its other activities. The manufacturing division develops experimental models, prototype equipment, and fabrication and assembly procedures.

Engineering services are directed largely to marketing and sales. Engineering field services are provided in places where many of a company's products are being used. Engineering services include providing test equipment, determining spare parts availability, and training customer personnel.

DEVELOPING AN ELECTRONIC DEVICE

Figure 45–4 shows six main steps in developing an idea into a practical electronic device. An idea conceived in the mind of someone is the first step. Ideas for developing new products or materials come from two main sources: (1) from within the company, its scientists, engineers, management personnel, and others, and (2) from a customer who has a need. A customer may have a problem and ask for help to solve it. This idea or need is usually expressed in a written report or proposal. This gives specifications of the new product or material and an estimate of the time needed to produce it. The research and development company and the customer come to an agreement about the factors and costs involved. Then the work begins.

Exploring Ideas. Research and development generally involves solving new problems. Thus, much exploratory work

Fig. 45–3. Flow diagram of the various processes and their interrelationships in the development of a new material for the electronic industries (*Melpar*)

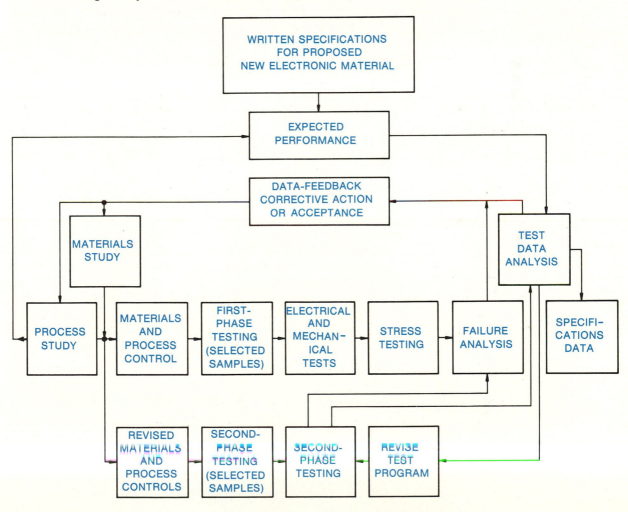

is done first. This exploratory work is the second step in the research and development process. Its main purpose is to find out if an idea is practical. In the electronic industry, the exploratory process is called *breadboard development.* This is a design and development process in which scientists and engineers test theories. They want to find out whether their ideas work electronically. Rough pencil sketches are used to record ideas about new circuits. Technical notes are written down to record the results. It may take 6 to 8 months before a satisfactory circuit is found and accepted. At this point, a customer may want to order two or three deliverable *prototypes,* or units.

Deliverable Prototypes. The engineering development process of making a deliverable prototype is the third step. The deliverable prototype is developed from the ideas found

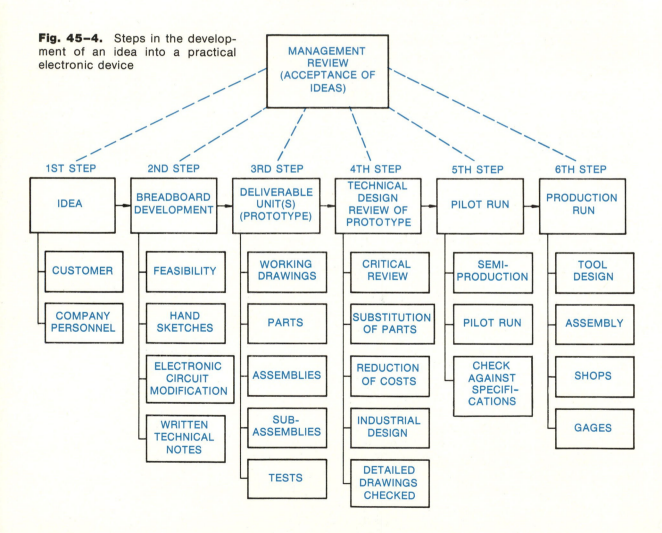

Fig. 45-4. Steps in the development of an idea into a practical electronic device

to be practical during the breadboard development. During this step, working drawings of the parts are made. These include the specifications and the designs for supporting parts, assemblies, and subassemblies. Tests are done during the development of a deliverable prototype to make sure that it continues to work and to improve the way it works.

Reviews. The fourth step in developing an electronic device from an idea is the *technical design review*. This involves taking a critical look at the whole process up to this point. Representatives from various divisions—such as manufacturing, engineering, purchasing, quality control and reliability, technical writing, and others—ask questions about the units already developed. This review is often called *value engineering*. This is a team approach to solving technological problems. Value engineering involves attempts to eliminate unneeded parts and to substitute new, lower-cost materials where possible. At this point, detailed drawings are reviewed. Any remaining mechanical and electrical design problems are solved.

Pilot Run. The fifth step is the *pilot run*. The pilot run is done to check a small number of the devices against the original specifications. The pilot run also provides a means of checking all tools and fixtures to be used in the manufacturing process. During the pilot run, the *process sheets* are brought up to date. These explain in detail the operations that must be performed to complete the work. They include details on the time of manufacture, number of parts, sequence of operations, and assembly operations. When all parts of this step have been completed, the device is released to production.

Production. Producing the device in the quantities needed is the sixth and last step in the development process. Manufacturing activities may include sheet-metal work, machining, welding, brazing, painting, and finishing. The model and tool shop has facilities for making models and building breadboards and experimental equipment. Wiring and assembly shops do wiring for all kinds of electronic equipment, including printed circuits and modular and chassis units. Some precise work is done in *clean rooms*. These have a controlled environment in which dust and other foreign particles are kept to a minimum and a constant temperature is maintained. The manufacturing division also controls the scheduling of all shop activities and shipping.

SELF-TEST

Test your knowledge by writing, on a separate sheet of paper, the word or words that most correctly complete the following statements:

1. Research is closely related to _____ .
2. A range of limits and capabilities is a _____ .
3. Development ideas come from company personnel and from _____ .
4. In the electronics industry, the exploratory process is called _____ .
5. The first deliverable units are called _____ .
6. Another term for technical design review is _____ .
7. A _____ explains in detail the operations in making a product.
8. A manufacturing site with a controlled environment is called a _____ .

FOR REVIEW AND DISCUSSION

1. Explain the relationship between research and development.
2. Name three steps in producing a new product.
3. List the steps in developing an electronic device.
4. What are the main sources of new ideas?

INDIVIDUAL-STUDY ACTIVITIES

1. Select a common household electronic device and try to trace its development to the present time.
2. Present an idea to your class for developing or modifying an electronic device to meet your needs. Your class or some interested student may want to follow the six steps suggested in the unit to develop your idea into a practical product.

Unit 46   Product Manufacturing

The manufacturing of electronic products is one of the major industries in the United States. In contrast to electronic research and development, electronic product manufacturing is done mostly by mass production (Fig. 46–1).

An electronic manufacturing company is often independently owned, with its own stockholders. It may also be an electronic division of a larger parent company. Electronic companies often have several divisions, such as chemical, laboratory, services, and engineering divisions. Each division has its own organization and functions. Cooperation between the divisions is directed by division managers and their assistants.

DEVELOPING THE PRODUCT

The six basic steps in developing an idea into a practical electronic device are almost the same in product manufacturing as in research and development. However, in product manufacturing, the sales department is often the first to be aware of the need for manufacturing a particular product or device. This information is vital feedback data from which management makes decisions on a particular line of products to be manufactured.

Product manufacturing generally follows these seven basic steps: (1) developing finished drawings, (2) ordering raw materials and parts, (3) planning and controlling the manufacturing process, (4) fabricating the parts, (5) assembling the parts, (6) testing the parts (assemblies or subassemblies) against specifications, and (7) inspecting and shipping the final product.

Two important forms have been developed to control product manufacturing. These are the manufacturing process sheets and the manufacturing process specifications.

Manufacturing Process Sheets. Figure 46–2 shows a manufacturing process sheet. Note that the sheet has much information about the part, such as name, number, standard time of manufacture, and routing.

The *standard time* is the actual time it takes to do an operation. The times for operations are estimated before any parts are manufactured. Later, these estimates are checked against the pilot run and corrected as necessary.

Manufacturing Process Specifications. These specifications describe briefly the steps to be followed during the

Fig. 46–1. Completing the wiring of motor armatures in a manufacturing plant (*Ametek, Inc.*)

Fig. 46–2. A manufacturing process sheet

MANUFACTURING PROCESS SHEET
(Sheet ____ of ____ Sheets)

Employee # _____

Raw Material Code _____

Part Number _____, Part Name _____

Job # _____ Sub-assembly _____ Quantity _____

Location		Tool	Routing	Standard Time			Parts Per Hour	Materials	Labor	Over-time Hours	Total
Operation Number	Group			Standard Unit	Hours/Unit	Total					
					Standard Cost Total ➤					

Processed by _____ Date _____

435

Fig. 46–3. Manufacturing process specifications for printed-circuit boards: dip-soldering operation

I. Purpose

To outline a process for dip soldering printed-circuit boards.

NOTE: Dip solder is used primarily for boards over 3 1/2" in width and for boards with standoffs or terminal lugs on the circuit side. Boards with heavy land areas are better suited for wave soldering and, whenever possible, should be wave soldered.

II. Equipment

Dip-soldering machine

Soldering flux formula no. 22

Holding fixtures (4 available — 2 for boards of 1 3/4" to 2 3/4" and 2 for boards 3" to 4 1/2"

63-37 solder

Flux remover type 6x

Cleaner

Thermometer

Scraper (stainless steel with plastic handle for removing dross)

III. Preparation for dip soldering

A. Turn thermostat knob on soldering pot to no. 5 (setting of 500°). This setting will heat the solder from 440 to 470°F. Caution: This is the correct temperature for dip soldering and should be checked with a thermometer to verify prior to soldering.

B. After checking the solder temperature, jog the sweep arm of the dip-soldering machine into a perpendicular position. Attach a holding fixture set to the correct width of the board to be soldered (or a dummy board of the exact width). Adjust the holding fixture to the correct height, without the board. The top of the solder should be even with the bottom of the notch in the holding fixture. With a dummy board, the board should rest on the top of the solder. Caution: Do not allow the board to submerge. Turn the machine on and return the board to its original position. The holding fixture should be to the right side of the machine.

IV. Operation

A. Adjust two fixtures to the correct width.

B. Insert boards and dip in soldering flux no. 22.

C. Clean dross from top of soldering pot with the scraper.

D. Hang the fixture and press the start button. Allow to complete one cycle.

E. Remove the board from the fixture and clean with cleaner. Reclean with flux remover type 6x.

V. Touch-up and inspection

A. Return the board to bench to be checked and touched up as required.

B. After check and touch-up, return it to the ventilation booth to be spray-cleaned with cleaner.

C. Submit boards to Quality Control.

After acceptance, boards must be placed in plastic bags.

VI. Maintenance of dip-solder machine

When the solder on the boards assumes a dull, dingy, or stringy (solder skips on the land pattern) appearance, change the solder in the pot.

Approvals:

Originator   Manufacturing Engineering

Production   Quality Control

manufacture of certain assemblies (Fig. 46–3). They tell the workers exactly what equipment is used, the specific operations in the process, and the maintenance needed to ensure quality. Manufacturing process specifications are kept up to date as changes and improvements are made.

QUALITY CONTROL

The purpose of *quality control* is to make sure that manufactured products meet specifications. Carrying out the sequence of operations listed on the manufacturing process and specifications sheets is a major factor in quality control. This ensures that each part will be treated in the same way. The inspection department is important in quality control. It is responsible for checking tolerances, specifications, and the like on parts as they move through the plant. The final inspection often includes touch-up and testing. Quality control is considered to be the responsibility of all persons in the company rather than of one person or department. Whatever the job, it should be done well. This is the foundation of any effective quality-control program.

COMPONENT MANUFACTURING

The component manufacturer plays a key role in the electronics industry. These companies employ anywhere from twenty to several thousand people. The range of components they make is as varied as the parts of electronic equipment. Some companies specialize in one item, such as a filter. Others make a variety of items, such as terminal boards, interlock receptacles, anode connections, test prods, and many others.

Several elements are common to the organization of all companies, whatever their sizes. The product manufacturing process always includes five steps: (1) sales, (2) engineering, (3) prototype development, (4) production, and (5) shipping. These steps are quite similar in both small component manufacturers and large product manufacturers.

SELF-TEST

Test your knowledge by writing, on a separate sheet of paper, the word or words that most correctly complete the following statements:

1. In a product manufacturing company, the _____ department provides vital feedback information to management.
2. A document that contains information about a part's name, number, and routing is the _____.
3. Specific manufacturing operations can be found in the _____.
4. A system designed to make sure products meet specifications during the manufacturing process is called _____.
5. Small electronic companies that manufacture selected electronic parts are called _____ manufacturers.

FOR REVIEW AND DISCUSSION

1. Describe the major organizational difference between a product manufacturer and a research and development company.
2. List the seven basic steps in product manufacturing.
3. What five elements are common to manufacturing companies, regardless of size?

INDIVIDUAL-STUDY ACTIVITIES

1. Obtain the organization chart of a medium-size electronics company. Discuss the functions involved in manufacturing an electronic product.
2. Develop an outline of a plan to mass-produce an inexpensive electronic device for each member of your class. Your class may wish to organize itself into an electronics "company."

Testing and Troubleshooting

Unit 47   Electronic Test Instruments

In previous units of this book, you have learned about one basic test instrument, the multimeter. You have also learned about instruments used to test capacitors and transistors. Many other testing and measuring instruments are used in the fields of electricity and electronics. Three of these are the oscilloscope, the signal generator, and the signal tracer. They are used to test the performance of certain circuit systems. They are also used to adjust several kinds of circuits so that the circuits will work in the most efficient way. The operation of the signal generator involves two very important circuit functions, which are also briefly discussed in this unit. These are oscillation and modulation.

OSCILLOSCOPE

The *oscilloscope*, or *scope* as it is often called, is one of the most important test instruments. It is used in all kinds of design, servicing, and maintenance activities (Fig. 47–1). The main purpose of the oscilloscope is to show a *waveform*, or picture, of a voltage (Fig. 47–2). This makes it possible to test circuits by seeing if the waveforms of their voltages at different points are correct.

The Cathode-ray Tube. The waveforms shown by an oscilloscope are displayed on the screen of a cathode-ray tube (CRT). This is in many ways similar to the picture tube of a black-and-white television set. An electron-gun assembly in the neck, or narrow end, of the tube produces a beam of electrons, sometimes called a *cathode ray* (Fig. 47–3 at A). When

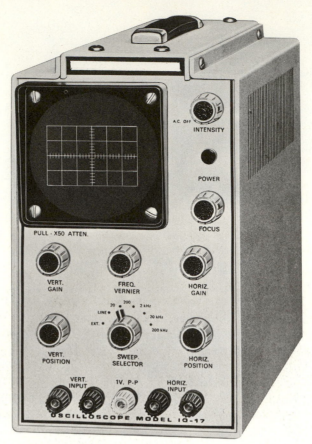

Fig. 47–1. Oscilloscope. Most of the controls (intensity, focus, and so on) seen in this photograph are found on all general-purpose oscilloscopes. (*Heath Co.*)

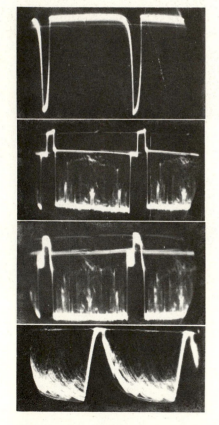

Fig. 47–2. Oscilloscope displays showing voltage waveforms present at different points of a television circuit

the electron beam strikes the phosphor coating on the screen of the tube, the phosphor *fluoresces*, or glows. If the beam is not moving horizontally or vertically, a tiny spot of light appears on the screen (Fig. 47–3 at B).

Electrostatic Deflection. The electron beam of the cathode-ray tube in an oscilloscope is moved by a process known as *electrostatic deflection.* Since the beam is made up of electrons, or negative charges, it can be made to move by the electrostatic forces of attraction and repulsion. This is done by applying voltages to the horizontal and the vertical deflection plates in the tube (see Fig. 47–3 at A). Examples of horizontal and vertical deflection of the beam are shown in Fig. 47–4 at A and B. If one horizontal and one vertical deflection plate are made equally positive or negative at the same time, the beam will move across the screen diagonally (Fig. 47–4 at C).

Sweep Circuit. When the vertical gain control of an oscilloscope is turned down fully and the horizontal gain control is

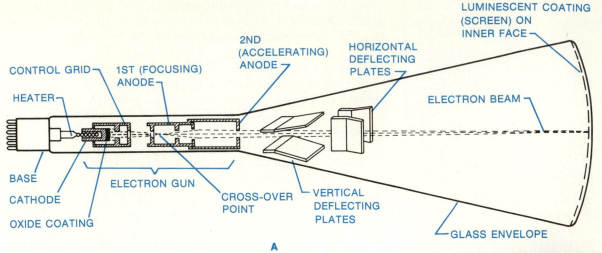

A

turned up, the electron beam sweeps, or moves, across the cathode-ray tube in a horizontal direction. The beam spot moves so quickly that, because of a characteristic of our eyes known as *persistence of vision,* it appears as an unbroken line (Fig. 47–5).

The horizontal sweep of the beam is produced by an oscilloscope circuit commonly called a *sweep generator.* This circuit generates a sawtooth-shaped output voltage. When this voltage is applied to the horizontal deflection plates, the beam is moved across the screen from left to right. It is then very quickly brought back to the starting position. The speed at which this action is repeated is determined by the frequency of the sweep generator. The frequency of the generator is varied by adjusting the time base, sweep frequency, or similarly named control of the oscilloscope.

SIGNAL GENERATOR

A *signal generator* is a test instrument used to supply output voltages of different frequencies (Fig. 47–6). These voltages,

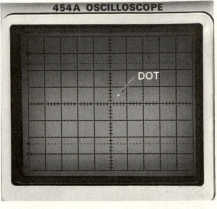

B

Fig. 47–3. The oscilloscope cathode-ray electron tube: (A) tube construction; (b) tiny spot, or dot, of light appearing upon the screen of a cathode-ray tube (*photo courtesy of Tektronix, Inc.*)

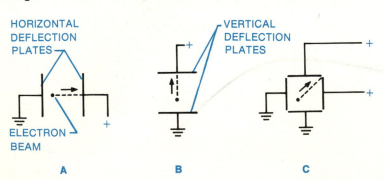

A   **B**   **C**

Fig. 47–4. Electrostatic deflection of the electron beam in a cathode-ray tube: (A) horizontal; (B) vertical; (C) diagonal

441

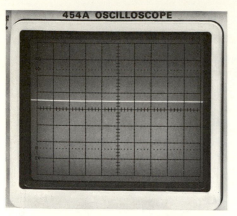

Fig. 47–5. Horizontal line, or trace, appearing on the screen of an oscilloscope cathode-ray tube (*Tektronix, Inc.*)

or signals, are then applied to certain circuits such as those in a radio receiver. This is done to test and adjust the circuits so that the most efficient operation can be obtained.

Oscillator Circuit. The heart of a typical signal generator is an oscillator circuit. Such a circuit produces an alternating current through the action of an inductance-capacitance circuit or a capacitance-resistance circuit.

A common inductance-capacitance oscillator circuit known as a *Hartley oscillator* is shown in Fig. 47–7. As direct current begins moving through the collector circuit of the transistor, it creates an expanding magnetic field around coil L1. This magnetic field induces a voltage across coil L2. The induced voltage excites the circuit consisting of L1, L2, and capacitor C1 into oscillation. This oscillation produces an alternating voltage across the output circuit. The combination of L1, L2, and C1 is commonly called a *tank circuit.*

While the tank circuit is oscillating, part of its energy in the form of a voltage is fed back through capacitor C2 to the input of the circuit. The input is at the base connection of the transistor. The feedback voltage is then amplified by the transistor. This resupplies the tank circuit with energy through the collector circuit. Because of this, the energy losses from resistance in the circuit are replaced. The circuit thus keeps oscillating. The frequency of oscillation is determined by the inductance of L1 and L2 and the capacitance of C1. Since C1 is a variable capacitor, the frequency of the output signal can be varied.

Audio-frequency and Radio-frequency Signal Generators. Audio-frequency (AF) signal generators produce output signals with a frequency range of from about 10 Hz to 100 kHz. Audio-frequency generators produce a sinusoidal, or sine-wave, output. Many also produce a square-wave output.

Fig. 47–6. An RF signal generator (*Precision Apparatus Division of Dynascan Corporation*)

Radio-frequency (RF) signal generators have a frequency range of from about 100 kHz to 100 MHz or more.

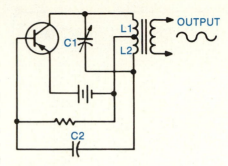

Fig. 47–7. Hartley oscillator circuit

Modulation. In the typical radio-frequency signal generator, the signal from a 400-Hz oscillator circuit in the generator can be used to modulate the output signal. *Modulation* is varying the amplitude or the frequency of a carrier wave. In this case, the carrier wave is the sine-wave output voltage. When the amplitude is varied, it is called *amplitude modulation* (AM). When the frequency is varied, it is called *frequency modulation* (FM).

Because of its modulated outputs, the radio-frequency signal generator is very often used as a miniature radio transmitter for testing the circuits of radio receivers. In such testing, the signal generator is connected to certain parts of a radio circuit. If the circuit is operating properly, the 400-Hz signal, or tone, with which the carrier signal is modulated is reproduced by the loudspeaker. Special kinds of radio-frequency signal generators are also commonly used for testing and adjusting television receiver circuits.

SIGNAL TRACER

The *signal tracer* is an instrument used to trace, or follow, a signal voltage through circuits such as those found in amplifiers and radio receivers (Fig. 47–8). This is done by connecting different points of the circuit under test to the input of the instrument. If the circuit is operating properly, the signal tracer will indicate this by an audio output signal heard through a loudspeaker or headphone.

When a signal tracer is used with a modulated radio-frequency signal, it must have a detector circuit. This circuit separates the audio signals from the radio-frequency carrier. This is necessary since a signal-tracer loudspeaker will not respond to signals above the audio range of frequencies.

Fig. 47–8. Signal tracer (*ELCO Electronic Instrument Company, Inc.*)

USING ELECTRONIC TEST INSTRUMENTS

The procedures given below are intended to provide basic information on operating the oscilloscope. The controls mentioned are those found on the typical, general-purpose instrument (Fig. 47–9). After learning to operate an oscilloscope in this way, you will be able to use the oscilloscope for various purposes. Many oscilloscopes have additional controls. Before operating an oscilloscope or other test equipment, you should carefully read its instruction manual. Basic information on the operation of a signal generator is also given in the following procedures.

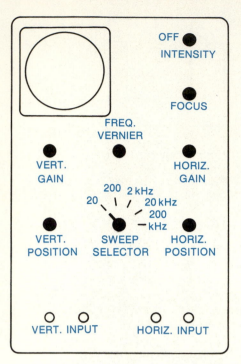

Fig. 47-9. Typical oscilloscope controls

Procedure

1. Turn the oscilloscope on. Adjust the following controls as stated:

Vertical gain	fully counterclockwise
Horizontal gain	fully clockwise
Intensity	approximate center of rotation

2. Adjust the vertical position control so that a horizontal trace, or line, appears through the center of the oscilloscope screen.

3. Adjust the intensity control so that the line can be clearly viewed without too much brightness. Then adjust the focus control so that the line is "clean," or well-defined.

4. Adjust the horizontal gain control to make the line just long enough to stretch to each side of the screen. This may also take an adjustment of the horizontal position control. You will note that the horizontal position control allows you to move the line to the right or to the left.

5. Turn the vertical gain control about one-fourth of its full clockwise range.

6. Connect an audio-frequency signal generator to the vertical input jacks of the oscilloscope. Adjust it to a sine-wave output at a frequency of 1,000 Hz.

7. Adjust the sweep selector control of the oscilloscope to the 2-kHz position. Then adjust the frequency vernier control so that a complete sine wave appears on the screen. Now adjust the vertical gain control to produce the height of waveform you want.

8. Adjust the signal generator to produce signals of other frequencies. As this is done, you will note that the sweep selector and/or the frequency vernier controls must be readjusted to produce one complete sine waveform on the screen.

9. Adjust the signal generator to produce a square-wave output. Observe these waveforms at different frequencies.

10. Connect a crystal microphone to the oscilloscope. Observe the voice waveforms generated by talking into the microphone.

11. Connect a 6-volt battery in series with a door buzzer. Connect the vertical input leads of the oscilloscope to the terminals of the buzzer.

12. Adjust the necessary controls of the oscilloscope so that a clear pulsating-current waveform appears on its screen. This waveform is produced by the rapid action of the buzzer contacts in making and breaking the circuit.

Test your knowledge by writing, on a separate sheet of paper, the word or words that most correctly complete the following statements:

1. The main purpose of an oscilloscope is to show a _____, or picture, of a voltage.
2. Waveforms shown by an oscilloscope are displayed on the screen of a _____ tube.
3. The electron beam in the cathode-ray tube is moved by a process known as _____ deflection. This is done by applying voltages to the _____ and the _____ deflection plates in the tube.
4. A signal generator is a test instrument used to supply output voltages of different _____.
5. The heart of a typical signal generator is an _____ circuit.
6. Two general kinds of signal generators are _____ signal generators and _____ signal generators.
7. When the amplitude of a carrier signal is varied by modulation, this is called _____ modulation. When the frequency of the carrier signal is varied by modulation, this is called _____ modulation.
8. An instrument used to trace a signal through circuits such as those in amplifiers and radio receivers is called a _____.

FOR REVIEW AND DISCUSSION

1. What is the main purpose of an oscilloscope?
2. How is a spot of light produced on the screen of a cathode-ray tube?
3. Explain the process of electrostatic deflection in the cathode-ray tube of an oscilloscope.
4. For what purpose is the sweep circuit used in an oscilloscope?
5. What is a signal generator? Why is it used?
6. Explain how a basic Hartley oscillator circuit works.
7. State the difference between audio- and radio-frequency signal generators.
8. Explain the difference between amplitude modulation and frequency modulation.
9. For what purpose is a 400-Hz oscillator circuit used in a typical radio-frequency signal generator?
10. What is a signal tracer? For what purpose is it used?
11. Why must a signal tracer that is tracing a radio-frequency signal have a detector circuit?

INDIVIDUAL-STUDY ACTIVITIES

1. Give a demonstration to your class on how an oscilloscope is used to show different kinds of waveforms.
2. Give a demonstration showing your class how an audio-frequency signal generator is operated. An audio amplifier can be connected to the signal generator to provide audible tones of its output.

Unit 48 Troubleshooting and Repair

Troubleshooting is the process of determining why a device or a circuit is not working as it should. To do this efficiently, you have to have a broad knowledge of how things work. Technicians who do troubleshooting, repair, or servicing must know a number of things. They must know what test instruments and procedures are needed to find a defect. They must also know what tools and materials are needed to repair it. Finally, they must have the ability to *diagnose.* This means that they must be able to recognize the symptoms of a defective device or circuit and then form a plan of action to repair it.

TROUBLESHOOTING PROCEDURES

The troubleshooting process for a circuit often takes the form of a step-by-step procedure of specific instructions. For example, the following are the steps for troubleshooting a radio circuit.

1. Check all wiring visually or by means of continuity tests.
2. Check battery and if necessary, replace it.
3. Check voltages at electrodes of transistors.
4. Test voice coil of loudspeaker with an ohmmeter.
5. Check volume-control potentiometer with an ohmmeter.
6. Clean contacts of headphone jack.
7. Replace detector diode with same or equivalent type.
8. Test transistors in driver and output stages and replace if necessary.

In the general troubleshooting of devices and circuits, there are usually no definite rules to be followed. In the typical case, technicians must apply their knowledge of theory, test instruments, tools, and materials to solve a problem. This often involves mechanical skill. Many electrical and electronic devices and circuits depend to some extent on certain mechanical operations. Several suggestions for troubleshooting and repair are given in the rest of this unit.

VISUAL INSPECTION

Unless the probable source of difficulty is known, troubleshooting should usually begin with a thorough visual inspection. Loose connections, broken wires, poor solder joints, burned-out resistors, or even a burned-out fuse or a tripped circuit breaker can very often be found in this way.

For example, if you find that a floor lamp is not working, you might replace the bulb. If this does not correct the trouble, you might replace the lamp switch and then the lamp socket. If the lamp still does not work, you would have wasted your time.

Perhaps if you had inspected the plug before replacing anything, you would have discovered that a wire had become unfastened from a plug terminal screw. With this corrected, the lamp might have worked. Therefore, in this case, a thorough visual inspection would have saved some time and possibly some expense. In more complicated troubleshooting and repair jobs, more time and more money can be wasted in procedures that would have been unnecessary if a visual inspection had been made first.

Vision also plays an important part in troubleshooting television circuits since the picture tube can be used as a test instrument. Many technicians are able to quickly localize defects in a television circuit by observing the lack of a picture or the distortions that appear in a picture or in a test pattern.

SMELL, HEARING, AND TOUCH

In addition to sight, the senses of smell, hearing, and touch can be valuable aids in troubleshooting. Overheated resistors, transformers, choke coils, motor windings, and insulation materials can very often be detected by their odor. Each of these conditions produces a characteristic odor that the experienced technician can use to locate defective components. When the overheating of any component or part in a circuit is noticed, the circuit should be turned off at once.

Overheated transformers, choke coils, and motor windings will sometimes produce a cooking, or frying, noise that can easily be heard. In high-voltage circuits such as those in television receivers, the arcing of current between certain wires and the chassis or between two wires produces a chirping noise. By tracing this noise, it is often possible to quickly locate the source of the arcing.

The sense of hearing can also be used in troubleshooting a circuit that has a loudspeaker. By using the loudspeaker as a test instrument, the lack of sound or the distortions of sound heard through it can indicate specific circuit defects. An experienced technician can quickly locate defects using this procedure.

Overheated components can often be located by touch. This is especially true of transformers, choke coils, electrolytic capacitors, and transistors. If any of these components feels too warm, there may be a defect in the component or in the circuit wiring.

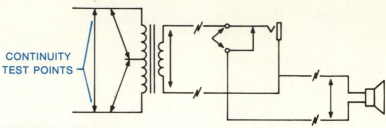

Fig. 48–1. Typical continuity test points used in checking the output transformer, headphone jack, and loudspeaker of a radio receiver

CONTINUITY TESTS

After the visual inspection and the smell, hearing, and touch tests have been done, the troubleshooting of many devices and circuits involves continuity tests made with an ohmmeter. Continuity tests should be made wherever an open-circuit condition is suspected (Fig. 48–1). The test can also be used to check for shorts. It is important to remember that an ohmmeter should never be used on a circuit to which a voltage is applied. Also, a component should be isolated by disconnecting one of its leads before continuity tests are done.

ELECTRON TUBES AND TRANSISTORS

Troubleshooting circuits with electron tubes usually begins with testing the tubes, unless a circuit defect is first noticed. The transistors in a circuit are generally not tested until there is definite reason to believe that at least one of them is defective.

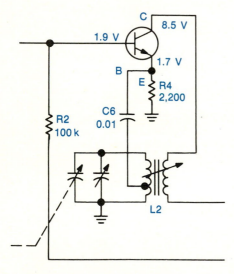

Fig. 48–2. Portion of a radio-receiver schematic diagram showing the dc voltages that should be present between the electrodes of a transistor and the common (ground) point of the circuit

VOLTAGE TESTS

After it has been determined that all power leads, fuses, switches, and electron tubes, if any, are in good condition, the next step in troubleshooting is to test the voltage at certain supply points. These include the secondary winding or windings of transformers, battery terminals, and rectifier outputs.

If the supply voltage is found to be normal, it is good practice to then measure the voltages at the electrodes of the transistors or the electron tubes. Unless these voltages are of the correct value, a circuit will fail to operate or will operate improperly.

The voltages that should be present between the electrodes of transistors or electron tubes and the common point of their circuit are usually shown on the schematic diagram of the circuit (Fig. 48–2). If you do not have this diagram, it is often helpful to take a look at the schematic diagram of another circuit in which identical (or similar) transistors or electron tubes are used.

RESISTANCE TESTS

A lack of voltage or an incorrect voltage at a given point of a circuit is very often caused by a defective resistor or capacitor. In order to localize this defect, it is helpful to make resistance tests of that part of the circuit from which voltage is obtained for the electrode in question (Fig. 48–3). In this way, it is possible to locate defective components that may be preventing the correct distribution of voltage in a circuit.

DIAGNOSIS OF CIRCUIT TROUBLES

Defects in a circuit, such as a broken connection, a loose solder joint, or a worn-out component, can usually be taken care of by simple repair or replacement. However, a repair job often involves a further diagnosis of the reason or reasons for a defect. For example, a resistor, a diode, a transistor, or a solenoid coil will become overheated and burn out because of a second defective component in the circuit. If the first component is replaced without further diagnosis, it is probable that the replacement component will also be damaged.

Let us suppose that the carbon-composition resistor in the rectifier circuit of an amplifier is burned out (Fig. 48–4). Let us further suppose that an inexperienced person immediately replaces the resistor and finds that the replacement resistor also quickly burns out. Obviously, then, it is necessary to diagnose the situation more carefully.

First of all, fundamental theory tells us that the resistor burned out because it was forced to carry an excessive current. From experience, it is reasonable to assume that an excessive current is not caused by a defect in the resistor itself. The conclusion must be that the excessive current was caused by some other defect in the circuit.

In the circuit in Fig. 48–4, two electrolytic filter capacitors, C1 and C2, are connected across the rectifier output. We know that electrolytic capacitors can become shorted. If C1 were shorted, perhaps one or both of the diodes would be damaged. However, because of the circuit arrangement, this would not cause excessive current through the resistor. On the other hand, if C2 were shorted, excessive current would definitely pass through the resistor. Therefore, the solution to the problem is replacing both C2 and the resistor.

The situation just described is a very basic troubleshooting and repair job. It is, however, a good example of how knowledge of theory, practical experience, and the ability to diagnose trouble can be used in solving a circuit problem. The efficient technician must constantly do this in finding circuit defects and correcting them.

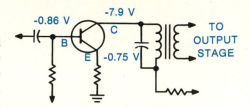

Fig. 48–3. Portion of a radio-receiver circuit showing the resistors that should be checked if correct voltages are not present at the electrodes of the transistor

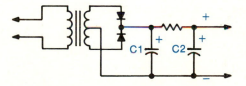

Fig. 48–4. Troubleshooting a rectifier filter circuit

LOGIC PROBE

INTEGRATED CIRCUIT BEING TESTED

Fig. 48–5. Using a logic probe to check a digital-clock integrated circuit

LOGIC PROBE

A *logic probe* is a test instrument used mostly to trouble-shoot different families, or kinds, of integrated circuits (ICs) (Fig. 48–5). The logic probe detects pulses of voltage. In digital work, these pulses of voltage are called *logic levels*, or *threshold levels.* The logic probe also stores and displays the logic level. This logic level is displayed on a light-emitting diode (LED) readout.

By flipping a switch, the logic probe can be set to detect threshold levels of 2.25 volts (logic "1") and 0.8 volts (logic "0") for a family of ICs known as DTL or TTL. These abbreviations refer to *diode-transistor logic* (DTL) or *transistor-transistor logic* (TTL or T²L). Switching to another position sets the probe to detect threshold levels for a family of ICs known as CMOS/HTL. These abbreviations refer to *complementary metallic oxide semiconductors* (CMOS) or *high-threshold logic* (HTL). In the CMOS/HTL ICs, the threshold levels for digital operation are determined by the *applied voltage* (Vcc). When a threshold level is greater than 70 percent of Vcc, it is logic "1." When the threshold level is less than 30 percent of Vcc, it is logic "0." Some logic probes detect pulse widths as narrow as 10 nanoseconds at a frequency of 50 MHz. These features make the logic probe an important troubleshooting instrument. The power supply of the circuit under test is also used to operate the logic probe. The following general procedures are suggested for using a logic probe:

1. Connect the clip leads to the power supply of the circuit.
2. Set the switch to the logic family to be tested (DTL/TTL or CMOS/HTL).
3. Set the MEMORY/PULSE switch to the PULSE position.
4. Touch the probe tip to a node, or terminal, on the IC to be analyzed. The three LEDs (high, low, and pulse) on the logic probe will provide an instant reading of the signal activity at the node.

If the logic probe is to store a pulse, touch the probe tip to the node under test. Then move the MEM/PULSE switch to the MEM position. The next pulse will activate the PULSE LED.

SELF-TEST

Test your knowledge by writing, on a separate sheet of paper, the word or words that most correctly complete the following statements:

1. The process of determining why a device or a circuit is not working as it should is called _____ .
2. In addition to sight, the senses of _____, _____, and _____ can be valuable aids in troubleshooting.
3. Overheated components can often be located by _____ .
4. The troubleshooting of many devices and circuits involves _____ tests.
5. The voltages that should be present between the electrodes of transistors and electron tubes are often shown on _____ diagrams.
6. Repair jobs often involve further diagnosis of the _____ for defects.
7. Integrated circuits can be analyzed with a test instrument called a _____ .
8. CMOS refers to a family of integrated circuits called _____ .

FOR REVIEW AND DISCUSSION

1. What is meant by the troubleshooting of a circuit?
2. Name some of the things technicians should know and be able to do in order to be efficient at troubleshooting and repair.
3. Tell why a thorough visual inspection is so important in the troubleshooting of many circuits.
4. How can the senses of smell, hearing, and touch be put to use in troubleshooting?
5. What circuit defects can be located by means of continuity tests?
6. Explain how resistance tests can be used to locate defects in a circuit.
7. What can happen if you replace a part without making a full diagnosis of the trouble?

INDIVIDUAL-STUDY ACTIVITY

Give a demonstration showing your class the correct procedures to be followed in troubleshooting a device, an appliance, or an electronic circuit.

Suggested Projects

The following projects provide several practical and inexpensive circuits to build. Each project has a schematic diagram and a parts list. Hints that will help in the construction and the operation of a project are also given when appropriate.

Chassis construction and other mounting and enclosure information has been omitted purposely. This will provide an opportunity to exercise creative design skills in determining the appearance of the finished product.

Project 1   LOW-VOLTAGE POWER SUPPLY

A variable-voltage (0 to 15 V) dc power supply that will provide an adequate amount of current to operate many solid-state products and experimental circuits.

PARTS LIST

P1, 3-conductor polarized connector with male contacts, rated: 15 A, 125 V ac (attachment plug)
C1, electrolytic capacitor, 25 µF, 25 WVDC
C2, electrolytic capacitor, 500 µF, 25 WVDC
CR1, CR2, CR3, and CR4, rectifier diodes, 1 A, 50 PIV (An equivalent four-diode bridge-rectifier unit can be substituted for the individual diodes.)
F1, fuse, AGC, $\frac{3}{4}$ A
M1, voltmeter, 0 to 25 V, dc (optional)
Q1, transistor, p-n-p, International Rectifier TR-01 or equivalent
R1, potentiometer, 500 kΩ, $\frac{1}{2}$ W
R2, resistor, carbon-composition or film, 1 kΩ, $\frac{1}{2}$ W
R3, resistor, carbon-composition or film, 1.5 kΩ, 2 W
S1, switch, spst (individual unit or part of R1)
T1, transformer, 117 V primary, 12.6 V secondary, 1.2 A

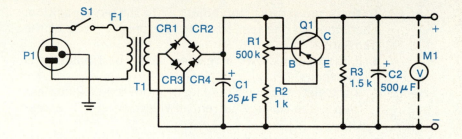

Project 2 TRANSISTOR TESTER

A circuit designed for testing typical p-n-p and n-p-n transistors. Provides tests for shorts, excessive leakage, and gain.

PARTS LIST

BT1, battery, 6 V or 4 size-C dry cells connected in series
M1, milliammeter, dc, 0 to 1 mA
R1 and R3, resistors, carbon-composition or film, 220 kΩ, ½ W
R2, potentiometer, 10 kΩ, ½ W
S1, switch, slide-type, tpdt (p-n-p or n-p-n selector control)
S2 and S3, switches, spst, push-button, normally open (N.O.) momentary-contact type (p-n-p and n-p-n gain test controls)

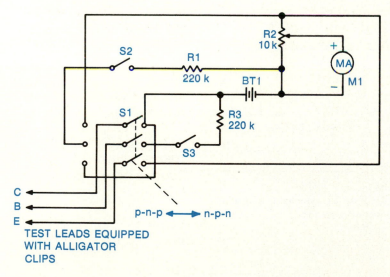

TEST LEADS EQUIPPED
WITH ALLIGATOR
CLIPS

Using the Tester

1. To calibrate the tester before testing any transistor, first adjust switch S1 to the p-n-p position. Then bring the alligator clips of test leads C and E into contact with each other. Next adjust R2 until the meter reads 1.

453

2. To test a p-n-p transistor, adjust S1 to the p-n-p position. Then connect the test leads to the proper leads of the transistor:
   a. If the meter shows a reading of 1, the transistor is shorted.
   b. If the meter shows a reading of more than 0.2 but less than 1, the leakage current of the transistor is excessive.
   c. To test for gain, depress S2. If the meter reading is significantly higher than the reading before S2 was depressed, the transistor is good.
3. To test an n-p-n transistor, adjust S1 to the n-p-n position. Then follow the procedures given for testing a p-n-p transistor, using S3 to provide an indication of gain.

Project 3 CONTINUITY TESTER

A useful product that can be used to determine broken or open connections and grounded circuits. Continuity (a complete circuit) is indicated by either a buzzer signal or an indicator lamp.

PARTS LIST

BT1, battery, 6 V or 4 size-C dry cells connected in series
DS1, pilot lamp, no. 47
LS1, buzzer, 6 V
S1, switch, toggle, spdt with center "off" position

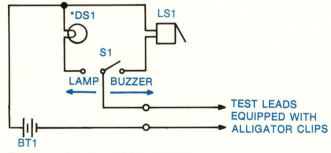

*Lamp can be mounted in a miniature bayonet-base socket or an indicator-light assembly.

Project 4 RESISTOR SUBSTITUTION BOX

This project gives a handy way of having available a variety of often-used resistors. This is very useful when working with experimental circuits or servicing equipment. A similar cir-

cuit made with capacitors can be easily attached to provide
for capacitor substitution.

PARTS LIST

R1 through R17, resistors, carbon-composition or film,
1 W, having the following resistance values:

3.3 ohms	680 ohms	47 kilohms
10 ohms	1,000 ohms	100 kilohms
27 ohms	1,800 ohms	220 kilohms
56 ohms	4,700 ohms	560 kilohms
100 ohms	10 kilohms	1 megohm
470 ohms	27 kilohms	

S1, switch, single-gang nonshorting rotary, 1-pole,
17-position, with pointer knob
switch plate, dial, shopmade. Dial should have resis-
tance values printed, typed, or lettered on its face to
show which resistor is connected to the substitution
box test leads.

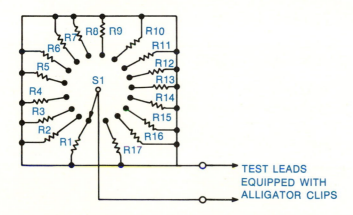

TEST LEADS
EQUIPPED WITH
ALLIGATOR CLIPS

Project 5   LAMP-DIMMER CONTROL

This circuit allows you to vary the light output of a 100-watt
incandescent lamp from zero to full brilliance. It is very convenient for controlling the operation of a study lamp or a
lamp that is used as a night light. This circuit can be used
with any 60- to 200-watt incandescent lamp. However, if you
expect to use a lamp larger than 100 watts, fuse F1 should be
increased to 2 amperes.

PARTS LIST

P1, 3-conductor polarized connector with male contacts,
rated: 15 A, 125 V ac
C1, capacitor, 0.0022 μF, 200 WVDC

DS1, neon lamp, NE-2
DS2, incandescent lamp, medium-base, 100 W, with socket
F1, fuse, AGC, 1 A
R1, resistor, carbon-composition or film, 330 kΩ, ½ W, 5 or 10% tolerance
R2, potentiometer, 2 megohms, ½ W
S1, switch, spst (individual unit or mounted on R2)
SCR1, silicon controlled rectifier, International Rectifier IR106B1 or equivalent

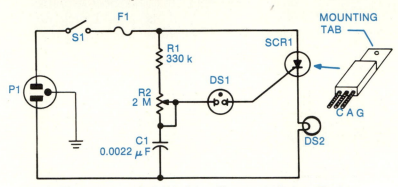

Note 1: The center terminal (A) of the silicon controlled rectifier is internally connected to the mounting tab of the device. Therefore, as a safety precaution, the mounting tab must be completely insulated from any metallic enclosure within which the circuit is placed.

Note 2: For the safest operation, the shaft of the potentiometer, R2, should be equipped with a plastic control knob.

Project 6   CODE-PRACTICE OSCILLATOR

A loudspeaker oscillator circuit that provides adequate volume for either individual or group code practice. It is useful in preparing for the amateur radio operators' international Morse code examination. The most commonly used characters of the international Morse code and their dot-dash equivalents are given in Fig. 37–3, page 344.

PARTS LIST

BT1, battery, 9 V
C1, capacitor, 0.05 µF, 200 WVDC
C2, electrolytic capacitor, 100 µF, 25 WVDC
LS1, loudspeaker, 8 Ω miniature
Q1, transistor, n-p-n, SK3005 or equivalent
Q2, transistor, p-n-p, SK3444/123A or equivalent

R1, resistor, carbon-composition or film, 100 kΩ, 0.5 W
R2, resistor, carbon-composition or film, 100 Ω, 0.5 W
R3, resistor, carbon-composition or film, 10 Ω, 0.5 W
S1, telegraph key

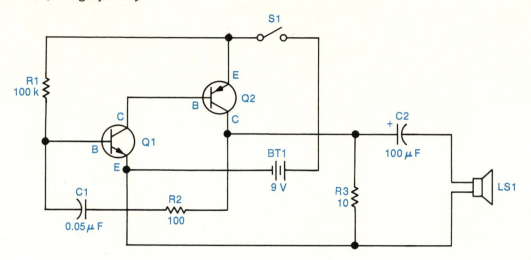

Project 7   MINI-ORGAN

A multivibrator circuit similar to that used in several models
of electronic organs. The fundamental multivibrator frequency is determined by the RC time constant of C1 and R1.
Changes of this frequency are produced when other capacitors (C2 through C8) are connected into the circuit by
switches S2 through S8. Any changes in the frequency provide a different tone.

PARTS LIST

BT1, battery, 6 V or 4 size-C dry cells connected in
    series
C1, capacitor, 0.005 µF, 200 WVDC
C2 to C8, capacitors, 0.02 µF, 200 WVDC
C9, capacitor, 0.001 µF, 200 WVDC
C10, capacitor, 0.2 µF, 200 WVDC
CR1, germanium diode, 1N54 or SK3087
LS1, loudspeaker, 2½ in. (63.5 mm), 8 Ω
Q1, transistor, n-p-n, 2N388 or SK3011
Q2, transistor, p-n-p, 2N408 or SK3003
R1, potentiometer, 1 MΩ, ½ W (tone control)
R2, resistor, carbon-composition or film, 1 MΩ, ½ W

S1 to S8, switches, spst, push-button, normally open
(N.O.) momentary-contact type
S9, switch, spst (individual unit or part of R1)

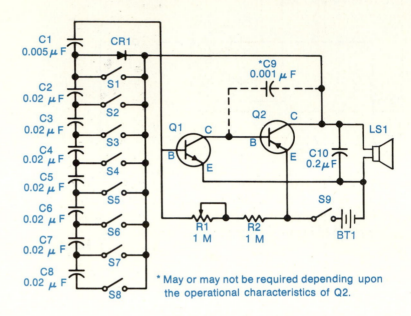

* May or may not be required depending upon the operational characteristics of Q2.

Project 8   LIGHT-CONTROLLED SWITCH

A relay-control circuit operated by changes in the intensity of the light that strikes the active surface of a photoconductive cell. The relay contacts are used as a switch to control the operation of a load, such as an incandescent lamp or a small motor.

PARTS LIST

BT1, battery, 9 V

K1, relay, miniature sensitive, spdt, 500-Ω coil, with
   contact rating 600 mA at 50 V ac/dc or 2 A at 117 V
   ac

Q1 and Q2, transistors, p-n-p, GE-2 or SK3004

R1, potentiometer, 100 kΩ, ½ W (sensitivity control)

R2, photoconductive cell, Clairex no. CL704M or
   equivalent

R3, resistor, carbon-composition or film, 10 kΩ, ½ W

S1, switch, spst (individual unit or part of R1)

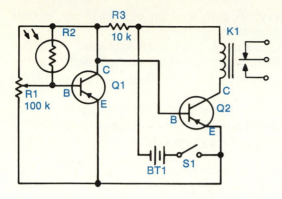

Project 9   ORNAMENTAL LAMP FLASHER

A circuit that provides the attractive effect of neon lamps flashing on and off at different times. Useful for lighting jack-o-lantern eyes, for Christmas decorations, and for other ornamental-lighting displays.

PARTS LIST

P1, 3-conductor polarized connector with male contacts,
*   rated: 15 A, 125 V ac*
C1 and C2, capacitors, 0.1 μF, 200 WVDC
C3, capacitor, 0.12 μF, 200 WVDC
CR1, rectifier diode, 1 A, 200 PIV
DS1 and DS2, neon lamps, NE-2
F1, fuse, AGC, ½ A
R1 and R2, resistors, carbon-composition or film,
*   10 MΩ, ½ W*
S1, switch, spst

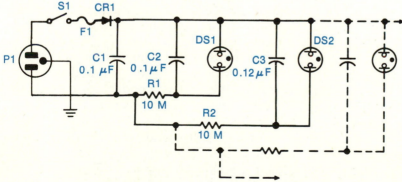

Dotted lines show circuit connections for additional lamp.
More lamps can be connected into the circuit in a similar way.

All resistors are 10 megohm and all lamps are type NE-2. Additional capacitors suggested are 0.15 μF, 0.18 μF, 0.22 μF, 0.27 μF, 0.33 μF, and so on.

Glossary

ammeter An instrument used to measure the amount of current in a circuit. An ac ammeter measures alternating current. A dc ammeter measures direct current.

ampere The unit of measure for a specific quantity of electrons. A current of 1 ampere represents the movement of 6,280,000,000, 000,000,000 electrons past a given point in a circuit in 1 second of time.

amplifier A device used to increase the amount of electric power output in a circuit by increasing either the input current or voltage, or both.

amplitude modulation A common method of radio broadcasting in which the amplitude, or strength, of a radio carrier wave is changed above and below a certain level. The changes in the strength of the carrier wave, when detected, represent the information that was transmitted.

anode The positive terminal of a cell or battery. In an electron tube, the anode, or plate, is that element within the tube to which a positive voltage is applied.

antenna The system of wires or other electrical conductors used to receive or to transmit radio waves.

arc, electric A visible glow of light that is formed under certain conditions when electrons move through gases or through the air space between two points in a circuit.

armature The moving part or parts of a magnetically operated device such as a motor, generator, buzzer, bell, relay, or loudspeaker.

atom The smallest part of a chemical element having the properties of the element.

automation The method of controlling the operation of machinery and equipment for the production of goods in such a way that the input and output of machines are controlled by the use of electronic circuits and devices.

battery A combination of two or more cells connected together.

breadboard A circuit arrangement in which electrical components are fastened together temporarily for testing or experimental work.

brush A carbon or metal object used to make contact with the commutator or slip rings of a motor or a generator.

capacitance The property of a capacitor that enables it to store an electric charge. The unit of measurement for capacitance is the farad.

capacitive reactance The opposition that a capacitor presents to alternating current or to a direct current that is changing in value. Capacitive reactance is measured in ohms.

capacitor A device capable of storing electric energy. It is constructed of two conductor materials separated by an insulator.

cathode The negative terminal of a cell or battery. The part of an electronic device from which electrons are emitted.

cell, voltaic A device that produces voltage by means of chemical action. A voltaic cell is made of two different kinds of conductor materials placed within a paste or a fluid (electrolyte) that also conducts electricity.

charge, negative The electrical property of an object containing more electrons than protons.

charge, positive The electrical property of an object containing more protons than electrons.

charge, battery A device that supplies dc voltage to charge a cell or a battery.

circuit A system of conductors and devices through which electrons can move. A complete circuit contains (1) conductors, (2) a switch, (3) a load, and (4) an electrical source.

circuit, open An incomplete or broken circuit.

circuit, parallel A circuit in which the loads are connected across two wires or conductors of the power line.

circuit, printed A method of circuit wiring in which the conductors are copper strips on a sheet of insulator material.

circuit, series A circuit in which the loads are connected into one wire or conductor of the power line in a "one after the other" fashion.

circuit, short A circuit containing a defect that causes electrons to follow a path of smallest resistance.

circular mil The unit of measure for the cross-sectional area of a round wire. A round wire 1 mil or 0.001 of an inch in diameter has a cross-sectional area of 1 circular mil. The circular mil area of any round wire is found by squaring the diameter of the wire given in mils.

coil A device made up of turns of insulated wire wrapped around a hollow or a solid-core form.

commutator A device used to reverse the direction of current. A motor or a generator commutator is made of a ring of insulated copper bars, or segments, mounted on a shaft. The bars are connected to the coils of wire wound into the armature core slots.

conductor A solid, a liquid, or a gas through which electrons pass easily.

continuity test The test made to determine whether a circuit or a part of a circuit provides a complete electrical path.

crystal cartridge The device, mounted in a record player pick-up arm, used to change the variations of the record grooves into an electric current. The needle is fastened to the cartridge.

crystal diode A device made of germanium or silicon used to rectify an alternating current.

current The movement of electrons through a conductor material.

current, alternating The movement of electrons through a conductor, first in one direction and then in the opposite direction.

current, direct The movement of electrons through a conductor in one direction only.

current, induced A current present in a circuit because of the voltage produced by means of electromagnetic induction.

detector An electronic device or a circuit that changes the form of a current. In radio and television receivers, a detector is used to separate the audio or video portions of a carrier wave into currents that can be changed to sound or to a picture.

diagram, pictorial A diagram that shows circuit parts by pictures or drawings.

diagram, schematic A wiring diagram that shows circuit parts by standard electrical symbols.

diathermy The process of heating with high-frequency current.

dielectric An insulation material.

electric car An automobile that gets its power from rechargeable batteries.

electricity, dynamic A form of energy present when electrons move through a circuit.

electricity, static A form of energy present within the space between two oppositely charged objects.

electrolysis The process of producing a chemical change by passing current through a conducting solution.

electrolyte A solution able to conduct current.

electromagnet A magnet produced by passing current through a conductor.

electromagnetism The magnetism produced by an electric current.

electromotive force The electrical force that causes electrons to move through a conductor. Electromotive force is commonly known as *voltage.*

electron A negatively charged particle that revolves about the nucleus of an atom.

electron gun A device capable of producing an electron beam.

electron tube A device consisting of two or more electrodes enclosed in a glass or metal envelope from which most of the air has been removed. Also commonly known as a *vacuum tube.*

electronics The study of electrons and how they move through space and through special conducting materials. The term *electronics* is usually applied to the study of how electrons flow in electron tubes and semiconductor materials.

electroplating An electrochemical process of plating with metal by passing current through a conducting solution.

energy The ability to do work.

farad The unit of measure for capacitance. A capacitor is said to have a capacitance of one farad when a voltage applied to it, changing at the rate of one volt per second, produces a current of one ampere in the capacitor circuit. Because the farad is such a very large unit, the term *microfarad,* or *one-millionth of a farad,* is most commonly used.

filament A threadlike conductor found inside electron tubes and light bulbs. The filament glows to give off light and heat.

fluorescent lamp A lamp in which light is produced by the action of ultraviolet rays striking a phosphor-coated surface.

frequency The number of cycles of an alternating current completed in a certain period of time—usually 1 second.

frequency modulation A common method of radio broadcasting in which the frequency of the radio carrier wave is changed above and below a certain given frequency. The changes in the frequency of the carrier wave, when detected, represent the information transmitted.

friction A force that opposes motion.

fuse A protective device made up of wire that melts and breaks when the current through it is more than the ampere rating of the fuse.

fuse puller A tonglike device used to remove cartridge fuses from their mounting clips.

galvanometer An instrument used to measure very small amounts of current.

generator A machine used for changing mechanical energy into electric energy.

grid The element in an electron tube that controls the flow of electrons from the cathode to the plate.

grid bias The controlling voltage applied to the grid of an electron tube.

ground An electrical connection between a circuit and the earth, or between a circuit and some metal object that takes the place of the earth.

ground-fault circuit interrupter An electronic device that opens a circuit automatically due to ground faults. This protects the user against serious injury.

ground wave A radio wave that travels on, through, or near the Earth.

headphones A device that changes audio-frequency current into sound. It is worn over the ears.

henry The unit of measure for self-inductance or mutual inductance. A coil is said to have an inductance of one henry when the current through it, changing at the rate of one ampere per second, produces a voltage of one volt across the terminals of the coil.

heterodyning The production of one frequency by combining two different frequencies. The frequency produced is equal to the difference between the two frequencies.

high-efficiency motors Electric motors designed to efficiently convert electric energy into mechanical energy by reducing stator and rotor losses.

high fidelity The faithful reproduction of sound by an electronic device such as a loudspeaker.

hydrometer A device used to test the specific gravity of acid in a storage-battery cell.

impedance The total opposition of a circuit to alternating current.

incandescent lamp A lamp that produces light by the heating of a filament to a white-hot temperature.

induced magnetism The production of magnetism by placing magnetic material within a magnetic field.

inductance The property of a coil or of any part of a circuit that causes it to oppose any change in the value of the current flowing through it. The unit of measure for inductance is the henry.

induction The process of magnetizing an object or of inducing a voltage in an object by placing it within a magnetic field.

inductive reactance The opposition that a coil presents to alternating current or to a direct current that is changing in value. Inductive reactance is measured in ohms.

insulator A material that will not conduct current. An object made of an insulator material that supports or separates conductors in a circuit.

ion A charged atom containing more or less than its normal number of electrons.

jack Any of several devices used to make a temporary and easily disconnected electrical contact in a circuit.

junction A point in a circuit to which two or more conductors are connected together.

key A hand-operated switch used for sending the international Morse code.

kilo A prefix meaning thousand.

kilohertz One thousand cycles per second.

lamp A device used to produce light.

laser An abbreviation for *light amplification by stimulated emission of radiation.* A laser produces an intense directional beam of light.

lead Either the beginning or the ending wire of an electrical device or an external wire connected to a device such as a resistor, a coil, or a capacitor.

lightwave communications A communication system that uses light pulses beamed by a laser through glass fibers.

load The device in a circuit operated by the current.

lodestone One of the common names for magnetite, a mineral that is a natural magnet.

logic probe A troubleshooting instrument for digital circuits.

loudspeaker A magnetically operated device that changes electric energy into sound.

lug An insulated strip of metal to which wires can be connected.

magnet An object with the property of magnetic polarity and that attracts magnetic materials.

magnetic field The space near a magnet in which magnetic forces are present.

meter A measuring instrument.

meter, watthour An instrument used to measure the amount of electricity in kilowatthours.

mica A mineral that is a good insulator of electricity and heat.

micro A prefix meaning one-millionth.

microphone A device used to change sound energy into electric energy.

modulation The process of changing the amplitude or the frequency of a carrier wave by means of an audio or a video signal.

motor A machine that changes electric energy into mechanical energy.

National Electrical Code A set of guidelines for the installation of electric wiring and apparatus sponsored by the National Fire Protection Association under the auspices of the American National Standards Institute.

negative The electrical property of an object that contains more than the normal number of electrons.

neon lamp A device that produces light by passing current through neon gas.

nonmagnetic material A material that cannot be magnetized and is neither attracted nor repelled by a magnet.

ocean thermal energy conversion system An energy system under study that is expected to make use of ocean water to collect and store energy from the sun.

ohm The unit of electrical resistance. A circuit has a resistance of one ohm when one volt produces a current of one ampere.

ohmmeter An instrument used to measure resistance.

Ohm's law A basic electrical law that states the relationship between current, voltage, and resistance in a dc circuit.

optical fiber A hair-thin ultratransparent filament of glass. Several fibers brought together form a lightwave communications cable.

oscillator An electronic device that generates an alternating current.

oscilloscope An instrument that produces the picture of a waveform upon the screen of a cathode-ray tube.

OSHA An abbreviation for the *Occupational Safety and Health Administration,* an agency of the U.S. Department of Labor. This agency issues safety and health standards related to working conditions.

output transformer A transformer placed between the output stage and the loudspeaker of a radio receiver, television receiver, audio amplifier, and so on.

pentode An electron tube with five elements.

phototube A two-element electron tube, the cathode of which emits electrons when light strikes it..

piezoelectricity The production of voltage by straining or pressing upon certain kinds of crystal substances.

plug A device inserted into a jack to make an electrical connection.

points, breaker A device that interrupts a continuous direct current. In an automobile, breaker points are part of the distributor unit.

points, contact A mechanism with special surfaces that make contact to complete, or separate to open, a certain part of a circuit. Contact points are found in switches, relays, circuit breakers, and so on.

polarity An electrical condition that determines the direction of current. In a circuit, electrons move from a point of negative polarity to a point of positive polarity.

polarization The process by which hydrogen gas is deposited on the positive electrode of a dry cell.

potentiometer A variable resistor.

power The rate, or speed, at which work is done.

primary cell A cell that cannot be recharged.

radio A general term used in connection with equipment that receives electromagnetic waves and changes them into sound energy. The process by which radio waves are sent through the air is called *radio transmission.*

receptacle A socket or an outlet into which a plug is inserted to make an electrical connection.

rectifier A device that changes alternating current into direct current.

regulator, generator An electromagnetically operated device used to control the current and the voltage output of a generator.

relay A switch operated by electromagnetism.

resistance The tendency of a device or a circuit to oppose the movement of current through it. The unit of resistance is the ohm.

resistor A device used to insert electrical resistance into a circuit.

resonance The condition of a tuning circuit when the capacitive reactance of the circuit capacitor is equal to the inductive reactance of the coil in the circuit.

rheostat A variable resistor usually used in circuits through which large amounts of current move.

rotor The revolving part of an induction motor.

scanning The process of moving an electron beam over the inner surface of the face of a cathode-ray tube.

selectivity The property of a tuning circuit that enables it to pass only one frequency at a time.

semiconductor A group of solid materials that are good conductors of current under one condition and poor conductors under another condition.

series-parallel circuit A circuit containing one or more combinations of series and parallel circuits.

signal generator An electronic device used to generate a signal to test and adjust a radio or a television circuit.

signal tracer An electronic device used to trace the progress of a signal in a radio or television circuit.

solar cell A cell, usually made of silicon, that produces current when light falls on it.

solenoid A coil of wire wrapped around a hollow form.

space-based power conversion system A solar energy system under study that can collect pollution-free energy from the sun and transmit it to Earth from a fixed orbit.

splice A form of permanent electrical connection—usually soldered.

switch A device used to complete or to open a circuit.

telegraph A system of communicating by means of coded signals.

telephone A system of communicating by means of audio signals transmitted over conducting wires or by radio waves.

teletypewriter An electrically operated typewriter that reproduces, in written form, the coded signals that it receives.

television Electronic apparatus that changes information transmitted by carrier waves into pictures and sound.

terminal A mechanical connecting point used to fasten wires to cells, batteries, switches, panels, and so on.

theory An idea that is thought to be true but has not yet been proved.

thermocouple A device consisting of two different kinds of metals joined at one end. When the junction is heated, a voltage is produced across the open ends of the metals.

thermostat A switch operated by the application of heat.

tool, alignment A tool made of insulating material used to adjust variable capacitors and coils in radio and television receivers.

tower, transmission A tower on which a radio or television transmitting antenna is mounted.

transformer A device that transfers electric energy from one coil to another by means of electromagnetic induction.

transistor A semiconductor device used to amplify electrical signals.

transmitter The electronic apparatus that sends radio waves through the air.

triode An electron tube with three electrodes.

tube, cathode-ray An electron tube containing an electron gun that can reproduce a signal in the form of a picture on its face.

tube, diode An electron tube with two elements.

tube tester A device used to test the condition of electron tubes.

tuner The circuit of a radio or television receiver that selects the carrier frequency desired.

volt The unit of electromotive force, or voltage.

voltage The electromotive force that causes electrons to move through a circuit.

voltmeter An instrument used to measure voltage.

watt The unit of electric power.

waves, radio A form of electromagnetic energy that transmits information through the air.

windmill A machine that converts wind energy into useful work.

wire, magnet Copper wire usually insulated with insulating varnish. Magnet wire is used for winding many kinds of coils.

wire stripper A tool used to remove the insulation from wire.

wire table Table of sizes and properties of round copper wire.

wire wrapping A process of making an electrical connection by tightly coiling wire around a metal terminal.

INDEX